W9-CNG-181

BUTCHERING

BUTCHERING

POULTRY · RABBIT
LAMB · GOAT · PORK

The Comprehensive Photographic Guide to Humane Slaughtering and Butchering

Adam Danforth

PHOTOGRAPHY BY KELLER + KELLER

*The mission of Storey Publishing is to serve our customers by
publishing practical information that encourages
personal independence in harmony with the environment.*

EDITED BY Carleen Madigan
ART DIRECTION AND BOOK DESIGN BY Carolyn Eckert
TEXT PRODUCTION BY Theresa Wiscovitch

BACK COVER AND INTERIOR PHOTOGRAPHS BY
© Keller + Keller Photography,
except pages 335 by © Adam Danforth,
127 by Carolyn Eckert, and 74–76 by Mars Vilaubi.
ILLUSTRATIONS BY © Karin Spijker, Draw & Digit

INDEXED BY Christine R. Lindemer, Boston Road
Communications

**IN ADDITION TO THOSE MENTIONED IN THE AUTHOR'S
ACKNOWLEDGMENTS,** the publisher would like to thank
the following people for their help: Wendy Warner at
Windward Farm and the Martin family at Elmartin Farm in
Cheshire, Massachusetts; Christopher Bonasia, Michael
Gallagher, and Ashley Amsden at Square Roots Farm in
Lanesborough, Massachusetts; Jake Levin from the Berkshire
Food Guild; everyone at The Meat Market in Great Barrington,
Massachusetts; John Fazio Farms in Modena, New York;
Brett McLeod; Shiloh Partin; Billy Barlow; Greg Stratton
and Michael Montgomery from Stratton's Custom Meats &
Smokehouse in Hoosick Falls, New York.

© 2014 by Adam A. Danforth

The information in this book is true and complete to the
best of our knowledge. All recommendations are made without
guarantee on the part of the author or Storey Publishing. The
author and publisher disclaim any liability in connection with the
use of this information.

Storey books are available for special premium and
promotional uses and for customized editions. For further
information, please call 1-800-793-9396.

Storey Publishing
210 MASS MoCA Way
North Adams, MA 01247
www.storey.com

Printed in China by R.R. Donnelley
10 9 8 7 6 5 4 3 2 1

**Be sure to read all instructions thoroughly before attempting any of the activities in this book. No book can replace the
guidance of an expert nor can it anticipate every situation that will arise. Always be vigilant when working with animals and use
extreme caution when employing potentially lethal implements.**

LIBRARY OF CONGRESS CATALOGING-IN-PUBLICATION DATA

Danforth, Adam.
 Butchering poultry, rabbit, lamb, goat, and pork: the comprehensive photographic guide to
humane slaughtering and butchering / by Adam Danforth.
  p. cm.
 Includes bibliographical references and index.
 ISBN 978-1-61212-182-6 (pbk. : alk. paper)
 ISBN 978-1-61212-188-8 (hardcover : alk. paper)
 ISBN 978-1-60342-931-3 (ebook)
 1. Slaughtering and slaughter-houses. 2. Game and game-birds, Dressing of.
 3. Meat cutting. 4. Meat—Preservation. 5. Meat animals. 6. Animal welfare. I. Title.
TS1960.D25 2014
664'.9029—dc23
            2013030702

DEDICATION

For Moxie. I hope you feel comfortably known.

CONTENTS

FOREWORD BY JOEL SALATIN

MOST AMERICANS TODAY FEAR FOOD because they don't know much about it. But as we learn more and more about the shortcomings of industrial food, concentrated animal feeding operations (CAFOS), and the shenanigans of the food processing industry, we yearn for an antidote but don't know where to turn.

Our grandmothers and grandfathers weren't afraid of food. They knew how to turn cucumbers into pickles (how many Americans don't even know pickles come from cucumbers?) and which vegetables could be root-cellared. Not too long ago, the shared agrarian understanding in the culture included knowing the difference between hay and straw, shoats and gilts, cows and heifers.

Today, farmers' market shoppers looking through farmers' scrap books routinely explain to their children that the cow with the horns is a bull. Of course, farmers know horns have nothing to do with sex and joke among themselves about how ignorant their customers are. When our farm began raising and butchering pastured poultry, every homemaker knew how to cut up a chicken. Today, most don't realize a chicken even has bones. I have to explain to them that chicken nuggets in the shape of Dino the Dinosaur is not a muscle group on a chicken.

Into this profound ignorance, timidity, and fear steps a delightful remedy from Storey Publishing and domestic artisan Adam Danforth. The book you hold in your hands is a recipe for self-reliance and faith rather than dependency and fear: faith in the ability of individuals — thousands of them — in their own backyards and homesteads to access nature's bounty with home-scale meat preparation.

Unlike formal butchery textbooks, this one assumes beginner understanding, rudimentary equipment, and do-it-yourself (DIY) labor. The dramatic visceral photographs captivate the imagination. They draw you into the topic rather than repel. Indeed, this is exactly the kind of information that empowers people to try new things, that dispels the fears and anxiety, and that propels all of us to reconnect with our ecological umbilical.

I especially appreciate Storey and Danforth encouraging backyard butchery because it is exactly the kind of democratized, decentralized food system our country desperately needs. The opaqueness and centralization of America's food system, from mono-speciated factory farms to mega-supermarkets, has birthed a brand new lexicon of pathogenicity, toxicity, and ills. While self-empowerment makes food regulators shudder ("What, turn a bunch of novices loose with butchering animals in their backyards? Goodness, they'll kill themselves!") those of us who have done this for generations and encouraged others to do it realize the benefits for food safety, nutrition, and taste.

This book is yet another indication that the burgeoning local food tsunami continues to gain strength. It started with a resurgence of culinary interest and the Food Channel, then moved into urban farming and local produce, and now can include meat, which accounts for 40 percent of the grocery dollar. If we're ever going to move our food system to a place of regeneration, accountability, integrity, and transparency, we have to tackle the issue of how meat is produced. This book, titled simply *Butchering Poultry, Rabbit, Lamb, Goat, and Pork,* is a launch pad for this next step in healing our collective foodscape.

If we wait for the government or land grant colleges or big food corporations to change, we'll be waiting a long time. Our culture's assault on humane livestock rearing, sanitary slaughtering, and processing with integrity can be corrected quickest when thousands of people, empowered by simple instruction, take control of their own meat and return to the social, small-scale artisanship of our forebears.

Using the best understanding of microbes, the latest knowledge regarding muscle development, the most modern infrastructure from cooling to knives, Danforth opens a world of can-do that invites the most timid onlooker to participate in this dramatic farm-to-plate choreography. It's a world of profound sacredness — the sacrifice of life to sustain life. Perhaps few things can express ideas more clearly than personally taking up a knife and carving a carcass. While that may sound repulsive to some, for many of us, it speaks to a deep yearning, a primal call, to rediscover the foundations of human existence and the integrity of ecological cycles.

Because it is arguably the most nutrient dense of all foods, meat formed the basis of all ancient diets. In modern times, we have refrigeration, hot water, stainless steel, and efficient packaging materials that make this ancient art of butchering more efficient and safe. Preserving this tradition and explaining it to what is now a nation of novices, Danforth's gift pushes us forward culturally and personally. Now that we have thousands of homesteaders growing critters, it's time to encourage this legion of food participants to process them. Thank you, Adam, for showing us the way.

INTRODUCTION

T **HERE IS NO BETTER LOCATION** to harvest an animal than the land on which it lives. You're present with the animal, in a calm, familiar environment, during its final moments. Natural surroundings reinforce the normalcy of one animal's sacrifice for another's existence. Earth cushions the hard fall of a stunned animal, preventing bruising. Blood fertilizes the field.

Honorable harvesting prioritizes the well-being of the animal; the process resembles nothing of the horror stories — and horrific realities — coming from inside improperly operated abattoirs or industrial-size slaughterhouses. When I'm harvesting an animal, I feel right knowing that I have done everything I can to ensure a natural (albeit domesticated) existence and painless departure for any animal I process. This book will provide you with the same assurances.

Slaughtering an animal is not for everyone, and trepidation when beginning an education in this kind of work is to be expected. In fact, I would encourage it. A bit of uncertainty will make you slow down and will encourage you to deeply consider the importance of what is about to happen. Speed is for the machines. Act with intent and go slowly. Remember: preventing mistakes by practicing often and working with precision will increase efficiency more than working quickly will. At home there is no need to rush, either in raising the animal or in harvesting it. Slow growth yields a more flavorful meat and cautious processing yields a nourishing product.

For people raising animals for their own consumption and for those looking to purchase and process animals that have been raised locally, this book is the key to your food freedom. Everything you need to know in order to successfully — and respectfully — slaughter the most common species found on a farm is contained in these pages.

You'll learn exactly how to prepare animals for slaughter, how to set up a slaughtering and butchering area, how to select the tools and equipment you'll need to ensure a successful slaughter, and, most importantly, how to stun and bleed animals with the certainty that they are experiencing the least pain and discomfort in these final moments of their lives.

Ensuring the animal's well-being during the slaughtering process isn't just about the ethics of humane animal handling; it's also about producing quality meat. Chapter 1, which focuses on meat science, explains exactly why providing better care, and slaughtering with respect, produces the best-quality meat. When aberrations occur — an unsuccessful stun, punctured viscera, or damaged meat — there is comfort in knowing that you did what you could to avoid them. The slaughtering methods in this book focus on ensuring the animal is insensible to pain, not only for the animal's well-being but also for your own.

Knowing the ins and outs of slaughtering is just the first stage, though; the bulk of this book focuses on butchering the resultant carcasses. For each animal, there is a wide range of approaches to breaking down the carcass and options for cutting. It behooves us to maximize the usage of each and every carcass we process, not only for our own return on investment but out of respect for these formerly living creatures. The many different butchering methods covered in this book ensure that you will find a system that works for what you want to eat and how you like to cook. Your favorite cuts are definitely in here, as well as many others that you've probably never encountered that might become new favorites. And, of course, there is detailed information on the ideal butchering setup for each species, along with the best options for equipment and how to care for it.

The methods I demonstrate, using common livestock, will allow you to harvest any land animal, domestic or otherwise. All animals share similar anatomical structures that have allowed us to saunter, trot, sprint, and exist outside of the oceans. All birds are anatomically similar. Bears are just like big, furry pigs. A giraffe is nothing more than a cow with very long shanks and the world's greatest neck roast. Don't let the skeletal maps and scientific names of muscles send your mind spiraling; they're not critical for learning the skill of butchering. You will become a great butcher by committing shapes to memory, by understanding the intersection of muscles and the planes of connective tissue, by knowing the conformation of joints, and simply by making the same cuts over and over and over again.

An entire chapter is devoted to food safety and what we can do, as processors of a perishable and edible product, to ensure that the meat we produce is nourishing. The setup you devise at home may never be USDA certified, but the same rigorous dedication to sanitation and safe handling will benefit everyone who consumes the meat you produce.

It's deeply satisfying to know that, at my hand, an animal suffered as little as possible on its way to feeding myself and those around me. This kind of involvement also enables me to produce meat that conforms to my standards: quality and cleanliness are paramount. When mistakes happen, during slaughter or butchering, I recognize them and make every effort to learn from such incidents. I don't rush to the end result, be it a carcass in the cooler or a roast coming out of the oven. I revel in the intricacies of a process that extends back in time for generations. Slaughtering and butchering my own meat connects me to cultures, regions, and generations I've never known. We all share the need to eat, and having a hand in how another living being is transformed and contributes to my own existence is profound and cathartic. I dearly hope that you draw the same inspiration from your experiences and that the information in this book helps guide you along the way.

Adam Danforth

1

FROM MUSCLE TO MEAT

PRIOR TO LANDING ON YOUR PLATE, the meat that you choose to eat began life as muscle, a highly organized and complex living system. Muscles, which are made up of tissues and fibers, perform many of the voluntary and involuntary actions in a body. Each muscle has a unique structure that depends not only on function and position but also on species and environment. As muscles are transformed into meat they undergo many physical and chemical changes. These changes are initiated by death but are influenced by many factors, including, for example, how the animal was handled before it was slaughtered and how quickly the carcass was cooled after slaughter. These factors and the chemical changes they cause can have an enormous effect on the palatability of the final cuts of meat.

As it turns out, the better you treat an animal while it is alive, the better the meat from that animal is. To create delicious meat, you should understand not only the physiology of muscles but also the types of favorable treatment that enable the production of a high-quality product.

Muscles

MUSCLES ARE ORGANIZED STRUCTURES that enable movement. The heart, a muscle, pumps blood through the body; muscles move food through the stages of digestion; muscles in the legs allow an animal to stand and walk. Each of these functions, enabled by the contraction of muscle, showcases one of the three different types of muscles: cardiac, smooth, and skeletal. But first, we must explore the basic muscle structure.

Muscle Structure

Muscles are made up of cells called *fibers*; these are slender cylindrical structures that enable contraction. Muscle fibers are organized in bundles that are stacked together in one direction and bound by sheaths of connective tissue. Envision holding a bundle of dry spaghetti; the spaghetti is the muscle fiber, and your hand is the connective tissue. This pattern of spaghetti-style bundling continues for many levels, as bundles upon bundles are grouped together, level after level, with the final bundle completing the full muscle. The connective tissue holding the full muscle together is the *silverskin* (technically called the epimysium). Along with more connective tissue, the space between the bundles is filled with blood vessels and fat deposits. The visual grain patterns we recognize in meat are actually midlevel bundles called *fascicles*. These are most notable in cuts in which the grain is prominent, such as the flank steak.

At the most basic level of the muscular structure are *sarcomeres*, long threads of linked proteins organized into bundles. These threads initiate muscle contraction from inside the muscle fibers. The main two proteins in sarcomeres, *myosin* and *actin*, make contraction and relaxation possible. They're linked in an overlapping pattern that allows them to slide past each other. When the muscle contracts they overlap more, shortening and getting closer together, and when the muscle relaxes they overlap less, lengthening the long threads they make. These protein actions change the shape of muscles; this is evident, for example, when you move your leg or flex your bicep. In short, when a muscle contracts the action originates in the myosin and actin proteins. The action of these two proteins, shortening or lengthening, causes a chain reaction that repeats upward through every bundle: the fibers shorten, the fascicles shorten, and finally the entire muscle shortens for contraction, and vice versa for relaxation.

Muscle Fibers

Fascicles are the smallest muscle fiber bundle that we can easily identify with the naked eye and with the palate. The size of a bundle and its interior fibers plays a large part in how we experience meat. The larger the fascicle, the easier it is to see and the tougher it is to cut through. This gives us the advantage of being able to identify tenderness visually. Fine-grained muscles are more tender than coarse-grained muscles; thus a tenderloin is easier to chew than a skirt steak. The reason for this is that our teeth do a poor job of cutting through bundles of fibers; they are much more effective at separating them from one another. (Imagine trying to chop your way through a truckload of logs instead of just pushing the logs to one side or another.)

In addition, muscle fibers typically toughen during cooking, drying up as the heat ruptures water-holding structures and causes evaporation. The result is a denser, more resistant structure. This is the reason we cut meat across the grain rather than with it. Cutting with the grain would leave stacks of dense, lengthy fiber bundles that we would struggle to split with our teeth. Instead, we let a knife do the work of shortening the fibers so our teeth can do the job of separating them, an effort that in some cases takes ten times less energy than splitting.

MUSCLE FUNCTION AND AGE DETERMINE SIZE

The size of muscle fibers is partly the result of muscle function. The more power a muscle needs, the shorter and fatter the sarcomeres within the fibers. More power requires more contractile proteins (actin and myosin). This in turn requires stuffing more proteins into the same connective tissue casing. (Imagine, for example, filling a balloon to capacity with water.) When more proteins are created to produce more power, this causes the sarcomeres to fatten. As sarcomeres fatten, so do the fibers, the bundles of fibers, and the bundles of bundles throughout the entire muscle. Muscles requiring short, powerful bursts of energy — such as those responsible for an animal's fight-or-flight response in reaction to sudden danger — have the thickest fibers. One example of this is the breast muscle of birds that fly only when threatened, such as chickens. (You may be saying to yourself, "But the breast meat of a chicken is so tender." Muscle fiber size is not the only factor in determining tenderness; see page 12.)

Age also contributes to muscle strength and therefore fiber size. In general, the older an animal is the larger it gets, and the longer it has lived the more activity the muscles have experienced. An animal does not grow new muscle fibers; rather, the fibers increase in size as they develop more contractile proteins. The muscles require more strength to support the growing size of the animal; as the animal ages, increased activity promotes muscle expansion. Larger muscles need more strength, provided by an increase in contractile proteins (actin and myosin). The more proteins inside a muscle fiber, the denser and wider the fiber, and the tougher it is to chew. This is one reason why older animals have tougher meat.

Connective Tissue

Connective tissue is made primarily of *collagen*, a substance that accounts for about one-third of the protein in the entire animal. Collagen is concentrated the most in ligaments, tendons, bones, and skin. The other notable component of connective tissue is *elastin*, which is named for its elastic properties and provides some of the stretch that connective tissue needs in order to change shape and move with the muscles and other body parts.

CONNECTIVE TISSUE AND STRUCTURE

The structure of all tissues within the body, muscles included, is enabled by connective tissue. Muscle fiber bundles, and the bundles of bundles, are all wrapped by thin layers of collagen-rich connective tissue. Within these bundles, numerous strands of connective tissue fill the spaces between fibers. These strands weave themselves together, as in a tapestry, to form a complex structure. The strands are connected through a process called *chemical cross-linking*. The interior and exterior networks of a muscle's connective tissue all converge at either end to form tendons. When muscles contract, fibers tug on

Diagram of muscle structure

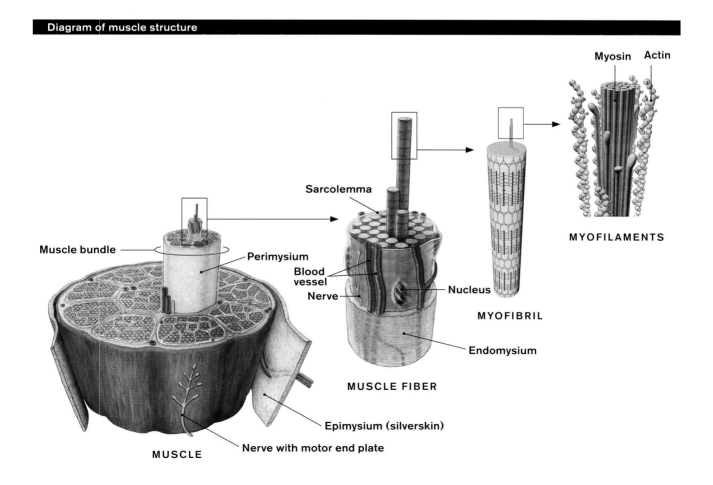

Myosin   Actin

MYOFILAMENTS

Sarcolemma

MYOFIBRIL

Muscle bundle
Perimysium
Blood vessel
Nerve
Nucleus
Endomysium

MUSCLE FIBER

Epimysium (silverskin)

MUSCLE
Nerve with motor end plate

their respective connective tissue sheaths, causing bundles of fibers to contract. Through a chain reaction across the bundles of bundles, the muscle pulls the tendons and causes skeletal movement.

COLLAGEN AND MUSCLE TENDERNESS

More than any other factor, the main property that governs muscle tenderness is the volume and strength of cross-links between collagen fibers. Just as with textiles, the more threads and connections you have, the stronger the fabric and the tougher it is to cut through. Many factors contribute to cross-link development, including not only the function of the muscle but also the animal's age, nutrition, and breed. The hardest-working muscles, and those that get the most exercise, require a dense network of collagen to provide adequate structure and functionality. Density is achieved through the development of intense cross-linked collagen fibers. As a rule, the closer to the ground a muscle is, the harder it works to provide support to the body. This is illustrated in the copious amounts of collagen found in meat from the lower limbs of all animals, including beef shanks, ham hocks, and chicken drumsticks.

BREAKING DOWN COLLAGEN BY COOKING

Fortunately, collagen and its cross-links can be broken down into gelatin through the application of heat and water, a process called *hydrolysis*. *Gelatin* is the sticky, unctuous substance that helps thicken liquids for sauces or desserts and provides the adhesive for traditional glues. In contrast to muscle fibers, which get drier and denser when cooked, collagen softens during a proper stewing, helping to turn otherwise tough cuts of meat like a beef shank into a succulent result. As a general rule, the tougher the collagen and the more cross-links it has developed, the longer it will take to break it down into gelatin. Thus, meat from an older animal will need more moisture and time to hydrolyze than meat from a younger animal of the same species.

Hydrolysis of collagen begins as the temperature rises above 122°F. The higher the temperature, the faster it happens. However, while higher temperatures increase the rate of hydrolysis, there is a trade-off. Once the temperature rises above 140°F the collagen also begins to shrink. The shrinkage begins to squeeze on the muscle fibers, causing them to expel liquid. The process is similar to twisting a wet towel: the more you twist, the more water flows out. The higher the temperature, and the quicker it rises, the faster and tighter collagen strands twist and squeeze out the moisture contained in muscle fibers. Hence, hydrolysis that occurs too quickly results in dense, dry meat.

Take a pork shoulder, for example. A pig's shoulder is heavily worked and therefore chock-full of extensively cross-linked connective tissue. Cooking this cut for a few hours at a high temperature, in moist or dry heat, will produce meat that is dense, dry, and a struggle to chew: the collagen has shrunk, squeezed out the liquid, and not been given adequate time to hydrolyze. Slow-cook it at a low temperature for many hours, and the muscle fibers will fall apart into the threads of meat characteristic of pulled pork: the collagen has been fully hydrolyzed, and the structure holding the fibers together turned into gelatin. With a longer cooking process, the transformation of collagen into gelatin provides a better mouthfeel. Slow-cooked meat has still been squeezed by the collagen, though, so it will still benefit from the application of moisture such as a barbecue sauce.

Allowing time for hydrolysis is pertinent only when dealing with tough cuts of meat in which there is a substantial amount of connective tissue. Tender cuts have weak collagen and in small amounts. The generally preferred method for tender meat is quick cooking because there is not enough collagen present to make chewing difficult. Further, keeping the internal cooking temperature of tender cuts from many animals to 140°F or lower avoids collagen shrinkage and the resultant moisture loss. This is why a lamb steak cooked to medium-well at 150°F or higher will be a denser, drier version of the same steak cooked to a 133°F medium-rare.

Collagen fiber

Young tropocollagen strands

Old tropocollagen strands

Tenderness is largely the result of cross-links between tropocollagen fibers within collagen; as the animal ages, more cross-links are formed, making the meat progressively less tender.

AGE AND ITS EFFECT ON COLLAGEN

As an animal ages, the volume of collagen decreases but the strength increases. Aging of the muscle causes the development of more chemical cross-links between the collagen fibers that remain. Muscles are exercised, fibers increase in density and girth, and the collagen fibers respond accordingly by increasing tensile strength through the addition of cross-links.

To avoid harvesting meat with tough collagen, those who raise animals for the meat industry slaughter most of those animals before they reach adulthood. For example, consider the difference between an eight-week-old broiler chicken and a two-year-old laying hen. The tender broiler, with its relatively weak collagen cross-linking, is suited to any kind of quick cooking; the layer, on the other hand, requires long, slow cooking in order to break down the strong collagen fibers.

Fats

Along with fibers and collagen, fat plays a distinctive role in our experience of consuming meat. Fats happen to be a unique form of connective tissue, primarily serving three purposes, in some cases simultaneously: to insulate the body, to protect the body and the internal organs, and to store energy. The latter function is responsible for many of the flavors that we associate with meat. To increase fat coverage, an animal does not add new fat cells but rather increases the volume of the cells already there.

FAT CELLS AND TASTE

Fat cells store energy in the form of fatty acids but also act as a repository of any substance that is fat-soluble. (Just as salt is water-soluble, any compound or substance that will dissolve in fat is fat-soluble.) So, while an animal gathers energy from its food, it also stores other fat-soluble compounds from the food within the fat cells. Which compounds are stored depends largely on species and diet, while the concentration of those compounds is mainly a result of age. The older an animal is, the more time it has spent storing fat-soluble compounds in its fat cells and the more flavors and flavor-enhancing components are released during cooking. This accounts for the typically stronger flavor and aroma of meat from older animals, as is the case with mutton.

An animal that is raised primarily on pasture, relying on a varied diet of foliage, both fermented and fresh, will process and store a diverse array of organic compounds and fatty acids. Upon cooking, these assorted odorous substances will strengthen the flavor of the meat. In contrast, an animal reared with a diet composed primarily of grain will have less diversity in its fat stores. It is for this reason that meat from grass-fed and pasture-raised animals has a stronger flavor than meat from grain-fed animals.

SATURATED AND UNSATURATED FAT

Within the world of animal fats there are two main categories: saturated and unsaturated. Fats are composed of carbon atoms linked together in chains. These carbon atoms like to bind with hydrogen, and in the case of a saturated fat, the carbon atoms bind with as many hydrogen atoms as possible. They are literally *saturated* with hydrogen bonds. Unsaturated fats are not. Instead, one or more carbon atoms are double-bonded to each other. Monounsaturated fats have a single double bond; polyunsaturated fats have more than one double bond. This double bond

TYPES OF FAT DEPOSITS

Within the body of an animal, there are four types of fat deposits:

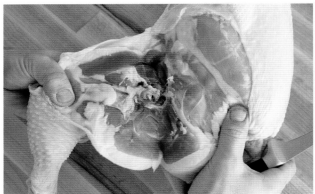

SUBCUTANEOUS FAT lies under the skin, as seen in this chicken.

VISCERAL FAT like this leaf lard lies inside the body cavity. It also surrounds kidneys and other organs (caul fat, for example).

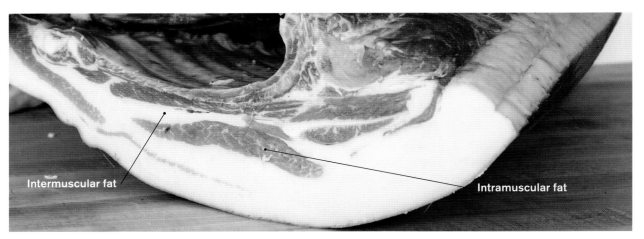

Intermuscular fat

Intramuscular fat

INTERMUSCULAR FAT is intermingled between muscles (as seen in this pork shoulder).
INTRAMUSCULAR FAT, or marbling, is interspersed within the connective tissue and fibrous bundles of muscles.

adds one or more kinks to the chain, changing its shape from clean and organized to a bit awry.

If you're good at organizing, packing a car, or stacking boxes, you know that things of a consistent shape fit tightly together. This is the case with saturated fats: the chains of evenly bonded molecules stack tightly together, forming stable fats that are solid at room temperature. Once there is a kink in the chain, those chains can't stack so closely, thus preventing them from forming tight, stable structures, often making them liquid at room temperature; the more kinks, the more unstable the structure.

Animal fats contain mainly saturated fats, making them solid at room temperature. But the amount of unsaturated fats certainly comes into play. Chicken and pork fat have higher levels of unsaturated fats than beef, sheep, and goat fat, making them less solid at room temperature. (Vegetable fats, like olive and canola oils, are mostly unsaturated, and that's why they're liquid.)

These kinks in the molecular chain affect not only the solidity of the fats but also the rate of rancidity. A double bond in the chain exposes carbon atoms to oxidizing elements, like oxygen and water. These elements

TYPES OF MUSCLES

The three types of muscles are defined according to both function and form:

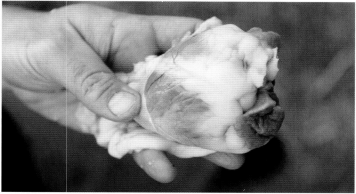

CARDIAC MUSCLE is found only in the heart. It's striated and involuntary.

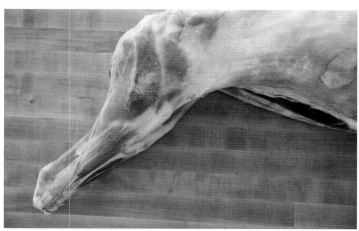

SKELETAL MUSCLE is attached to the skeleton and enables its movement and is the predominant type in the body, as well as in butchery. It's striated and voluntary.

SMOOTH MUSCLE comprises all the remaining involuntary muscles such as the walls of blood vessels, the gastrointestinal system, and the stomach lining. It's nonstriated and involuntary.

Because poultry and pork have higher levels of unsaturated fats, their carcasses can't be aged as long as beef, sheep, or goat carcasses can, and the cuts from these animals tend to spoil sooner, whether fresh or frozen.

react with the exposed carbon atom, often disrupting the chain, causing it to break apart. This is the basic action of rancidity or oxidation — the breakdown of molecular chains into smaller fragments. The more kinks in the chain, as with polyunsaturated fats, the more susceptible they are to rancidity. The more unsaturated fats in the animal, the more prone the carcass and its resultant cuts are to rancidity and oxidation. Because poultry and pork have higher levels of unsaturated fats, their carcasses can't be aged as long as beef, sheep, or goat carcasses can, and the cuts from these animals tend to spoil sooner, whether fresh or frozen. (Liquid fats have a tendency to go rancid even more quickly, especially nut oils that are high in polyunsaturated fats. This is why refrigeration of these oils is always recommended.)

FAT AND MEAT TENDERNESS

Fat is also a contributor to tenderness, but marginally so compared to fibers and collagen. Within muscles, the presence of intramuscular fat between bundles and fibers disrupts the mesh of collagen fibers and cross-links, making holes in the otherwise taut tapestry. This helps to weaken the solidity of a muscle's connective tissue, increasing tenderness when we chew. Unlike fibers,

which dry out, fats also melt during cooking. Melted fats turn to liquid, adding a necessary moisture to the drying fibers. This also lubricates the fibrous bundles, aiding our teeth's efforts to separate them, resulting in a more tender bite. Additionally, as fats melt under heat they release the aromatic fat-soluble compounds stored within the cells, contributing to the olfactory experience and increasing the perception of flavor.

Types of Muscles

The function of a muscle can be either *voluntary* or *involuntary*: voluntary contractions are performed with intention (the movement of an arm or the focusing of eyes), whereas involuntary contractions happen without conscious control (the drawing in of air to breathe or the pulsing of a blood vessel). The form of a muscle can be *striated* or *smooth*. Striated muscles contain parallel fibers, organized side by side, providing the grain and appearance we associate with meat. Smooth muscles have fibers organized in complex sheets and are associated with blood vessels, organs, and other internal functions.

Types of Muscle Fibers

The main function of muscles is movement, and animal movement can be separated into two basic types: quick bursts or slow and steady. The different types of movement are achieved by two types of muscle fibers: a fast-twitch fiber, which provides the sudden contraction needed for bursts of energy, and a slow-twitch fiber, which provides the endurance essential for sustained activities like standing or chewing.

FAST-TWITCH FIBERS

Fast-twitch fibers are strong, as their contractions require power to achieve the appropriate speed of action. They are short, fat, and hard to chew, in accordance with the aforementioned effects of strength on fiber size. Moreover, fast-twitch fibers do not need the energy from fat stores to operate, making the muscles dominated by these fibers very lean. Yet fast-twitch muscles are called into action only periodically, which makes for a sparse and weak network of collagen. So while the fibers are harder to chew, with meager fat, the muscle overall is in most cases considered relatively tender. A well-known example is the chicken breast, a tender, lean muscle composed of fast-twitch fibers that illustrates how connective tissue is the most influencing factor in tenderness.

SLOW-TWITCH FIBERS

Slow-twitch fibers are long and thin, requiring less power but more endurance than fast-twitch fibers. Their stamina is provided through a combination of extensive connective tissue and large intramuscular fat stores. The connective tissue provides the rigidity and support required to sustain lengthy periods of contraction and activity, while the fat stores provide the energy necessary to operate for extended periods. Muscles with slow-twitch fibers are normally considered tough. Despite the tender narrow fibers and the generous fat stores, the abundant connective tissue is the determining factor creating chewiness. Hardworking muscles in the shoulders of quadrupeds or in the legs of poultry are typical examples of how the concentration of connective tissue determines tenderness. Under the proper cooking conditions these muscles

Chicken breasts are an example of lean, fast-twitch fibers.

Chicken legs, and their slow-twitch fibers, are deep in color and require connective tissue and fat deposits for endurance.

provide some of the most unctuous results. The exception to this is a muscle like the tenderloin. It is a rare combination of typical slow-twitch muscle fibers, but since it is rarely used in the body of quadrupeds, it has very little connective tissue, making it the most tender of all cuts.

Each muscle in the body is a mixture of fast- and slow-twitch muscles. Fast-twitch muscles are distinctly paler than slow-twitch muscles, resulting in varied colorations within the muscles of an animal. An easy example is the typically white chicken breast and the dark-colored thighs. The breast, comprised mostly of fast-twitch fibers, explodes with energy when the chicken takes a sudden but short flight. The legs, which are slow-twitch dominated, require the stamina for daylong posture and movement.

As a general rule, muscles nearer to the skin are expected to respond less frequently than those farther away and are therefore lighter in tone and leaner in composition. Muscles residing closer to the skeleton are the ones handling the bulk of the posturing and movement; these muscles will be darker in tone and contain higher concentrations of slow-twitch fibers, connective tissue, and fat. So, while that chicken breast may be lightly colored, lean, and tender, the absence of connective tissue and fat means that it can easily be turned into a dense, dry cut, a lackluster comparison to a properly cooked chicken leg in which the collagen has been hydrolyzed into gelatin and the fats have flooded the meat with fragrant and tasty compounds.

The Color of Muscles

The coloration of muscles does not, as some people think, come from blood. In fact, there is no blood within the muscle fibers themselves. Muscles do need blood to operate, though, and vessels running among the connective tissue and fat that sits between the fibrous bundles carry blood into the muscles. Muscle fibers need oxygen, and to get it from the bloodstream to the fibers requires a local delivery system of sorts. In comes *myoglobin*, a protein capable of making the trip between the bloodstream and the muscle fiber while carrying a load of oxygen. While an animal is alive, oxygen exchanges are constantly happening according to the demands of a muscle. The more a muscle works, the more oxygen it needs and the more myoglobin is there to make it happen.

THE THREE STATES OF MYOGLOBIN

Myoglobin happens to be the main party responsible for muscle coloration and is able to change color according to its state. Myoglobin has three states: deoxygenated (not

DEOXYGENATED. The absence of oxygen within a vacuum-sealed package results in the purplish hue of deoxygenated myoglobin.

OXYGENATED. Red meat "blooms" to its characteristic red coloration as myoglobin bonds to oxygen in the air.

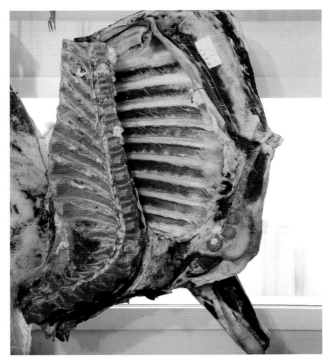

OXIDIZED. Extended exposure to the elements results in metmyoglobin, causing red meat to shift to an unattractive brownish-red hue.

carrying oxygen), oxygenated (carrying oxygen), and oxidized (exposed to external elements).

DEOXYGENATED. This first state occurs when myoglobin is not holding on to any oxygen. After an animal is dead there is no more circulation of oxygen-rich blood and no need for muscle fibers to operate: no more oxygen for the myoglobin to carry around. So the default state of myoglobin in meat is the deoxygenated state, the result of the myoglobin's last delivery of oxygen to an eager muscle fiber. Meat in this first state is purplish-red. This may sound familiar to anyone who has sliced open a raw steak or separated a bunch of ground beef to find the interior tinged with a purple hue; red meat cuts that have been vacuum sealed also tend to show deoxygenated coloration. What you see is myoglobin without access to oxygen — that is, until you give it access to the air around it.

OXYGENATED. The second state of myoglobin occurs when it picks up some oxygen. When myoglobin is oxygenated it is called oxymyoglobin, which is bright red. Oxygenation can happen from anywhere oxygen resides: living muscles gather oxygen from the bloodstream; meat picks up oxygen during exposure to air. Myoglobin binds with oxygen whenever it has the opportunity, so when you cut open that raw steak, break apart a pile of ground lamb, or open a vacuum-sealed packaged of red meat, the hue shifts from purplish-red to cherry red, a process known as "blooming."

OXIDIZED. Myoglobin may like to carry around oxygen, but it is not all that great at holding on to it. The bond between myoglobin and oxygen is highly unstable and susceptible to being broken through exposure to things like bacteria, enzymes, light, or even more oxygen. When oxygen breaks away, it often steals an electron, leaving myoglobin in an oxidized state called *metmyoglobin*. Metmyoglobin is brownish-red and responsible for the unattractive hue of meat left in compromising conditions for too long — on the counter, in the fridge, under heat lamps — but it does not always indicate spoilage or rancidity.

No muscle is exclusively one type of fiber, so the shades of muscle color are a result of the mixture of slow- and fast-twitch fibers. This is easier to identify within sedentary animals, which have higher concentrations of fast-twitch fibers and intermittently used muscles, making the darker hues more evident. Pigs are not grazing animals and move much less than lambs do; hence, pork is lighter in color than lamb. Within grazing animals and other animals whose bodies are in constant motion, a third type of muscle fiber develops: the intermediate fiber.

> The size of the animal, its age, and its demands for movement account for some of the difference in the volume of myoglobin among species. Beef, which comes from large animals that cover a great deal of ground, has more myoglobin than either pork or lamb, making it exceptionally red.

The intermediate fiber is a fast-twitch fiber with aerobic capability; it therefore needs oxygen and contains myoglobin. A combination of intermediate and slow-twitch fibers mainly composes the muscles of red meat animals like cows, lambs, goats, and deer. It is this overall coloration of red meat that makes differentiation between slow-twitch and fast-twitch fibers difficult to discern in many species.

The size of the animal and demands for movement also account for differences in the volume of myoglobin among species. Beef, which comes from large animals that cover a great deal of ground, has more myoglobin than either pork or lamb, making it exceptionally red. This is also evident when you look at the muscle-color differences between a chicken, a largely sedentary bird that does little flying, and a duck, which migrates twice a year and spends most of its time moving in water. Again, the breast muscle is easiest to compare: the chicken breast is lightly colored and dominated by fast-twitch muscles, whereas the duck breast is dark red and dominated by intermediate and slow-twitch muscle fibers. Finally, as an animal ages the volume and the concentration of myoglobin increase, resulting in richer coloration.

Cured meats retain their attractive red hue because of a strong bond between myoglobin and nitric oxide.

COLOR AND THE FRESHNESS OF MEAT

The final color of meat is a mixture of purple, red, and brown — the colors of myoglobin molecules in all three states. Shifts between states, caused by oxygen handling and electron theft, are constantly happening, but the dominant condition will determine the color. A vacuum-sealed beefsteak will be purplish because there is not much oxygen inside the bag. Open it up, though, and the steak will bloom to red within a few minutes as the myoglobin is exposed to the oxygen in the air. If the meat is left out longer, the myoglobin loses hold of the oxygen and turns brown. We have come to associate color with freshness, as with the bright red of beef, but in many instances the presence of brownish hues is not an indication of rancidity or spoilage.

PRESERVING THE COLOR OF MEAT

There are two conditions in which we intentionally interfere with myoglobin to produce desirable colors under undesirable conditions: when we cure meat, and when fresh meat is packaged for sale.

CHEMICAL PRESERVATION. Cured products often use nitrites, a chemical added to meats to prevent bacterial growth and control the final color. During the curing process, nitrites convert to nitric oxide, a compound that can bind with myoglobin, effectively taking the place of oxygen. This turns myoglobin pink and is why bacon, hams, and other cured meats have a pinkish hue. The myoglobin's bond with nitric oxide is much more stable than its bond with oxygen, so the color has better staying power.

PRESERVATION WITH CARBON DIOXIDE. Bright red meat sells better than brown meat does, but keeping it red is a challenge. (Remember, myoglobin loses oxygen easily and turns brown.) To keep meat red, processors pump carbon dioxide into packages of meat for sale. Carbon dioxide reacts with myoglobin in the same manner that nitric oxide does: it replaces oxygen, forms a stable bond, and turns the myoglobin reddish-pink. The color will not be as pronounced as the cherry red of the true oxymyoglobin, but it will certainly look better than the metmyoglobin brown and is therefore able to stay appealing in a meat case for a longer period.

Turning Muscles into Meat

TRANSFORMING MUSCLES INTO MEAT is a complex process that involves more than just proper slaughter and butchering. Fortunately, the same conditions that are most humane for the animal also produce the highest-quality meat products. How the animal is reared, fed, handled during transport, and treated prior to slaughter all play significant roles in the resulting meat. So do many postmortem conditions like storage, humidity, temperature, and especially time.

The effect of death on a body is nothing short of epic. The sudden standstill of the circulatory system halts all oxygen delivery, leaving aerobic cells without sustenance and causing immense cellular damage and chemical transformation. Enzymes and bacteria freely roam the defenseless carcass, mercilessly breaking down structural components. Fatty and amino acids are left to disintegrate. It may seem chaotic, but under proper conditions this natural process transforms previously living tissue into palatable meat.

The First 24 Hours Postmortem

The events that occur in the first 24 hours after death can determine the quality of all meat coming from a carcass. Thankfully, the conditions that can cause those events are largely within our control. Living muscles are constantly using energy, typically in the form of glycogen. When muscles process glycogen, the by-product is lactic acid. The more a muscle works, the more lactic acid is produced. (Lactic acid is responsible for the muscle burn associated with intense exertion.) Oxygen from the circulatory system maintains the level of lactic acid, removing it when too much accumulates. After death, muscles continue to generate lactic acid, but with no blood circulation the acid builds up, changing the pH level from a nearly neutral state (around 7.0) to a slightly acidic state (near 5.7). This is beneficial, as the mild acidity retards microbial activity, including that of enzymes and bacteria. In addition, the drop in pH causes proteins to slightly unravel (a process called denaturing), releasing some fluid and further moistening the meat.

RIGOR MORTIS

Muscles contract and relax in response to chemical reactions. Under normal conditions, the drop in pH is a slow decline over the course of many hours. Muscle fibers continue to contract and relax after death, but as the pH drops the chemicals that allow the fibers to relax become sparse. Eventually the muscles contract and never relax again, causing the stiffening known as *rigor mortis*. There are three phases to rigor mortis:

- **THE DELAY PHASE** — muscles continue to contract and relax after death.

- **THE ONSET PHASE** — muscles begin to lose the ability to relax and start permanently contracting.

- **THE COMPLETION PHASE** — all muscles have tightened to a fully contracted state.

Rigor mortis effectively exhausts muscles of their ability to contract and relax; after rigor, muscle functioning is completely finished. But, to become palatable meat, more time is needed and more chemical reactions must take place.

The effects of rigor mortis are counteracted by enzymes and time. Once an animal dies, enzymes are let loose, and with nothing regulating their activity, they begin to attack at random. Their targets include

Soon after death, muscles begin to permanently contract into a state known as rigor mortis.

the contractile proteins that keep muscle fibers in a contracted state (remember actin and myosin from page 6). Given enough time, these enzymes will have done enough dismantling to undo the effects of rigor mortis, producing tender meat. This is all part of the process of aging, discussed in detail below.

The Effects of Animal Stress on Meat Quality

Although normal changes in pH levels are advantageous, aberrations in pH levels can cause unsavory effects. Stress is the main culprit. Animals exposed to stressful conditions for extended periods prior to slaughter (e.g., uncomfortable temperatures, jarring noises, unstable surfaces for standing, confrontations in the pen) will become exhausted. After they die, their muscles will not produce enough lactic acid to induce the proper drop in pH levels, resulting in a high final pH (above 6.0, depending on the species). Meat with such a high pH is referred to as dark, dry, and firm (DDF) or "dark cutter." Without the acidity needed to curb microbial activity, spoilage quickens: proteins tighten, rather than denature, and strengthen their bond with water, causing the firm texture and dryness. Red meat is especially susceptible to DDF, with most cases occurring in beef.

Similarly, a sudden fall in the pH of muscles after death also results in poor-quality meat. When an animal is threatened, surprised, or excited its fight-or-flight response results in a flood of adrenaline entering its muscles. If this happens within a half hour prior to slaughtering — for example, because of fighting with other animals, being mishandled, or experiencing loud noises on the kill floor — the effects of the adrenaline will continue to persist after death.

An adrenaline rush spurs accelerated processing of glycogen and increased lactic acid production. A sudden shift of pH from neutral (7.0) to acidic (less than 5.8), when combined with warm muscles, causes the muscle fibers to unravel excessively. The effects are numerous, but the outcome is called pale, soft, and exudate (PSE) meat. In short, it's wet, mushy meat that dries out easily. As the muscle fibers unravel, they lose much of their ability to hold water, resulting in excessive moisture or drip loss and a compromised structure; at the same time, the myoglobin changes in structure and reflects light, causing a paler-than-normal color. PSE meat caused by pre-slaughter stress can come from any animal, but the most common occurrences are in pork because of the genetic disposition of pigs.

The Importance of Proper Storage after Slaughter

Production of high-quality meat requires careful planning and attention through each stage of life, slaughter, and storage. Even when all preceding conditions are carried out to the ideal, improper storage procedures have the potential to compromise the final quality of meat. Carcasses are chilled soon after slaughter to prevent the growth of dangerous microbes that can cause spoilage and illness; the quicker the chill, the less bacterial growth.

Chill the carcass too quickly, though, and the consequence is incredibly tough meat caused by a process called *cold shortening*. This occurs when the temperature of the meat drops below 59°F before the onset phase of rigor mortis. At temperatures below 59°F, muscles contract to abnormal extremes. This causes shortening of the muscle fibers (and therefore the muscles), in some cases to less than 50 percent of the original length. Therefore, when the final contraction of rigor mortis sets in, the result is a magnified tightening of the muscles, causing irredeemable toughness of the meat.

A similar condition, called *thaw shortening*, happens when meat from a freshly slaughtered carcass is frozen prior to the onset of rigor mortis. In this case the meat never goes through rigor mortis, therefore never exhausting the muscles of their ability to contract. When the meat thaws, the muscles come back to life (in a manner of speaking), contracting in the same manner as with cold shortening and ending with similar results.

Aging

PALATABLE MEAT IS THE RESULT of properly aging the carcass to counteract the effects of rigor mortis after the animal has been responsibly slaughtered. There is a sliver of time after slaughter in which the muscles are still relaxed and, if cooked, may provide a tender bite. This period is impossibly short for small animals like chickens and rabbits; for larger ones, the result of cooking the meat at this stage would be a watered-down, mildly flavored version of what you would expect. Soon after this sliver of time, rigor mortis sets in; attempting to butcher, cook, freeze, or do anything else with the meat during the period of rigor mortis would result in total failure. At this point the opposing muscles (e.g., hamstring/quadriceps, bicep/triceps) in the carcass are contracting, ending in the unusual posture associated with stiffened roadkill. The contractile proteins actin and myosin have permanently bonded to each other. This is where aging comes into play: meat will tenderize and improve itself given the right conditions and proper time.

Enzymes at Work

Aging does many things, but its main effect is to undo the results of rigor mortis. After death, naturally occurring enzymes within the meat go rogue, since the body systems that kept them in line shut down. The main agents are two types of enzymes called *calpains* and *cathepsins*. They indiscriminately attack proteins and, in a process called *proteolysis*, break them down into fragments, disrupting the structures responsible for keeping muscle fibers in a contracted state.

Given time, the enzymes dismantle enough contractile structures in the muscle to undo the structure of contraction and the effects of rigor mortis, essentially relaxing the muscles and increasing tenderness. Moreover, the cathepsins take apart cross-links and fibers within the connective tissue, another benefit for tenderness. The result is collagen that hydrolyzes more easily during cooking, producing more gelatin and thereby increasing succulence. The weakened collagen structures also prevent excessive squeezing on muscle fibers during cooking, stemming excessive moisture loss. The by-product of all this enzymatic tenderization is a wide range of broken-up proteins and molecules: a great thing for our palate. Formerly bland fats, proteins, and other molecules are all transformed into fragmented and intensely flavored compounds. These account for many of the sweet, savory, and aromatic attributes associated with aged meat.

The longer a butcher waits to process the meat, the more structures collapse from enzymatic activity and the more tender the product. The meat is quite literally rotting, but under very controlled conditions. The bustle of enzymes increases with temperature, doubling every 18°F, but the development of harmful microbes also spikes. So, while we could speed up aging by raising the temperature, it behooves us to bide our time and keep the meat between 32°F and 38°F. At these temperatures, the enzyme mayhem continues, albeit sluggishly, and the meat is safe. If the temperature of the meat goes below 28°F, which runs the risk of freezing, proteolysis slows down to an inconsequential crawl.

> The longer a butcher waits to process the meat, the more muscle structures collapse from enzymatic activity and the more tender the product.

Hanging carcasses in a temperature-controlled environment allows for extended aging.

HOW LONG TO AGE

All animals benefit from controlled aging to counteract rigor mortis, though the length of the aging process is different for each species and is further influenced by ambient temperatures, airflow, humidity, and personal taste. General rules are that smaller animals age more quickly than larger ones, younger animals age more quickly than older ones, and white meat ages more quickly than red. Animals left skin-on will benefit from a period of aging within an airflow-controlled environment. This will help the carcass "set up," tightening the skin and reducing some of the surface moisture. Leaving the skin on is especially helpful with poultry intended for roasting.

TYPE OF MEAT	AGING TIME	NOTES
Chicken	1 to 2 days	
Turkey (domesticated)	1 to 4 days	Larger birds will benefit from more time.
Duck (domesticated)	1 to 3 days	Aging time varies according to personal preference.
Rabbit (domesticated)	1 to 2 days	
Pork (skin-on)	3 to 5 days	Wet aging* should be used for longer periods, in order to avoid development of rancid flavors in unsaturated fats.
Pork (skinned)	2 to 3 days	The exposed fat of the carcass will oxidize much more quickly than that of a skin-on carcass, so aging needs to be limited. Wet aging should be used for longer periods.
Lamb and goat (under 12 months)	3 to 5 days	Individual loins and legs can be aged for a much longer period, up to 21 days. Lean carcasses can be draped with cheesecloth, muslin, or caul fat to prevent excessive drying.
Sheep and goat (over 12 months)	3 to 10 days	See above.
Veal	10 to 14 days	Lean carcasses can be draped with cheesecloth, muslin, or caul fat to prevent excessive drying.
Beef	10 to 28 days	Middle-meat primals can be aged for several additional weeks, according to personal preference.**

*See page 21 for more information on wet aging.

**Middle-meat beef primals — rib, short loin, and sirloin — can continue to develop past other primals. The decision to age anything for more than 28 days, however, will be based on flavor and texture development rather than tenderness, as the enzyme action is minimal at that point.

Aging in the Open

All aged meat will increase in tenderness, but there are other beneficial repercussions, depending on airflow and the ambient humidity of where the meat is stored. In one method, called *dry aging*, water evaporates from the meat, sometimes reducing the original weight by as much as 20 percent. With the water gone, the muscle fibers shrink, and so does the overall size of the meat. This also concentrates the tasty, water-soluble protein fragments, strengthening the flavor of the meat.

During the dry-aging process, meat is kept at the proper temperatures while humidity and airflow are controlled. Humidity is held at 70 to 80 percent, allowing the meat to dry out gradually. If the humidity is too low, the meat will lose moisture too quickly, resulting in dried, unpalatable meat; if the humidity is too high, moisture remains on the meat surface, promoting rancidity and microbial development. Air circulation is also critical to maintaining humidity equilibrium and promoting evaporation. To allow air access to all parts of the meat, meat processors usually hang carcasses from rails and place cuts on perforated shelves, while high-velocity fans work to keep the currents continuous.

A dry-aged carcass or primal cut will have a hardened, blackened exterior that is very likely to be dotted with patches of white mold. All mold patches must be removed and discarded, exposing the underlying nutty, aromatic meat. Between the loss of meat from trimming and the loss of weight through evaporation, the edible portion of a dry-aged primal may be 70 percent of its original weight. This makes dry aging an expensive process: it requires equipment, ample space, and lots of time for hanging and trimming, and it ends with a considerable loss of salable weight. Yet the result, with its unique taste, will fetch high prices and yield flavorful results, making up for the product loss.

Aging in a Bag

These days, dry aging is rarely done within the commercial meatpacking industry. Carcasses are typically hung for the minimum amount of time to allow rigor mortis to resolve, after which they are broken down into primal cuts. Primals are then vacuum-packed and shipped to customers within refrigerated containers.

While the meat is bagged and in transit, the enzymes do the work of dismantling proteins and tenderizing muscle. Upon arrival to a customer, bagged primals can continue to be stored and aged further or butchered into cuts. This approach is called *wet aging*, and the results in tenderness are pretty much the same as in dry-aging. In wet aging, the meat is aged within a hermetically sealed environment, staving off microbes and preventing any oxidation or moisture loss from the product. There is no need for controlled humidity or airflow, just temperature, so equipment and space costs are lower. Furthermore, there is minimal loss of product, thus maximizing salable weight. For these reasons, the adoption of wet-aging has been widespread within the commercial industries.

The downside is the resulting lack of flavor enhancement or development. The meat ages in a bag, spending days or weeks sitting in a collection of its own juices and blood. It picks up the flavors of these juices and blood: *serumy*, *metallic*, and *irony* are all adjectives used to describe the profile of wet-aged meat. Despite this, the benefits of minimal weight loss and convenient handling have made wet aging the standard in modern meatpacking, an industry focused on speed and volume.

FOOD SAFETY

THERE ARE MANY POSSIBLE REASONS that may have motivated you to begin slaughtering animals — animal welfare, cost reductions, custom cutting, to name a few — but one you may not have considered is the reduction of exposure to "industrial-strength" pathogens. Increased amounts of antibiotics in livestock, used for a myriad of reasons including promoting growth and preventing infections (rather than treating infections), have given rise to new strains of pathogens. Many of these new strains are showing resistance to the drugs we use to combat them. As the pathogens move from animals to humans, through our consumption of infected meat, their drug resistance poses a serious risk to all of us in the general public.

Moreover, the need to meet the demand for feeding billions of people has motivated agriculture industries worldwide to rely on more and more automation and technology. The rise of automated systems has made the potential for spreading foodborne pathogens greater than ever. Ground beef, one of the products most susceptible to contamination, serves as a good example. A single pound of the stuff, sourced from your local supermarket, may come from hundreds (if not thousands) of different animals. Within these types of systems a single contaminated animal can cause widespread havoc. The consumption of such contaminated foods leads to infection, and the virulent strains responsible for these infections are becoming harder and harder to combat as more and more show signs of drug resistance.

Cleanliness during processing is paramount, not just for industrial meatpackers but also for home butchers. As butchers, we can help prevent transmission of foodborne pathogens, starting with mindfulness toward cleanliness. More than half of all foodborne illnesses in the United States result from contamination within the household, so it behooves us to take the appropriate steps to prevent contamination at home and in our work. Education and awareness are key to food safety, and as a home butcher you must make a commitment to reducing the potential for contamination.

Pathogens and Contamination

THE MOST FUNDAMENTAL CONCEPT to understand about food safety is that all food, without exception, is contaminated. No food is naturally sterile. No matter what we do, everything we eat is crawling with microbes. It just so happens that most of those microbes are not harmful to us, and the ones that are harmful may not be pervasive enough to overcome average immunity defenses. Our job, as butchers, is to do whatever we can to reduce contamination. The most powerful measure we can take is also the easiest: washing our hands. We'll cover more details later, but in essence, if you do nothing else to stem contamination (though you should take other steps as well), wash your hands; wash them often and wash them thoroughly.

The vast majority of foodborne illnesses stem from contamination caused by the people who have handled the food. For meat, contamination starts at the slaughterhouse. In most cases the muscles of an animal can be considered sterile prior to handling. With few exceptions, contamination is limited to the surface of the meat. The interior of a muscle is sterile until it's punctured or ground. The moment we cut into a muscle, we introduce

surface contaminants; this may occur during cutting, tenderizing, probing for temperatures, or other kinds of meat handling. Simply put, everything we do escalates the level of contamination in the food. It is up to us to minimize this escalation.

Microbes, Good and Bad

Microbes are everywhere: on our desk, on our food, on our skin, even existing symbiotically inside our bodies. Most microbes are benign; some are even beneficial, like those residing in our digestive tract that aid in nutrient absorption. Microbes that pose a health risk to humans are considered pathogens (i.e., disease-causing agents); diseases contracted from food are caused by foodborne pathogens. Some pathogens can cause infection with only a single entity, as is the case with parasitic worms, while others require a colony of microbes to make an impact. We often consume small amounts of foodborne pathogens, mostly bacteria, but the immune system of a normal human is quite capable of handling these minor invasions. And although all people are vulnerable to contracting serious illness through foodborne pathogens, it's those with weak or compromised immune systems — infants, children, the elderly, and those with immune disorders like AIDS — who hold the highest risk of infection.

Foodborne Pathogens

Although there are five main types of foodborne pathogens (see page 27), the types can be split into two categories according to how they infect people: invasive and noninvasive.

INVASIVE INFECTIONS are caused by pathogens that penetrate the walls of the intestines, the stomach, or other areas of the digestive tract. The invasive microbe may also produce toxins or other means of infection. All five pathogens have a species capable of invasive infection.

NONINVASIVE INFECTIONS result from the body's reaction to toxins produced by the pathogen. These toxins may be excreted while the pathogen is inside our gut or may be produced in the food prior to our consuming it. As an example, take food poisoning, a seemingly catchall term for foodborne illnesses. In food poisoning, rather than waiting to produce toxins until it is inside our body, the food is adulterated with copious amounts of toxins before we eat it. It is for this reason that the onset of food poisoning is quicker than other conditions — the bacteria have already produced their toxins and therefore do not have to multiply and produce poisons while residing in the gut.

Causes of Contamination

There are many ways for food to become dangerously contaminated, but the most common involves either human feces or bodily fluids. Poor hygiene is the main culprit in the spread of foodborne pathogens. Food handlers (especially those in food service positions) who do not maintain strict hygiene habits have incredible potential for promoting widespread infections, often unintentionally and without awareness. An inadvertent cough or sneeze can spread millions of microbes. When produced by someone working near a vat of ground beef, a single sneeze can send microbes nationwide within distributed packages.

Animal feces are also a potent contaminant. Slaughterhouses are the most common site for contamination of meat via animal feces. The interior cavity and musculature of an animal is sterile, but the hide and feet abound with microbes. Because of this, removing the hide cautiously and hygienically is the single most important step to achieving a clean carcass. Species that are processed with the skin intact, such as pigs and poultry, are especially susceptible to contamination, because skin has some of the highest concentrations of contaminants. The communal scalding tubs used in cleaning skin-on carcasses, for example, are a veritable breeding ground for spreading pathogens from one infected carcass to another. Contaminated meat can cross-contaminate any other meat it comes in contact with, which is why the high-volume operations of modern meatpackers carry an extreme risk of large-scale contamination: one bad apple will spoil the bunch (and anyone who consumes it).

Conditions for Contamination

In most cases pathogens need to multiply in order to be effective, and for them to multiply the conditions need to be right. After we ingest foodborne pathogens, there is a period in which they multiply inside our gastrointestinal

The vast majority of foodborne illnesses stem from contamination caused by the people who have handled the food.

system, building their population large enough to produce symptoms of infection. This period between ingestion and infection is called the *incubation period*, and the length of time can be dramatically different depending on the pathogen. For some bacteria, incubation can take as little as a couple of hours; for hepatitis, a common viral foodborne pathogen, the average incubation period is 28 days.

While many pathogens operate uniquely, they generally reproduce under similar conditions: warm temperatures, ample access to nutrients, and a slightly acidic pH. Both cooked and raw foods can provide the right nutrients and an acceptable pH condition, so the real variable for microbial growth on food is temperature. The range of temperatures that pathogens can actively persist in, often referred to as the "danger zone," is between 40°F and 140°F. (It is probably no coincidence that most pathogens multiply exceedingly fast at the same temperature as

CONTAMINATION BEYOND MEAT

It is often a misconception that meat is more susceptible to pathogenic contamination than other food. The main issue with contamination is feces, and feces can be anywhere. Vegetables carry a high risk. In fact, some of the largest outbreaks of foodborne illness in the United States have come from vegetables. Farms may inadvertently spread infected manure as fertilizer on crops, leading to surface contamination on produce. An infected wild animal defecating in a field can also be a source of infection. In the home, proper washing of produce is paramount, especially for salads and other raw preparations.

our body, 98.6°F.) Changes in the population of a pathogen can happen surprisingly quickly given the right conditions. Bacteria, for example, multiply exponentially (see Exponential Growth of Bacteria, page 28).

A general food service industry rule is that any animal protein exposed to temperatures within the danger zone for a cumulative amount of four hours should be considered unsafe for consumption. Unfortunately, it is rarely that simple. There are other influencing factors, but the higher the ambient temperature, the quicker the growth of microbes. Four hours may safely apply to temperatures on the lower end of the danger zone, but leaving a perishable product out in 100°F heat for four hours is a recipe for rancidity. As a specific example, leaving deli meat on the counter while you eat lunch during the winter may be fine. Leave it out a few times over the course of a hot summer week, and you'd probably want to throw it out instead of feeding it to your children.

Once food is contaminated, nothing can bring it back to a state that is safe for consumption. Therefore, we take measures to prevent contamination early on, to thwart not only the microbes that cause spoilage but also the proliferation of pathogens. Holding food at a temperature below 40°F, the lower threshold of the danger zone, by

Although roasts like these should come up to room temperature before they're cooked, leaving them in the "danger zone" — between 40°F and 140°F — for more than four hours creates a breeding ground for pathogens.

refrigerating or freezing is one method. Cooking food to a temperature above the upper threshold of the danger zone is another way to prevent the growth of dangerous microbes. While you may aim for an internal temperature lower than 140°F for a rib-eye steak, the exterior is undoubtedly heated well above that, making it impossible for surface contaminants to grow.

Types of Pathogens

FOODBORNE ILLNESSES ARE THE RESULT of an infection from one of five types of pathogens: viruses, bacteria, protists, parasitic worms, and prions. Incidents in the United States are common; estimates show that 9.4 million people become ill every year through foodborne contraction of one or more pathogens. The most common type is viruses, responsible for nearly 5.5 million cases of foodborne illness (59 percent). Bacteria account for about 3.6 million cases (39 percent), and protists and parasitic worm infections for around 200,000 (2 percent). Prions, while considered a foodborne pathogen, have yet to be specifically linked to any cases of human illness. While most cases are easily remedied, around 2,700 Americans die every year from foodborne illnesses. Though they account for less than half of the cases of illness, bacteria are by far the leading culprit in the deaths, claiming more than 65 percent of the lives lost. Protists — specifically *Toxoplasma gondii* (discussed further on page 30) — are responsible for another 25 percent or so of the deaths. Viruses, while being the largest infector, are the least deadly, causing 12 percent of foodborne deaths.

Bacteria

Bacteria, causing the most deadly forms of foodborne illnesses, catch all the headlines. Most people have heard of *Escherichia coli* (commonly referred to as *E. coli*) and salmonella infections, which in some cases have caused massive recalls of ground beef, eggs, spinach, and even peanut butter.

SPOILAGE BACTERIA

Bacteria that spoil food can be considered the "miscreants of rot." While often disgusting, rancid food is actually the work of a benign group of bacteria. Rotting meat is far more capable than other foods of harboring a menagerie of malodorous compounds, because the proteins present in meat are themselves capable of being broken down into mixtures reminiscent of eggs, fish, skunks, and more. The "smell test" is a good method of determining rancidity created by spoilage bacteria. Consuming the bacteria, and the resultant rank food, may trigger some nausea, but the bacteria responsible for rancidity will not cause any extended periods of illness. Nonetheless, the conditions that promoted the development of spoilage bacteria are identical to those needed by other, more malignant strains of bacteria. Therefore, the main concern with rancid food is the potential presence of dangerous pathogens in addition to the identifiable actions of spoilage bacteria.

INFECTIOUS BACTERIA

Bacteria that infect food are either invasive or noninvasive. The two most famous, *E. coli* and salmonella, are both invasive. Once consumed, invasive bacteria penetrate the walls of our digestive system and begin to wreak havoc on our system. Most of these bacteria have both benign and malignant strains. Symbiotic forms of *E. coli* exist in our digestive tract; without them, we would fail to absorb many of the necessary nutrients from the food we eat.

Most types of infectious bacteria produce dangerous toxins that cause disease or illness. After ingestion, the bacteria feverishly multiply, quickly increasing their population inside our digestive tract. Before the infection is evident, the increase in concentration of bacteria in our feces dramatically raises the likelihood of transmission to another host. Soon the bacteria begin excreting into our systems a toxin that is the actual cause of the illness. Signs of infection from invasive bacteria will most often appear before signs from infection from noninvasive forms because of the damage the invasive bacteria are doing to our tissues in combination with the toxins being produced.

POISONOUS BACTERIA

Often the onset of illness from ingesting food previously poisoned by bacteria will be quicker than that of illness caused by invasive or noninvasive species. Unlike infectious bacteria, poisonous bacteria have already produced the toxins required for illness or disease. Active bacteria rapidly excrete toxins in the food. These may sometimes leave a sour or off-putting taste, but often the presence of toxins is unknown to the eater until it is too late, with symptoms manifesting soon after consumption. Raw meats and dairy are highly susceptible to poisonous bacteria because the environments in which they are processed have high transmissions of fecal matter.

In many cases, cooking can destroy poisonous bacteria. However, while the bacteria themselves will die off at certain temperatures, the toxins they have produced are left behind and ingesting those toxins inevitably results in an ill reaction. Therefore, you should discard without question any food that you suspect of having become poisoned.

Viruses

As microorganisms, viruses are a hardy bunch. They are not considered to be alive, because they do not procreate or operate independently. Therefore, they do not die but rather have to be inactivated to prevent infection. Heat is the most common method of inactivating viruses; cold temperatures do nothing. Unlike bacteria, viruses rely on living host cells, such as our own, to reproduce. Viruses attach to cells and use the interior functions of the cell to create duplicates of themselves. After copies are made, the host cell often dies and the copies of the virus flood the host and continue the cycle. Without living host cells to invade, viruses can lie dormant on food for extended periods. The population of the virus will never increase prior to consumption, but in many cases contamination with small populations of a virus is enough to produce illness after consumption.

COMMON FOODBORNE VIRUSES

The most common foodborne viral pathogen, responsible for 58 percent of all foodborne illnesses in the United States, is the norovirus, named after Norwalk, Ohio, the location of an outbreak in 1968 that led to its discovery. Rotavirus and hepatitis are two other virus families that are a common cause of foodborne illnesses. Both norovirus and rotavirus have similar symptoms and cause gastroenteritis, more commonly known as the "stomach flu" or "stomach virus." The symptoms — vomiting and diarrhea — also happen to be the symptoms of many other foodborne illnesses, the similarity of which results in many misdiagnoses.

TRANSMISSION

Viruses are spread primarily through fecal-oral transmission, and in most cases from one infected individual to another. In a food service establishment a single infected employee can contaminate surfaces and foods throughout the workday, further increasing the localized cases and chances of a larger outbreak.

Other common culprits for passing a persistent virus are filter-feeding mollusks, like mussels and especially

EXPONENTIAL GROWTH OF BACTERIA

Bacteria reproduce by cell division, meaning that one cell splits and becomes two self-sustaining single-cell organisms, each of those two cells then split, and so on. Through this method, bacteria populations can grow frighteningly fast. Their growth is exponential — the population doubles with every cycle of reproduction.

Exponential growth is challenging to comprehend, but a good example is to track the growth of a single bacterium. The sequence of growth would be: 1, 2, 4, 8, 16, 32, 64, 128, 256, 512, 1,024, and after another 10 cycles 1,048,576. Thus, in just 20 growth cycles, 1 cell becomes more than 1 million cells. The rate of a growth cycle is based on a balance between temperature and pH levels (assuming that nutrients are available for the foodborne pathogens, which is usually the case). Depending on the type of bacterium and the balance between temperature and pH, a growth cycle could range from as long as 12 hours to as short as 30 minutes. A 30-minute growth cycle means that a single bacterium can produce more than 1 million bacteria in around 10 hours! In some cases as few as 500 bacteria are enough to cause infection, illustrating the priority of curbing growth in any way we can.

Methods abound that attempt to reduce the growth of bacterial colonies. Drying, curing, salting, and adding preservatives are measures that all aim to curb bacterial growth. High temperatures, especially those outside the danger zone, are also known to kill off bacteria; heat is therefore the basis of pasteurization and many techniques of sterilization.

oysters, which are often eaten raw. Contamination occurs when the water surrounding the beds of growing mollusks becomes infected. Mollusks filter the water around them, absorbing nutrients but also viruses and other waterborne pathogens, harboring them until we consume the mollusks. Widespread cases of infection have resulted from ships discharging untreated sewage near oyster or mollusk farms. Other methods of transmission can include contaminated public bodies of water (e.g., pools, ponds, bathhouses), though any contact with infected individuals, especially those with poor hygiene, carries a risk of further transference.

Parasitic Worms

Foodborne pathogens are most commonly found on the surface of meat that has been contaminated. Parasitic worms are the exception: they burrow inside the tissue of animals. Like other foodborne pathogens they are either invasive (i.e., they penetrate the gut walls and travel elsewhere in the body) or noninvasive (i.e., they live in the gut, reproducing or causing other damage).

TRICHINELLA SPIRALIS

The most notorious foodborne parasite is *Trichinella spiralis*, a roundworm responsible for the disease trichinosis and the main concern for those processing land animals for food. The trichina worms proliferate through the ingestion of infected muscle tissue. The larvae burrow into the flesh of the host animal, forming cysts and causing initial infection. There they lie dormant, waiting for the animal to die, after which another animal consumes the infected flesh. Once consumed, the larvae mature in the intestines and grow into adult worms. The adult worms reproduce, forming more larvae that, again, burrow into the flesh and await death and transference.

Pigs are viewed as the main risk animal on a farm because of their omnivorous diet, but these days cases of trichinosis are exceedingly rare. Since the discovery of *T. spiralis*, farmers are taking the appropriate steps to rid pork of infection, and in the United States there are fewer than 10 reported cases of infection from commercial meat a year. In fact, most cases are from hunters, because the meat of many carnivorous game animals holds a high risk of infection. Even so, the reported cases are only in the dozens.

TAPEWORMS

Other worms are also capable of inflicting damage on livestock. Two tapeworms, *Taenia solium* and *Taenia*

The most common foodborne viral pathogen, responsible for 58 percent of all foodborne illnesses in the United States, is the norovirus.

saginata, can infect pigs and cattle, respectively. These worms have a similar life cycle to that of *Trichinella spiralis*, with one major difference: humans are the only host in which these worms can mature and reproduce. Therefore, humans are also the only living beings that can spread the larvae, which is done largely by fecal transmission. Cattle and pigs that are raised around poorly treated human waste or given food that is handled by infected hosts with poor hygiene are the main hosts for the larvae. Cases in the United States are extremely rare, mainly because of food safety regulations.

PREVENTING INFECTION

Fortunately, killing parasitic larvae is relatively easy: it is accomplished by exposing meat to cold or heat. Most commercial pork is frozen prior to sale, killing off the larvae before it reaches the customer. For decades, the U.S. government recommended absurdly high cooking temperatures for pork in order to kill parasitic larvae, suggesting that all cuts be cooked to an internal temperature of 165°F. Only recently has the recommendation changed to the saner target of 145°F. More information about freezing or cooking pork to kill *T. spiralis* can be found on page 35.

Protists

Most people are familiar with certain forms of *protists*, single-cell organisms that are most often found in bodies of water. Algae, molds, fungi, and amoebas are all part of the protist family. Rivers, watering holes, swimming pools, and the like are all teeming with protists, most of which pose no risk to human health. Pathogenic protists, on the other hand, pose high risks to our well-being. They account for only 2 percent of documented foodborne illnesses in the United States, but the death rate from protist-related illnesses is staggering — protists cause almost 25 percent of foodborne related deaths. (Compare that to salmonella, which infects more than 1 million people a year, around 11 percent of the cases of foodborne illnesses, and results in a similar number of deaths.)

TOXOPLASMA GONDII

There are a host of pathogenic foodborne protists, but the substantial death rate is primarily due to the lethality of a single entity: *Toxoplasma gondii*. The U.S. Centers for Disease Control estimates that around 60 million people in the United States carry the *T. gondii* parasite. That is a shocking number, especially when you consider the low numbers of reported illnesses from the parasite. The reason for this low ratio of infection to illness is that the aver-

age immune system is capable of keeping the parasite at bay. Those who exhibit signs of toxoplasmosis (the disease caused by the parasite) are often misdiagnosed, because the symptoms are like those of the flu and may resolve themselves before the person seeks medical help. People with compromised or weak immune systems — infants, children, pregnant women, and those with immunity disorders — are more susceptible to illness, and potentially death, from toxoplasmosis.

PROTIST LIFE CYCLE

The life cycle of protists, including their proliferation, depends on fecal-oral transmission. Water, the most commonly contaminated substance, allows for the parasitic eggs, called oocysts, to travel and find a host. Once ingested, the oocysts mature into parasites in the gut of the host animal, causing illness and disease while also producing more oocysts that can be passed through the feces of the host, continuing the cycle. Humans can contract protist infections through ingesting oocysts in bodies of water,

AVOIDING CROSS-CONTAMINATION

Food, especially meat, is an inherently clean substance. By and large, the pathogens that we encounter through consumption are not harboring on food as a permanent home: they're merely squatting temporarily until they find residence in a human host. The squatting and the taking up of a new residence is called *cross-contamination*, the process by which one substance is soiled by another already-infected substance. A doctor sanitizes a scalpel not to keep it clean, but rather to prevent the introduction of new microorganisms. We do the same with our tools and surfaces when we process meat and should also adhere to a similar approach when dealing with food storage.

Cross-contamination occurs constantly. Everything acts as a method

of travel for microorganisms en route to better lands where they can proliferate at will. For some, humans are the greener pasture, and the microorganisms arrive through pathways that may involve doorknobs, toilets, utensils, or simply being around another human. When we handle or process food, cross-contamination is a serious concern.

Certain substances are more likely to contain specific pathogens than others. An easy example is chicken, which should always have a dedicated cutting surface. The basic reason for this is that meat from chickens has a high risk of containing one of many strains of salmonella. Consequently, we take steps to prevent cross-contamination by processing different foods on separate surfaces.

Chicken is an easy talking point, but the reality is that all food holds a similar risk of contamination. Fecal matter is the main culprit, and any surface is capable of spreading it. For this reason, we need to take safety measures during the processing of any type of meat. This includes carcasses, even of the same species. You may start the day breaking down a beef forequarter. When you're done and you plan on breaking down the hindquarter of the same animal, it behooves you to scrape and sanitize the surface prior to making that first cut. The same applies to switching between any home and kitchen processing tasks, whether they involve butchering different species or even chopping vegetables.

on poorly washed vegetables, or in the meat of infected animals that contain larvae in cysts. Notable protist pathogens that spread through these methods are *Giardia intestinalis* (the cause of giardiasis), *Cyclospora cayetanensis*, and *Cryptosporidium parvum*.

Toxoplasma gondii follows some of these laws of transference, but its life cycle has some bizarre twists. The definitive host of the parasite — meaning the animal in which the parasite can mature and reproduce — is the cat, either domestic or wild. Inside cats, the *T. gondii* larvae mature and produce more oocysts that are spread through the cat's feces. Cats contract the parasite from ingesting the meat of animals, primarily rodents, that are infected with oocysts and larvae-containing cysts. Because the feces of cats are the primary method of transference, cat owners, and the animals on farms with cats, face a higher risk of infection. Simply changing the litter box can spread the parasite; herbivores eating grass that was contaminated by a resident (or feral or wild) cat may result in contaminated meat, and the contamination is then passed to anyone who consumes it.

PREVENTING TRANSMISSION

Heat is the primary method of killing the oocysts and larvae of protists. Oocysts from some of the more rare pathogenic protists are exceedingly difficult to destroy at temperatures less than 200°F — far beyond the temperature you'd want to cook your meat to. Fortunately, the most common protist, *T. gondii*, can be destroyed at acceptable cooking temperatures. A link to the guidelines for cooking temperatures to kill off protists can be found on page 440.

Prions

The most obscure foodborne pathogen is the prion. Many aspects of its origins and methods of transmission are unknown, despite extensive testing and research. What we do know is that the structure of a prion is incredibly simple: it's a misshapen protein. And we also know that when this pathogenic protein gets into a host body it causes other prion proteins within the nervous system and brain to fold abnormally, leading to neurodegenerative disorders. The most famous prion is that which causes bovine spongiform encephalopathy (BSE), commonly known as mad cow disease.

How to destroy prions is another mystery. So far, all of the methods that have been tried — including incineration, chemicals, ultraviolet radiation, and ionizing radiation — seem to be ineffective. Thus, there is little we can do as meat processors to prevent prion transference. Any animals that show signs of neurological disease must be reported to local authorities, so you may consider testing for BSE in non-ambulatory animals older than 30 months that are destined for a compost heap rather than a slaughterhouse in order to prevent potential transfer. From a food safety perspective, few other measures help. Sanitizing is futile. Fortunately, the spread of BSE is rare in animals and any connection to human diseases is even more rare.

Preventing the Spread of Pathogens

KNOWING WHICH PATHOGENS are of concern is an important step toward reducing the spread of foodborne illness, but memorizing scientific names and symptoms will do nothing to stop the spread of pathogens unless you take the proper precautions each and every time you handle and process food. Remember that food is largely uncontaminated until it comes into human contact, and that there are steps you can take to limit contamination.

Personal Hygiene

The first and foremost step to preventing the spread of pathogens is good personal hygiene. Almost all of the 31 foodborne pathogens identified to date can be spread through fecal-oral transmission. Although repulsive to consider, the truth is that we're constantly consuming food that's laced with traces of feces, mostly from other humans. As soon as you accept that reality, though, you'll realize that proper hygiene is the key to food safety.

REGULAR, THOROUGH HAND WASHING

Good hygiene starts with clean hands. Washing your hands needs to be a regular activity throughout any day of processing food. It may sound surprising, but many people don't know how to properly wash their hands. A quick rinse and a splat of soap won't cut it. Alcohol-based and other hand-sanitizing solutions will work against bacteria, but their efficacy against viruses is low, so don't rely on them exclusively.

There is no quick fix: the proper washing technique for stemming the spread of pathogens involves time and attention. This is how you properly wash your hands:

- Start with hot water, not blistering hot but rather around 110°F, or as hot as possible while still allowing a regular flow over the hands.

- Use an antibacterial soap. Look for one without environmentally damaging ingredients like triclosan.

- Spread the soap between both hands, creating a lather.

- Scrub the front and back of the hands, as well as along the thumbs, between the fingers, and over the wrists. Spend 15 to 20 seconds doing this.

- Use a nail brush to clean under nails, especially at the beginning of any processing and after exposure to more concentrated contaminants like manure, hide, and digestive waste.

- Rinse thoroughly with hot, clean water.

- Dry with a clean towel, either paper or fabric.

Hand washing must take place *prior* to processing food, but each of the following situations should be *followed* with a thorough hand washing:

- Soiling hands (e.g., with manure, intestinal fluids, urine, dirt, hair).

- Handling waste and garbage (e.g., inedibles, entrails, hides).

- Going to the bathroom.

- Blowing your nose; sneezing or coughing into your hand.

- Switching species of meat or carcasses.

- Handling chemicals (e.g., sanitizers, cleaners, disinfectants).

Any sink that can be operated without hands, in this case with foot pedals, makes an ideal hand-washing station. An automatic soap dispenser would make it even better.

Hands that have open wounds should be cleaned thoroughly, bandaged, and then covered with latex or nitrile gloves. People who show signs of illness should not, under any conditions, be allowed to process food; the risk of contamination is far too great.

It is important to note that hand washing is not, in general, a method for killing pathogens. The soap acts as a surfactant, dislodging dirt, grime, and germs from the surface of your hands. After you thoroughly rinse them off, the pathogens go down the drain of your sink. Foot- or knee-controlled faucets and automatic soap dispensers, while optional, are helpful in ensuring that clean hands are not recontaminated while washing or rinsing.

The whole hand-washing process should take at least 30 seconds — far longer than most people typically spend scrubbing their hands. This may sound excessive, but it's really not. Strict hygiene policies will prevent potentially fatal illnesses from spreading while you're processing food. It's not enough to wash your hands properly at the beginning of processing and then briefly rinse them throughout the workday. Every time you wash, you should follow the procedure outlined above. Accidents do happen, but having a habit of good hygiene as standard policy will show that you are committed to creating clean food.

SANITARY CLOTHING

Washing your hands is certainly the most important step in achieving proper hygiene, but it doesn't end there. All clothes worn during the processing of food should begin clean and be kept sanitary. During slaughtering, it's helpful

> The successive application of vinegar and hydrogen peroxide has been proven to be effective at killing just shy of 100 percent of surface pathogens on contact.

to wear a rubber apron that allows for repeated spraying and cleaning after any soiling from manure, blood, or any other contaminants. It's necessary to also wear an apron while butchering; have a few on hand that you can change out during the day. You should wear a hat to restrain hair. Your sleeves should be rolled up, and their edges kept clean. Fabric towels should begin clean and be replaced during the course of the day as they become soiled. Avoid placing towels on anything but sanitized surfaces.

ACCIDENTS HAPPEN

While the phrase "safety first" may echo in your head while you're handling sharp knives, saws, and other dangerous tools, accidents will always happen — no matter how careful or how experienced you are. You will notice that I'm wearing a nitrile glove in some photos of the rabbit and poultry slaughter. No stranger to safe knife handling, I still managed to nick my forefinger and thumb while cutting an opening between the gambrel and the shank of a rabbit. The responsible action is to tend to your wounds before continuing: clean, bandage, and cover them, so as to avoid any contamination of the otherwise clean carcasses you're preparing for consumption.

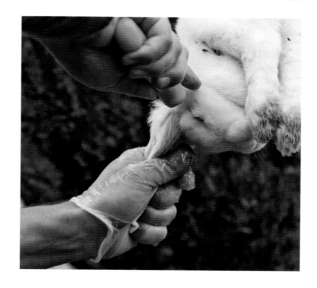

Workspace Hygiene: Tools and Surfaces

All tools should be sanitized at the beginning of any processing, and should continue to be sanitized throughout the day as they become soiled. Dropping a tool, placing it on an unsanitary surface, and using the same tool on different carcasses are all ways that tools become soiled. Tools that penetrate the surface of meat, such as boning hooks and mechanical tenderizers, are especially prone to spreading pathogens because they introduce microbes to interior areas that are often not cooked to the same temperatures as the exterior.

SANITIZING

QUICK DIP. Professional processing environments often employ the quick-dip method of sanitizing. This frequently includes a vat of water kept between 180°F and 190°F. After thoroughly rinsing the tool under hot water with soap — dirt, grime, manure, and the like need to be initially removed — you dip it into the scalding water for a brief period, killing any potential pathogens on contact. The quick dip is an excellent method and can be easily replicated with a pot of water on a hot plate or, depending on the capability of the unit, with a slow cooker set to high. Ensure that the water is maintaining the correct temperature by testing it periodically with a reliable and accurate thermometer.

A bench scraper is the ideal tool for keeping a cutting surface clear of debris and fat while you're working.

SANITIZING SOLUTION. Rinsing tools in a sink with soap and then dipping them in a sanitizing solution is another method. Sanitizing solutions can be specialized blends of chemicals or simply a mixture of bleach and water. Many chemical sanitizers require exposure for at least a few minutes; make sure you read through the directions for any commercial products to ensure proper usage.

Bleach and water is one easy and cheap solution for sanitizing everything from tools to surfaces. A mild mixture of one tablespoon of bleach per gallon of water will adequately sanitize any surface exposed to it for a minimum of two minutes. To properly sanitize tools, fully submerge them in the solution for the required two minutes. Surfaces should be liberally sprayed and left to sit. You don't need to rinse off this solution; in fact you shouldn't rinse, or you risk recontaminating the surface. Wipe surfaces with a clean towel and allow them to air-dry, after which the chlorine in the solution will evaporate.

There are many other household solutions that can also aid in disinfecting surfaces. Vinegar, alcohol, hydrogen peroxide, lemon juice, and salt are all commonly found in homemade cleaners. In general, they do work for clearing away some bacteria (some work better than others), though none of them is especially effective at disposing of viruses, especially the norovirus, which is the leading cause of foodborne illness. Because of this, they're not recommended as a solution for sanitizing food-preparation tools and surfaces.

One exception is the successive application of vinegar and hydrogen peroxide, which has been proven to be effective at killing just shy of 100 percent of surface pathogens on contact. Use a vinegar that is 5 percent acetic acid (regular white vinegar) and a 3 percent hydrogen peroxide solution (the basic drugstore concentration). Rather than mix the two, apply each one separately — the order doesn't matter — and then allow it to sit for a few seconds before wiping clean. This safe, effective, and cheap combination can be used for tools, surfaces, or food products like fruit.

Surfaces should be cleaned regularly throughout the workday, especially when switching between carcasses. Prior to sanitizing, use a bench scraper to clear away meat bits, residues, and fats, which inevitably accumulate during butchering. Further cleaning, with scouring pads and a surfactant, will help prepare heavily soiled surfaces, which should be relatively clean before sanitizing. Small cutting boards can be submerged; large cutting surfaces should be sanitized by applying a liberal amount of cleaning solution to the entire surface and allowing it to sit for the appropriate amount of time.

Cold Storage

SAFE FOOD STARTS with proper hygiene during slaughter and butchering, but the necessary precautions do not end there. Remember that all food is contaminated to some degree. Cold storage procedures, during every phase from slaughter to consumption, are equally important.

The one and only goal when storing meat or any other perishable product is to limit the growth of microbes. These include spoilage bacteria as well as the plethora of pathogens that may already be on the food but in benign numbers. The pathogens that are affected by cold temperatures are bacteria and parasitic worms. Bacteria are prolific at temperatures above 40°F; their multiplication rate increases with the temperature until the heat becomes high enough to kill them, which is usually around 140°F. This is the basis of the so-called danger zone: perishable foods left exposed in temperatures of 40° to 140°F for extended periods will undoubtedly spoil and increase the risk of dangerous infection.

Refrigeration

Refrigeration and freezing are the most common methods of storing perishable foods. The moderately low temperature of refrigerators does not halt all bacterial activity; it merely slows down the growth of bacteria to buy us more time to consume it. As you probably know, food will certainly go bad in the fridge if you leave it there long enough. As you lower the temperature, bacterial activity continues to slow down until it basically halts, which is around 0°F, the operating temperatures of most home freezers. At these low temperatures food will not spoil from bacteria, but it will still go rancid. The bacteria do not die but are preserved, just as the food is, becoming active once again with the rising temperatures of thawing and cooking.

TEMPERATURE

If the temperature of a piece of meat is kept low enough for long enough, parasitic worms are killed off. This is the general approach of the commercial industry for ridding pork of *Trichinella spiralis*; most commercial pork purchased in the United States has already been frozen to eliminate the risk of infection.

To be effective, all areas of a refrigerator — top shelves, bottom shelves, and drawers — need to be kept below 40°F.

Many consumer refrigerators will show a mean temperature below 40°F when in fact some areas are well above that. The only way to be certain that a refrigerator is cooling sufficiently is to place several small glasses of water in various areas of the fridge, allow them 24 hours to stabilize in temperature, and take a reading from each glass, using an accurate digital thermometer. Adjust the fridge as necessary to ensure that the overall temperature is safe.

PROPER ORGANIZATION OF FOOD

Organization is also an important tool for eliminating some of the classic cases of cross-contamination in the refrigerator. Open containers of lettuce sitting beneath packages of raw fish or other meats, juices of thawing meats spilling out into vegetable drawers, marinades splashing on surfaces — these are all examples of how germs can easily spread in the refrigerator. All meats should be stored on the bottom shelf and usually covered. (There are, however, times when meats should be left uncovered and exposed to refrigerated air, such as when you want to dry out the skin of a whole chicken that is destined for roasting.) When defrosting frozen meats, place them in a container or on a plate to contain drips. Finally, avoid storing raw meats together with cooked or ready-to-eat products.

Freezers

Freezers are often as inaccurate as refrigerators are. Many home models will claim to maintain a 0°F temperature when in fact they fluctuate. The lower you keep the temperature, the better kept the contents will be while minimizing the effects of inevitable fluctuations. Stand-alone freezers are far more effective than those integrated with refrigerators, though they require space and additional costs. (For more on freezing, see chapter 14.)

Bacteria are prolific at temperatures above 40°F . . . until the heat becomes high enough to kill them.

Killing Pathogens with Heat

IT'S A RELIEF TO KNOW that most of the food-borne pathogens we're concerned with, except of course for those pesky prions, can be killed with heat. This doesn't mean that every piece of cooked meat is safe to eat, but if you follow the recommended guidelines for cooking, in most cases you'll reduce the chances of infection to a negligible number. To achieve the desired results, you must follow the guidelines for time and temperatures, and use an accurate digital thermometer to ensure safe cooking.

Cooking Whole Cuts

It's relatively easy to gauge the cooking time for a whole, unpunctured muscle, because the interior can be considered sterile, with the exception of pork's proclivity toward parasitic worms. (Commercial pork can be considered parasite-free, and whole, intact pork muscles should be treated similarly to that of any other animal; homegrown pork that has not been frozen to ensure destruction of any parasitic worms should be handled according to internal temperature requirements, not surface temperature.) Such muscles include roasts, steaks, chops, shanks, and any other cuts in which the only area of the meat that has been exposed is the exterior surface. The surface of the meat can therefore be held at a certain temperature for a given time to kill off any residing microbes. In some instances, this may be far less time than what it takes to cook: exposure to temperatures over 160°F will destroy or inactivate pathogens in less than a second. The key is to ensure that the actual surface of the meat is achieving the target temperature.

When searing on the stovetop, this won't be such an issue — any browning done in a pan occurs at temperatures that immediately dispatch microbes — but oven-roasting can be more complicated. It's not enough to set the oven to 350°F and consider it a done deal, because ambient heat does not equal surface temperature. For oven-roasting, a digital probe thermometer comes in handy. Every so often, stick it just under the surface of the meat and take a reading until you are certain you are hitting the target temperature. And remember to sanitize the thermometer before each use since you are puncturing the meat.

Poultry, whether cooked whole or in parts (e.g., thighs and breasts), has its own considerations for times and temperatures to reduce potential pathogens. Those values can be found by following the link on page 440.

More Susceptible Products

Meats that have been ground, mechanically tenderized, or injected with marinades or brines — along with home-grown pork, wild game, and other meats susceptible to internal infection by parasitic worms — need to be handled differently than whole, intact muscles do. The interior of such meats is just as likely to be infected as the exterior is. Hence, we need to heat all parts of these meats to temperatures high enough to achieve the desired deadly effect.

For example, the exterior of a steak needs to be exposed to a temperature of 145°F for a minimum of three minutes. If the meat you're cooking is a hamburger, the interior of that burger needs to be held at 145°F for a minimum of three minutes. If you prefer your hamburger medium-rare, which corresponds to a temperature of about 135°F — the U.S. Food and Drug Administration and the U.S. Department of Agriculture will claim that 145°F is a medium-rare target temperature, but if you follow that guideline you will quickly realize that the temperature is too high — then it's recommended to hold that internal temperature for 36 minutes. Without some fancy modern cooking equipment, this is impossible; it is for this reason that you assume the risk of eating so-called undercooked meats. We've all done it, and it's not to say we won't again, but it is good to understand what the risks are before we assume them.

This meat is ready to grind. Grinding greatly increases the surface area of meat, making ground meats especially susceptible to contamination, thus requiring thorough cooking.

3

TOOLS & EQUIPMENT

TO EXCEL AT BUTCHERING, you need only a few basic, modestly priced tools. This chapter covers those tools and the other equipment required for nearly every butchering task described in this book. The main sections include an alphabetical list of the basic tools, an explanation of what makes each tool ideal for butchering, and advice on which to choose. Further on, the chapter covers additional tools and equipment, some task-specific, that allow for more customized or advanced preparations. You may keep these specialized tools around for occasional use, but they won't end up sitting on your block like the everyday ones will.

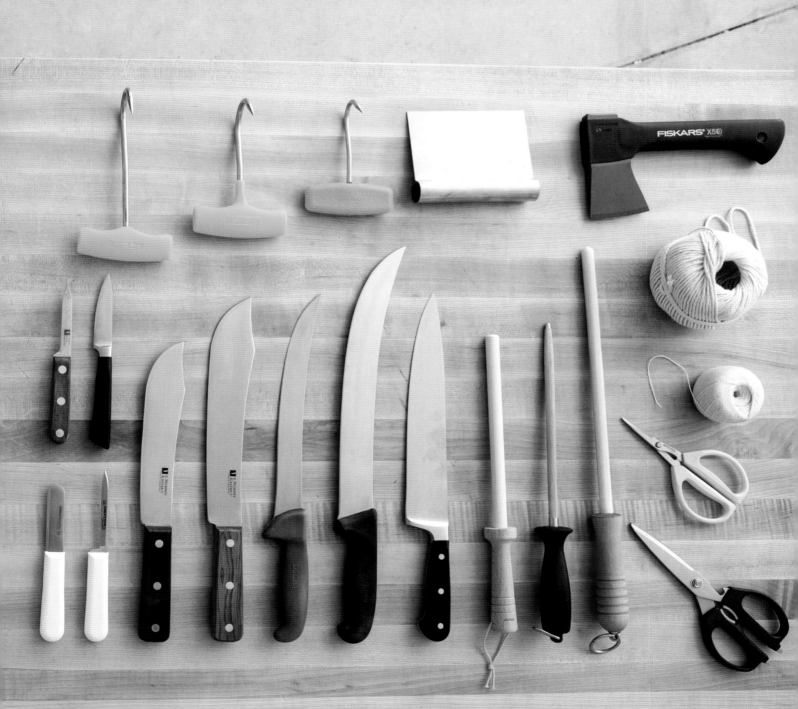

Knives

KNIVES ARE AVAILABLE IN MANY DESIGNS and are marketed under many names, so choosing the right knife can be a daunting task. When beginning to shop for a knife, do so in person in a store that will allow you to handle the knives prior to purchase. The knife you choose should feel good in your hand since you will be gripping it tightly for long periods. Test it out using both of the grips described on page 54. Make sure it is easy to grip and feels balanced when you hold it; also make sure that the handle is made of a material that is easily sanitized. Wooden handles often benefit from a quick sanding with a high-grit sandpaper to remove finishes and provide a textured surface for easier handling, especially because they inevitably get covered with fat and other slippery residues. All knife blades should be made from a high-carbon stainless steel, which resists rust and stain while maintaining a sharp edge.

FLEXIBILITY

FLEX BLADES
Allow for more nuanced cutting; great for shaping, denuding, and seaming; used mostly by experienced butchers.

SEMI-STIFF (semi-flex) blades
Allow for enough bend to keep the edge close to a bone or table, especially helpful when boning a loin, separating a top blade steak from the scapula, etc. Great for jointing.

STIFF BLADES
Allow for the most efficient transfer of energy into cutting. Precise overall; no risk of wandering when trying to make straight cuts. Great for jointing.

SHAPE

CURVED BLADES
Best for accessing the small spaces in joints or hiding.

STRAIGHT BLADES
Better for precision cutting, trimming, and portioning.

Blade Characteristics

Knife blades are available as flex, semi-stiff (sometimes known as semi-flex), or stiff. Semi-stiff and stiff blades are the two most common options for butchery and are best for those beginning to learn the nuances of cutting, following the curvature of bones, and making their way through joints. Flex blades, with their tendency to wander and change angles with less force, require deft handwork and are more often found on the blocks of experienced butchers. Semi-stiff blades allow for enough bend to keep the edge close to a bone or table, especially helpful when performing tasks like boning a loin or separating a top blade steak from the scapula. Stiff blades allow for the most efficient transfer of effort and energy into cutting, as none is lost in blade flexibility. The lack of flex also makes them more precise overall; there is no risk of wandering when trying to make straight cuts. Both are great for jointing, though the small amount of play in the semi-stiff blade can be forgiving when trying to access tight spaces.

Most knives are available in two shapes: straight and curved. Both shapes perform well, so the decision of which to use is going to be largely one of personal preference. Straight knives can feel like a more natural extension of your hand and tend to be better for precision cutting. Curved knives have a spine that slopes away from the edge. The point of the curved blade is also accentuated and designed to allow for access to small spaces, helpful for tasks like jointing a chicken or separating vertebrae. Furthermore, a curved blade works well with incremental cutting actions, such as removing beef hide, peeling the fatback off a pork loin, or removing connective tissue from curved surfaces.

The most common combination is a semi-stiff curved knife for boning and several stiff straight knives for trimming and cutting, and I would certainly encourage you to start there.

Types of Knives

Knives come in myriad shapes and sizes, with specialized designs intended to improve performance for various tasks. Having a small arsenal of knives at your disposal is fun, but for butchering you actually need only two or three of the right knives, and this is where I would start.

BONING KNIFE

Thin-bladed boning knives range from 5 to 7 inches. The boning knife will be your main knife, used for everything from separating primals, subprimals, and the like to boning, jointing, and trimming connective tissue. Start with

BONING KNIVES

Flex blade

Semi-flex blade

Stiff blade

a smaller blade length and, once you're comfortable with the relationship between your knife and non-knife hands, consider using a longer blade.

BREAKING KNIFE

A breaking knife is any long-bladed knife that's used to break down primals into subprimals and beyond, and to perform many trimming tasks. Make sure it's 8 to 10 inches long, with a stiff blade that is either straight or slightly curved. A breaking knife may be found under many other names, including *butcher knife* and *sticking knife*; find one based on size and shape rather than name.

BUTCHER KNIFE OR CIMETER

The term *butcher knife* tends to be a catchall for any long-bladed knife, but in essence a butcher knife is around 12 inches long and has a straight, stiff blade. The main function of this knife is to slice cleanly through large cuts of meat, leaving no marks from dragging blades or sawing motions, for example, slicing boneless steaks or completing the separation of a beef round and sirloin after the bone has been split. It will be the knife that needs sharpening the least since it will not come into contact with things like bones, joints, and the toughest connective tissues. Keep it sharp so that your cuts are as precise as possible. An alternative option for a butcher knife is a cimeter (also spelled scimitar), which, although its shape is slightly different, can perform all the same tasks.

Knife Accessories
HONING ROD/KNIFE STEEL

The primary objective of knife care is keeping the edge sharp; often that has to do more with honing (realigning microserrations in the blade) than with actual sharpening (removal of metal from the blade edge). The rods on the market are made from many different materials: polished steel, grooved steel, ceramic, borosilicate (Pyrex), and steel with an embedded diamond surface. The only one that won't remove material from your edge is the polished steel. Grooved rods, while wildly popular, can cause deformation of microserrations on the blade edge rather than promoting alignment. Ceramic, borosilicate, and diamond rods all hone while, due to their extreme hardness, also removing small amounts of material. Thus these materials allow for minor sharpening and upkeep along with honing.

For the price, a ceramic rod with a fine grit (i.e., a grit of 1,000 or greater) is the best bet, but for pure honing without the risk of edge damage through minor sharpening, a polished steel rod is ideal. You may also want to consider owning multiple rods in different materials: a polished steel for daily use and a fine- or coarse-grit rod for minor upkeep. Look for a rod that is at least 10 inches long, choosing a longer one for easier use with your butcher knife or cimeter. If you plan on attaching it to your scabbard chain (described on page 44), make sure it has a metal ring on the end of the handle.

Bone saw

BUTCHERING ESSENTIALS

Twine

Cleaver

Butcher knife

Breaking knife

Boning knives

(flex) (semi-flex)

Mallet

Bench scraper

Shears

Honing rod (steel)

Honing rod (ceramic)

Aligning the edge of your blade while honing requires very little force.

SHARPENING

When honing no longer produces a sharp blade edge, it is time to sharpen your knife. Ask your favorite restaurant where they sharpen their knives or find a professional in your area who offers sharpening services to reputable culinary establishments. You can also sharpen knives at home with great results using diamond, oil, or water stones. (I recommend any of the sharpening systems made by Edge Pro; see page 440.)

SCABBARD

Working with multiple knives on a tabletop can be dangerous. Commonly during butchering a loose knife falls out of sight within a pile of meat, presenting a serious danger when you try to move the meat and find yourself gripping a blade. Furthermore, loose knives may fall on the floor, damaging blade edges while becoming contaminated. Using a scabbard is one easy way to keep your knives easily accessible while protecting yourself and those working around you. It will also help you keep your knives sharp and clean.

Scabbards, and the chains that hold them up, are available in two materials: plastic and aluminum. Plastic scabbards are a one-piece construction and hold up to three knives and a meat hook. They are heat-resistant and therefore make sanitizing easy when working with knife sterilization boxes. An aluminum scabbard is a multipiece construction that comes apart for easy cleaning and sanitizing. The soft metal helps prevent dulling of knife edges, and the large size allows for up to five knives. Either plastic or aluminum will do the job just fine, but choose the aluminum if you plan on regularly wanting more than three knives at your disposal.

Other Cutting Tools
CLEAVER

Cleavers are traditional butchery tools but are not used frequently in modern meat cutting. For most bones you'll pull out a handsaw or go to your bandsaw if you have one, because the results are cleaner than those from using a cleaver and don't carry the risk of leaving bone shards in your product. But this doesn't mean that cleavers have no place. (One thing to consider is the cathartic effect of whacking your way through soft bones.)

Cleaver options run the full gamut of style, size, weight, and cost, but the most important factors are weight and the comfort in your hand. You want a cleaver that has heft, because its job has to do with impact, not

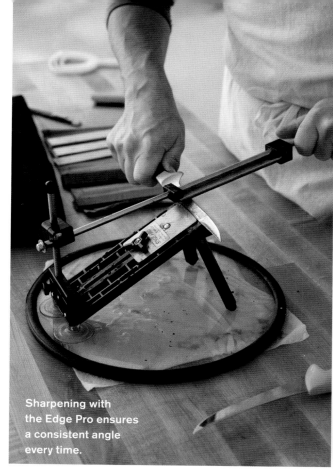

Sharpening with the Edge Pro ensures a consistent angle every time.

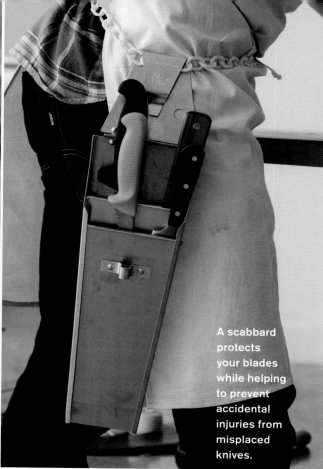

A scabbard protects your blades while helping to prevent accidental injuries from misplaced knives.

precision. It should be top-heavy, like an ax, so that when you swing it the weight of its broad blade, not your exertion, creates the momentum. In contrast to many other tools of the trade, antique cleavers can be just as effective as new ones and often are more affordable for the weight.

BONE SAW/HANDSAW

The blades of bone saws, especially the teeth, are designed to easily and efficiently do one thing: cut through bone — not meat. The quality of cuts that can be made with a bone saw far exceeds those made by other saws you may have on hand — such as a hacksaw, a carpentry saw, or a bow saw. A strong stainless-steel or aluminum frame provides the necessary rigidity; the thin profile enables splitting within difficult spaces and also prevents obstruction. The saw handle should be made from a material that is easily sanitized while being comfortable to grip. Blades should also be made from stainless steel and, when loaded, have a tension tight enough to prevent wandering while cutting.

Your choice of saw length should be based on the largest animal you will be working with. You should have enough room to freely move the saw in large strokes without the tip or handle having to enter the space you're splitting. For example, a 17- to 19-inch saw will work well for breaking down pigs (not including the larger boars). But trying to split a beef round from the sirloin using a 17-inch saw would be difficult, because the strokes would be short and the blade length would not fully traverse the area that needs to be split. For that job, use a 25-inch saw or larger, depending on what is also comfortable to hold and manage when sawing. Further details on using a saw can be found on page 55.

SHEARS

A good set of shears is an aid in processing small species, such as chickens or rabbits. Look for a pair that has blades made from high-carbon stainless steel and a comfortable grip that is easily sanitized. Two-piece constructions that separate easily can simplify cleaning. The more heavy-duty shears, like handheld bolt cutters or those sold for pruning and cutting through small branches, usually include a notch that makes cutting through round bones a cleaner task than when a standard poultry shear is used.

Tabletop Equipment

IN ADDITION TO CHOOSING equipment that slices and severs, a butcher needs to keep a selection of other tools intended to aid the job of cutting meat on a table or hook. Every item on this list is essential for my process, and I suspect you will find value in them as well.

Bench Scraper

Your work surface will inevitably get gummed up with fat residues, small meat scraps, bone dust, and other butchery debris. Keeping your workspace clean is part of conscientious meat cutting, and using a bench scraper regularly while working is the best way to clear your surface of such detritus. This tool works equally well with large butcher blocks or small cutting boards. Find one made of metal with a comfortable handle that can be sanitized easily.

Bone Dust Scraper

Anytime you saw through a bone-in cut of meat — such as bone-in pork chops, crosscut beef shanks, or lamb leg steaks — some debris is left on the cut face. This is especially evident when working with a bandsaw, which usually combines the bone dust with fat and moisture to form an unpalatable pasty mess. While you can use the edge of your knife to wipe the dust away, the lateral movement is not good for edge sharpness, especially around bones, and you run the risk of damaging the cleanly cut face of meat. It is better to use a bone dust scraper, a simple, cheap tool that does an effective job, leaving the face of your meat presentation-ready. Scrapers come in either plastic or metal, and both are equally effective.

Boning Hook

A boning hook not only is an essential tool for breaking down a carcass on the hook but also provides great benefits for table butchery as well. Its sharp, precise tip makes pulling and handling pieces of meat easier than gripping the carcass, and the length of the hook positions

A bone dust scraper is essential for removing the unpalatable debris left after portioning a bone-in cut.

It's important to choose a work surface that's big enough to accommodate the largest carcass you plan on butchering.

the non-knife hand away from apparent danger. A hook can also be used to help clean bones when dealing with marrow or when frenching. Find one with a comfortable handle and a hook length of at least 5 inches. Dulled or bent tips can be sharpened with a rasp or a grinder.

Butcher's Twine

Butcher's twine is a versatile tool; having some around, not just for tying off roasts and trussing chickens, will be helpful. Twines come in various ply ratings — such as 16-, 24-, or 30-ply — which indicate the number of strands twisted together to form the twine; the higher the number, the stronger and thicker the twine. Get 100 percent cotton, food-grade twine that is 16- to 24-ply.

Cutting Board/Surface Material

There are two main requirements for a good work surface: it should be easy to sanitize, and it shouldn't cause damage to your knife edge. The two main options are wood or plastic (polyethylene) and, while the commercial industry tends to use plastic, wood is my preferred option, for several reasons. Wood is a self-healing material, meaning that over time as you make shallow cuts and grooves, it expands to fill the space, leaving fewer crevices for bacteria to reside in and develop. Dry wood also has natural antimicrobial properties due to the capillary action of its fibers, which absorb surface bacteria and kill them. Plas-

tic, in contrast, is easily damaged from knives and saws, and the damage is permanent. Repeated use leaves deep grooves that are difficult to sanitize, even with soaking. Wood is also repairable: you can sand down the surface time and time again (until it is too thin to work on). Plastic, once badly damaged, needs to be thrown away and replaced. So, while both materials can be sanitized using the same solutions, the efficacy of sanitizing plastic is largely based on the effects of previous usage.

KINDS AND CUTS OF WOOD

Woods, the ones ideal for butcher blocks, are also gentler on knife edges than plastic. The softer the wood, the easier it is on your knife (though the downside is a shorter life through more wear). A balance between softness and durability is what makes maple the most popular material for butcher blocks.

Wooden boards are made from three different cuts: end-grain, edge-grain, and flat-grain. The terms are the same as with a piece of plank wood: *end-grain* is wood standing on end, *edge-grain* is the wood on its edge, and *flat-grain* is a board sitting flat.

END-GRAIN is an ideal option for knife care, because the fibers of the wood are aligned vertically, letting the knife edge fall in between them. Among the three types of cuts, it is the strongest, most resistant to wear, and longest lasting; fittingly, it is also the most expensive.

Cut-resistant gloves (left) can help protect your hands from injury. Nitrile gloves (right) are the best choice for covering hands that have cuts or open wounds.

EDGE-GRAIN, the next-best option, is moderately durable, and although the fibers run horizontally, they are vertically stacked and good for knives.

FLAT-GRAIN will not work for butchering use, because its durability is low. Just envision standing on a plank of wood laid flat: it bends, compared to when it is on its side or end. Flat-grain is the cheapest option, and for good reason: it will need to be replaced regularly.

A large wood butcher-block tabletop is the best surface to work on, but for many folks portable cutting boards are the only option. Choose a board that is as large as possible for the working space you have and for the manageability of moving and cleaning. Or combine multiple boards to make a mobile tabletop that is easy to clean. End-grain boards should be at least 2.5 inches thick, and edge-grain boards 1.5 inches thick. The thicker they are, the longer they'll last and the better they'll resist warping (though with proper maintenance no board should warp). Avoid culinary cutting boards that have a juice groove on the outer edge, because the groove will prevent easily scraping the board clean and will inevitably collect debris.

Kitchen Scale

A kitchen scale is an indispensable piece of equipment when processing meat. Using a scale improves the outcome of many tasks: you can label freezer-bound packages of meat with the weight; you can weigh portioned cuts to allow for adjustments that ensure better accuracy; and you can follow recipes based on ratios and weights, like the ones in this book, with certainty. Make sure your scale is digital and provides a readout in both the metric system (grams) and the imperial system (ounces). It should have a capacity of at least 5 pounds (ideally more), be battery-powered, and be easily portable.

Spray Bottle

Keep a quality spray bottle around, filled with a sanitizing solution, to encourage frequent cleaning of surfaces and tools. (For recommendations of sanitizing solutions, see page 34.

Towels

Towels are essential during processing. Keep a stack around, leaving some clean for tasks like drying hands and sanitized tools or surfaces, and some that can be soiled in tasks like wiping down saw blades and bone dust scrapers, and in intermittent clearing of surfaces.

Clothing

THE APPROPRIATE CLOTHING for butchering will not only keep you clean and protect you from injury but also improve the safety of the meat you cut by preventing cross-contamination.

Aprons and Butcher Coats

Protective clothing keeps you clean while reducing contamination of the product. Butcher coats, or frocks, are the most effective protective garments. Aprons can be worn on the outside of a coat and should be switched out once soiled. Aprons should also be changed when switching from working with poultry to working with other proteins. Choose clothing that fits snugly and reduces the amount of excess fabric that can easily get caught in machinery or soil.

Cut-Resistant Gloves

Wearing cut-resistant gloves on your non-knife hand can help you avoid minor cuts. Although you can still inflict a serious injury even while wearing gloves, they can reduce the damage. Gloves may be especially helpful when you're beginning to butcher and learning the relationship between your blade length and non-knife-hand positioning. You should change your gloves when they become heavily soiled or when you switch from working with poultry to working with any other protein or vice versa.

Nitrile Gloves

Always have a box of nitrile gloves on-site. If your hands have small nicks or any other kind of open wound, you should wear nitrile gloves to prevent contamination. Furthermore, wearing nitrile on the outside of cut-resistant gloves can reduce the need to swap the protective gloves when they become soiled or when you switch between different proteins. Choose a size that fits snugly but provides some room for the natural swelling of hands while working; they should never be tight at the wrist or feel like they're constricting blood flow to the hands.

Protective Aprons

Fast movements and sharp blades carry a high risk of injury. Some injuries may simply be nicks or shallow cuts, whereas others could be serious abdominal lacerations; you never know until it has already happened. One slip toward your stomach while knifing your way through cartilage could result in the equivalent of self-disembowelment. So, although wearing some protective gear is merely recommended, wearing something that protects your vital organs is essential.

There are three adequate options for abdominal protection, all of which come in the form of bibs or aprons: fabric, plastic, and chain mail. The fabric and chain mail exceed the protection of a plastic belly guard because they are able to cover more of the body without restricting motion, and it is certainly a benefit to have your groin and area blood vessels protected. The fabric from which butcher's bibs and aprons are made consists of lightweight cut-resistant materials, similar to what's found in cut-resistant gloves. Plastic is available in the form of a belly guard, which while not a proper apron wraps around the abdomen. Chain mail, or metal mesh, provides the best protection but is heavy and often expensive. Nonetheless, while 5 to 10 pounds (depending on the length of your apron) can certainly take its toll during a long day of butchering, the extra weight is well worth the protection. If you choose to use chain mail, consider wearing two fabric aprons, one underneath and one in front.

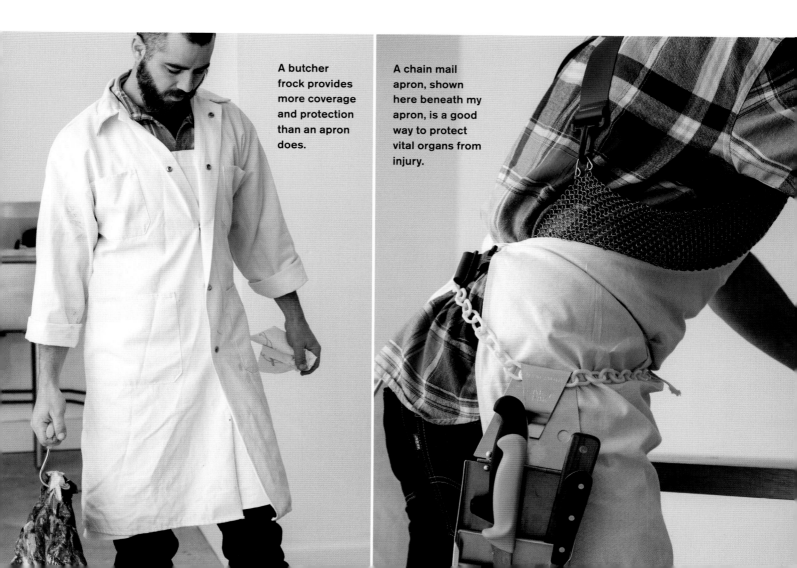

A butcher frock provides more coverage and protection than an apron does.

A chain mail apron, shown here beneath my apron, is a good way to protect vital organs from injury.

Optional Equipment

AS YOUR BUTCHERING SKILLS DEVELOP you may recognize a need for specialized pieces of equipment. Each of the following items provides a very specific function and is essential if you plan on doing things such as grinding or making cased sausages.

Bandsaw

A bandsaw is a powerful and versatile piece of equipment to have in a meat-processing shop. It makes quick work of many repetitive tasks like cutting steaks or crosscutting beef shanks, and can produce consistent results with the aid of measurement tools. The downside is that bandsaws are expensive, require heavy cleaning, and like all other machines need periodic upkeep. Everything that can be done with a bandsaw can, in theory, be done with a handsaw; the bandsaw may be quicker, cleaner, and easier, but it is not always better.

Brine Injector

Many wet-cure recipes suggest using a brine injector. These large syringes allow you to inject brine into the center of a muscle, where the salts and spices can equalize without waiting for penetration from the exterior. The time required to adequately brine meat is dramatically reduced when injecting. Look for a brine injector that's made out of stainless steel and comes with multiple needles.

Grinder

Ground meat is a common by-product of any whole-animal butchery, because it is one of the most effective ways to maximize carcass utilization, turning unpalatable bits into hamburgers, sausages, or other value-added products. Meat grinders come in all forms, from hand-crank models to kitchen mixer attachments to high-horsepower machines.

MANUAL GRINDERS are a challenge to operate for large quantities of meat; they're meant for the hobbyist.

GRINDER ATTACHMENTS for kitchen appliances provide adequate functionality for grinding several pounds of meat at a time, though their small capacity requires frequent cleaning for proper operation. They also require

Choose between a grinder attachment (right) or a dedicated grinder (middle) when deciding to grind meat at home. A basic sausage stuffer (left) is another option for the home butchering table.

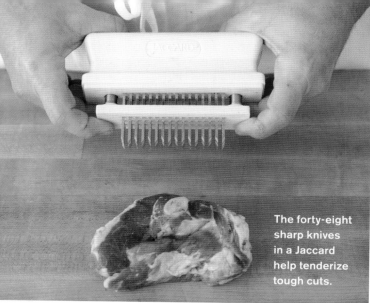

The forty-eight sharp knives in a Jaccard help tenderize tough cuts.

A mallet is good for pounding thin meats or for maintaining accuracy with a cleaver, as shown here.

you to have another appliance around, so unless you plan on having a kitchen mixer dedicated to processing meat (which could be great for making many value-added products), this won't even be an option.

DEDICATED GRINDERS are built for the sole purpose of grinding meat and perform the best of the three options. They range from home models with low horsepower to commercial machines with integrated paddle mixers and other features.

A meat grinder should have swappable plates for adjusting the size of the grind. The moving parts should be made from a durable material and should be easily dismantled for cleaning. All-metal construction is ideal, not only for durability but also because a chilled metal grinder stays cold longer than a grinder of another material, which helps prevent fat smearing and poor grind performance. Protective features can prevent injury from clothes or hands getting caught in the auger.

Jaccard Meat Tenderizer

Tough cuts of meat can benefit from tenderization (more on page 68); one useful tool for this is a Jaccard meat tenderizer. It uses dozens of sharp, flat knives to penetrate the muscle, severing and shortening the fibers, giving the impression of more succulent meat by making it easier to bite through. Along with helping tenderize the muscle, the Jaccard's minor penetrations can also help distribute marinades by providing multiple access channels into the interior of the meat.

Mallet

A mallet, whether made from wood or metal, should have a good weight and be head-heavy. You can pound out cutlets and flatten medallions with the side of a cleaver, but

having a mallet around will make the task much easier and cleaner. Some models have two sides: one flat and one textured, the latter being for tenderization and the former for flattening. I recommend using a Jaccard for tenderizing and a mallet solely for flattening or striking. Find a mallet that feels comfortable in your hand and has a large surface area to the head so that your strikes won't have to be accurate to achieve good results.

You can even use a mallet with a cleaver, striking the spine of a cleaver when it's in place to avoid inaccurate swinging. In this case look for a wooden or rubber mallet.

Meat Lugs

For home butchering, you'll need multiple containers to hold everything from offal to trim to inedibles. While you can use any plastic or metal container, lugs designed specifically for the task of meat handling provide some readily apparent benefits: they are constructed from heavy-duty and stain-resistant plastic that resists warping and cracking from exposure to extreme temperatures, optional lids allow for stacking, and interiors are designed with sanitization and cleaning in mind. Lugs come in a variety of sizes to fit the volume of meat you're working with, so purchase the largest size that works for your processing.

Sausage Stuffer

A sausage stuffer is necessary only if you plan on making cased sausages (not just patties) or quickly stuffing ground meat into bags. Hand-operated stuffers, while requiring more effort, carry the benefit of not heating the sausage mix during the stuffing process, which helps maintain the texture and quality of the product. Find one that has a capacity of 3 to 5 pounds and has interchangeable tubes for different casings.

BUTCHERING METHODS

FAMILIARITY WITH MUSCLE STRUCTURES and the skeletal map, knowledge about food safety and storage, and a collection of well-chosen tools will start you off on the right foot, but the real foundation to butchery as a craft is the mastery of methods. Understanding the conformation of muscle groups, the grain direction, the interworking of a joint, the connection points for seams: this knowledge is gained only through hands-on repetition. There is no substitute. The more pigs you break down, the better you'll become at boning the ham and removing the fat cap, as long as you aim to improve and take note of new details with each cut. A carcass is chock-full of cues that you will have to read to know where and how to cut. Every time you cut, you can learn something new; greater experience leads to a heightened awareness of the underlying intricacies, carrying you down the endless path of method and skill refinement.

We must all start somewhere, though, and this chapter provides the information you need for a foundational understanding of butchering methods, and then some. This is not beginner's knowledge per se; there is something to learn here for new and old butchers alike. These methods form the backbone of butchery. It is critical to have a solid grasp of the information in this chapter before you attempt any meat fabrication.

Foundations

EARLY BUTCHERING WAS, no doubt, less sophisticated than today's approach, which relies heavily on extensive research and analysis of carcasses. The basis of the craft, however — using a tool to segment a carcass into convenient edible parts — is still the functional foundation. Hence, before making the first cut, you must understand the basic foundations — skills that have been handed down through the centuries and are still practiced by modern butchers.

Holding a Boning Knife

There are two ways to hold a boning knife that you'll use during most of the butchering process: the pistol grip and the basic grip.

PISTOL GRIP

The pistol grip is the primary grip you'll use with the boning knife: invert the knife downward and hold the handle with a closed fist, the same way you would hold a gun (hence the name). The knife edge may face in either direction, depending on whether you're drawing the knife toward you or pushing it away from you; the former is the more common usage. The pistol grip allows you to hold on to the knife tightly, allowing the use of greater force to make your way through cartilage, fragile bone, ligaments, and other materials with resistance. Be especially careful when drawing the knife toward you: after cutting through some part of a carcass, it's not uncommon to accidentally jerk the knife toward your body, hence the need for protective aprons (see page 49). In most cases when this book suggests the use of a boning knife, this is the grip you should use.

BASIC GRIP

The basic grip is how most folks hold a knife when they pick it up. Any knife in your arsenal can be held this way: boning, butcher, cimeter, cleaver, you name it. The basic grip is used for cuts requiring accuracy and finesse, like navigating the shape of vertebrae while boning a loin. To further increase control, extend your forefinger onto the spine of the knife. Switch to this grip when the pistol grip feels crude or prevents the right angle of approach.

Pistol grip

Basic grip

Positioning the blade in the saw so that the teeth are facing toward you will give you more control, especially when you're working with smaller cuts.

Finesse with a Bone Saw

The use of a boning saw may seem straightforward enough, since its form is familiar and in many instances its operation mimics the similar tools found in a garage: load the blade with the teeth facing away from the handle and saw with an action that couples force with outgoing strokes. This works for sawing large round bones such as femurs or arm bones, crosscutting vertebrae (separating shoulders and legs from loins), or dividing other rigid structures.

Finer, more fragile bones present a different kind of challenge, one that is overcome with strategy rather than brawn. Examples are sawing through the ribs of a lamb rack, removing the chine bone from a suckling pig loin, or producing veal loin chops from a short loin. The use of force behind the teeth of the saw blade will disrupt these structures, making the bone edges jagged and the cuts sloppy. Furthermore, sawing through the more petite meat structures requires securing them to the table, most

often with the opposite hand. Maintaining a steady hand is a challenge in itself when plunging a saw forward, teeth dragging across bone. The solution to these predicaments may not seem obvious at first but becomes clear with very little practice: reverse the saw blade, teeth facing the handle, and allow the weight of the saw to be the only force behind the teeth. You have much greater control when drawing the saw toward you than when pushing it away, and this will allow for the deft work required to make clean cuts with fragile bones.

However, don't just plop the entire weight of the saw onto the bone and start sawing, expecting fine results. Instead, hold the saw so that the teeth are just touching the bone — they're sharp enough to cut, as long as you don't get impatient and try to muscle your way through — and then maintain this level of contact as you make smooth strokes with the saw. Don't rush it: just let the saw do all the work of cutting; your work is to maintain light pressure.

Honing Rod

The frequent use of a honing rod is one of the keys to maintaining knife sharpness and ensuring clean cuts. The honing rod realigns the misshapen edge of the blade; this requires only light pressure, just a bit more than touching, and only a few good passes to do the trick. If you find that using more pressure is required to garner results, then the knife needs to be sharpened, not honed. There are two approaches to using a honing rod: tabletop or freehand.

FIND YOUR ANGLE

Start by placing the rod tip down and perpendicular to table. (A towel underneath the tip can prevent the rod from slipping or forming divots in wood work surfaces.) Place the heel of the knife, nearest to the handle, at the top of the rod. Aim to hone an angle around 22 degrees. This is easiest to achieve by starting at 90 degrees, directly perpendicular to the rod. Split that in half, giving you 45 degrees, and then split that in half again, giving you about 22 degrees. (If you know the specific angle of your knife edge, estimate that instead.)

TABLETOP METHOD

For beginning butchers, the tabletop method is a great way to start. The rod is secure on the table, eliminating one of the variables (rod movement); this allows you to focus on factors like angle and pressure.

While maintaining your angle, draw the knife toward you while moving down the rod. Repeat with the opposite side.

FREEHAND USE

Once you've become comfortable with honing, integrating freehand use of the rod will be much easier than using the tabletop. This method uses the same angle and pressure, but the motion is performed while holding the rod in one hand.

Hold the rod, in the hand oppoosite your knife, at a slight upward angle. Place the heel of the knife, near the handle, at the tip of the rod, choosing an angle near 20 degrees. Draw the knife toward you, moving down the rod while curving your hand motion inward, ensuring that the entire edge makes contact with the rod. Maintain the angle and a light pressure throughout. Repeat with the opposite side of the knife, working on either the top or underside of the rod, moving the knife away from you (as shown) or drawing it toward you.

Trimming

AS A CARCASS GIVES WAY TO THE BLADE and muscles separate, falling from the bones, you'll come to realize that dissection is only the first step to producing palatable cuts of meat, a roughing out of the final product. Muscles are wrapped in fascial sheaths that connect with one another and converge to form tendon attachments to bone; an animal's glandular system is scattered throughout; halted suddenly, blood within the circulatory system coagulates as pockets in areas that went undrained; muscles require reshaping for culinary preparation. These, and many more, are all examples of how trimming is the true underpinning process of butchery.

Tenets of Trimming

A butcher is constantly trimming. It's the main action of cutting and is guided by a couple of ground rules.

TRIM AS LITTLE AS POSSIBLE

The first rule of trimming is to trim only what needs to be trimmed: nothing more, and certainly nothing less. Financially, trim represents lower value than whole muscle products. Ground meats and sausages are superb ways to garner value from edible trim, but, in general, meat is more lucrative when sold in a solid form, as portioned cuts or roasts. This is especially true when fabricating high-value primals like loins.

Trimming, like all other components of butchery, is a learned skill that gets refined through repetition and practice. When you first start butchering, you'll inevitably

The grain direction of muscles is important to recognize, especially when you're trimming, so that the meat can maintain a smooth surface and attractive appearance.

find yourself removing large swaths of meat during the trim process. This is okay. Focus on the priority, which is removing inedibles or shaping the cut mildly. (These priorities are further discussed on page 60.) Don't focus too much on the finer details of muscle curvature and final presentation. As your knife handling improves, your actions will become more precise. You will undoubtedly notice a natural reduction in the amount of meat you remove with each trimming motion, and it is then, and only then, that you should shift your focus to the next couple of ground rules. Otherwise it'll be like trying to cut a perfect circle out of a square piece of paper — you'll end up with a large amount of scrap paper (the trim pile) and a small circle (the piece of meat).

TRIM WITH THE GRAIN

When possible, trim in the direction of the muscle grain. Every muscle has a distinct although not always obvious grain direction. (Refer to page 6 for information about how muscle fibers create grain direction.) Think about

Trim with the grain whenever possible.

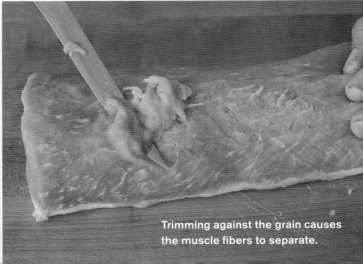

Trimming against the grain causes the muscle fibers to separate.

Trimming from Start to Finish

In the following sequence, a boned lamb rack is trimmed for palatability and presentation: beneath the thin layer of fat that sits atop the loin muscle is a swath of silverskin that is best removed. First, we remove the excess fat, then the silverskin, trimming with the muscle grain the entire time. The excess fat can be trimmed and tied back onto the roast, a process called barding: the fat renders while cooking, basting the lean roast while adding flavor.

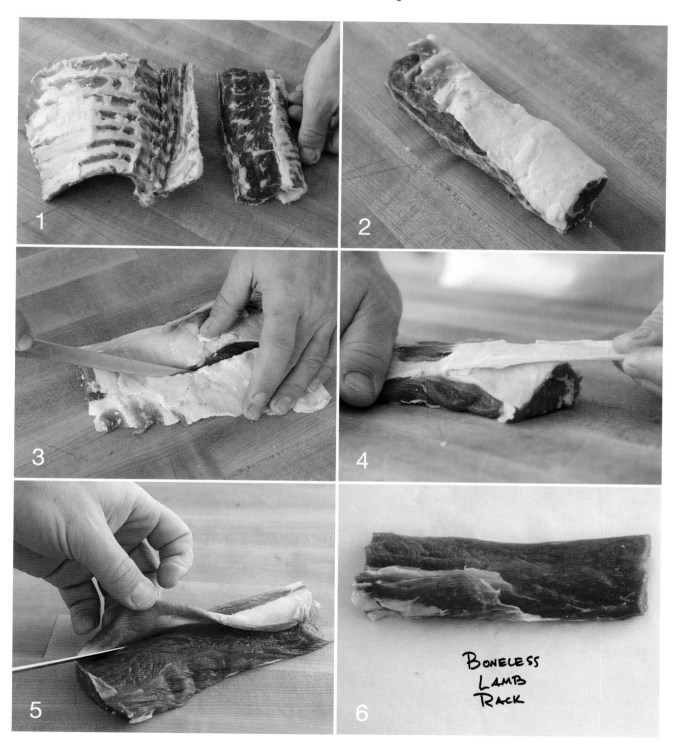

1

2

3

4

5

6

BONELESS
LAMB
RACK

those fibers like you would strands of hair. When you run your hand across the back of a dog, from head to tail, the hair lies flat and you can feel the fibers side by side. Switch directions and you have the canine equivalent of a bad hair day.

Muscle grain behaves in the same way. As a knife cuts meat, it severs individual muscle fibers. Cutting with the grain direction ensures that fibers remain tightly packed and maintain a uniform surface. Each fiber is sliced and laid next to another, in effect smoothing the surface. If the knife travels against the direction of muscle fibers (against the grain) it causes them to separate and lift, creating an uneven surface. This is most obvious with dense, thin-fibered muscles like the tenderloin, but it becomes evident in all muscles once you are aware of it. Additionally, the sharpness of your knife can dramatically affect the separation of the fibers. If you need to cut against the grain, use a recently honed knife and move slowly as you make the cut.

Removing Inedible Trim

The main priority when trimming is removing inedibles: the unpalatable pieces of meat and other bits found throughout the carcass. This may include anything from shriveled exterior surfaces to pus-filled glands.

ABSCESSES

Gland-looking growths that occur within a muscle are most likely abscesses and will need to be removed. An abscess is a sign of infection and, to be safe, a large area surrounding the abscess should be removed along with it. If you discover abscesses or other signs of infection or illness and are concerned about the safety of the meat, ask a local veterinarian to inspect the carcass.

BRUISES

During slaughtering, the process of stunning and the subsequent fall to the ground (along with some of the agitation that may lead up to slaughter) can result in bruised tissue. After an animal is skinned, the carcass should be inspected for bruising and any damaged tissue removed.

CONTAMINATED TISSUE

Between the time the animal was alive and the time its carcass landed on your butcher block, the meat you're cutting was exposed to contaminants. Areas close to where the hide was opened or the digestive tract removed are prone to contamination during slaughter. Furthermore, transportation of a carcass will often result in dirt or grime on the exterior, necessitating a careful inspection prior to aging or cutting.

DRIED TISSUE

Aging whole carcasses or extended storage in refrigerated spaces with ample airflow will cause surface drying of the meat. Discoloration and a hardened texture are both sure indicators of where to trim.

GLANDS

All animals have glands that need to be removed and discarded. Two glands that are exceptions to this rule are the thymus and pancreas, also known as sweetbreads. To be safe, if you are unable to identify a given gland, trim it away and dispose of it. All glands are located outside muscle, often among fat or connective tissue. The most common places to find large glands are at the convergence of large areas of the body — for example, at the leg and cavity, at the neck and shoulder. Smaller glands may show up in any other area of the body.

Separating Inedibles

Keep a separate container available for inedibles when you're butchering. Clearly differentiate it and keep it on

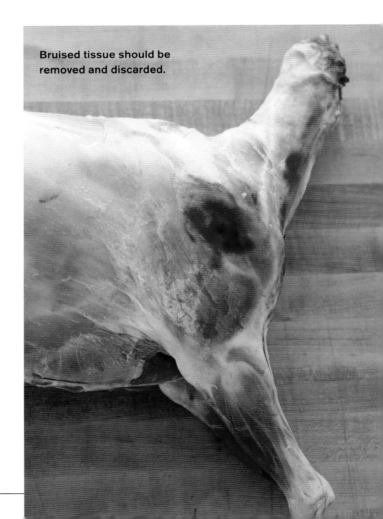

Bruised tissue should be removed and discarded.

the floor, on a pushcart, or on some other surface — never on the tabletop! — to help you avoid mistaking it for edible trim, especially when you're working in a team environment. Useful ways of marking containers of inedibles include using clear labeling on the outside and lid; buying color-coded containers; or, if you are not composting or rendering, just using a common trash can. Never keep inedible trim on the work surface; immediately dispose of it. Tightly integrating this habit into your workflow will dramatically reduce the risk of contamination and spoilage. (A single sour-tasting gland can ruin an entire batch of ground meat.)

Removing Unpalatable Trim

Fascia, ligaments, and tendons are unpalatable, but this does not mean they are without purpose in other culinary pursuits. They can be used to strengthen stocks, create a consommé raft, and add body to sauces.

FASCIA

All muscles are covered by a continuous layer of connective tissue called fascia. It exists throughout the entire mammalian anatomy, providing structure and support to muscles, organs, and the body overall. In some instances its presence is negligible, while in others it takes the form of a thick, fibrous, impenetrable layer called *silverskin*. Because it's difficult to chew and digest, silverskin always needs to be removed — a process called *denuding*, or peeling, the muscle. (See page 62 for more detail.) Thinner membranes of fascia are often left untrimmed, since they may not detract from the tenderness of the meat or they may melt into gelatin during cooking.

TENDONS

The fascial covering of a muscle or muscles converges to form a tendon, the attachment to bone. Tendons, rarely consumed in the United States, should be removed and reserved for other uses. Since they are composed mainly of collagen, tendons can be added as a natural thickener to stocks, soups, or any slowly cooked moist-heat dish. Tendons are often confused with ligaments, which are also collagen bands; however, ligaments connect bone to bone; no muscle connection is involved.

Edible Trim from Shaping

Save the edible trim: its utilization is one of the tenets of a zero-waste approach to whole-carcass fabrication. Edible trim is the by-product of the second major trimming priority: shaping.

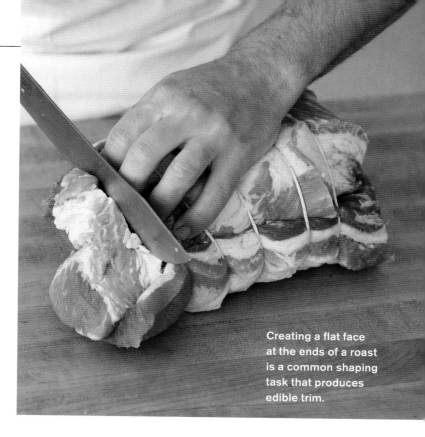

Creating a flat face at the ends of a roast is a common shaping task that produces edible trim.

The natural form of muscles is not always advantageous to the methods we use to make them edible. The tenderloin is a classic example, with its elongated tail that tapers as it extends from the butt end; another is a boneless sirloin, with its undulating shape after the pelvis is removed. These do not make for ideal shapes when aiming for cooking evenness or other culinary concerns. It is for this reason, among others, that we trim for shape. (Another way to affect shape is tying, discussed on page 74.)

Portioning cuts is a form of trimming for shape. It takes experience and skill to maximize portions from an oblong muscle. Cutting 1.25-inch pork chops from a loin primal is one thing; lopping off 8-ounce boneless portions of top round is a far greater challenge. Portioning requires accurate estimation along with knowledge of the ideal shapes and orientations to maximize repetitive shapes (the overall yield) from an uneven object. Along the way, the process produces edible trim. (See page 69 for more on how to portion.)

Edible trim to be used for grinding should be placed in a clearly labeled container separate from those containing tendons and fascia. Excess fat should be cut away from the lean meat and kept in a third container, allowing for control over the fat content in ground meat or for use in rendering.

Denuding

Denuding is the action of removing the exterior surface of a muscle or group of muscles. The word is most often applied to the removal of silverskin, and there are a couple of different methods. Here is an example of one method, showing how to denude a pork tenderloin while also demonstrating how to test for grain direction. Some muscles – loin, sirloin flap, and clod heart, to name a few – are better denuded by removal of the entire silverskin in one piece rather than in stages, as with the tenderloin. This is often true for thick silverskins that cover large areas of a muscle.

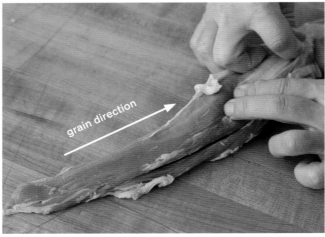

1. BEGIN BY DETERMINING THE GRAIN DIRECTION.
Scrape the blade along the surface in both directions and notice which one provides resistance, maybe with some fibers lifting from the surface. Thinly trim a small area of the surface and run your fingers along it, feeling for fibers. The direction without resistance – in which scraping lifts no fibers and you feel no fibers when running your fingers along it – is the direction in which you want to cut.

On the top side of the tenderloin, this direction is from the tapered tail toward the butt end. The underside will always be the opposite, so in this case from the butt end toward the tail. Once you discover the grain direction for a muscle, it will be the same for other species. For example, pork, lamb, goat, and beef tenderloins all come from the same muscle group – *psoas major* and *psoas minor* – making the grain directions identical.

2. CUT ALONG THE GRAIN.
Cutting direction follows grain direction: we now know to cut from the tail toward the butt end. Find the end of the silverskin and slide the knife underneath a portion of it. Secure the tenderloin by placing a hand on the muscle behind the knife. Angle your knife edge slightly toward the silverskin. Cut the length of the silverskin, moving your hand to stabilize the muscle as needed. Sever the strip of silverskin at its origin and then repeat until all the silverskin is removed.

Seam Cutting

Separating muscles along natural seams makes for more efficient cutting and maximizes the carcass yield. It also requires less effort, making for less fatigue. With a little force applied and the right connections severed, muscles tend to fall away from each other. Gravity may be all the force that is needed, though often it will require some muscle of your own.

The action, regardless of whether you are working on the hook or on a table, is basically the same. In both instances a boning hook is a useful tool for increasing the force applied with minimal exertion.

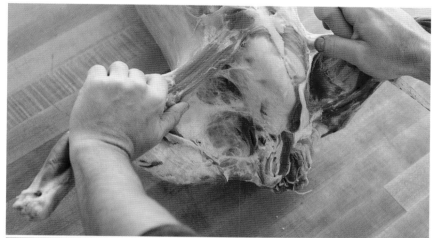

1. **FIND THE SEAM.** Start by finding a seam of fat and connective tissue that joins two muscles. This can be done at known junctions between muscles, as with this lamb shoulder, or by exploring with your hands, looking for areas where muscles meet. You can work your fingers into the seam to test for the ease of separation; some muscles, like those in the flank, can be separated mostly without the use of a knife.

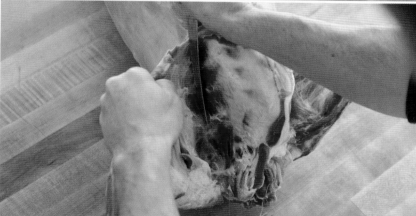

2. **SEVER ATTACHMENTS.** Begin severing the attachments between muscles using a knife, staying in the seam. Keep the cuts shallow: you're mainly trying to peel the muscles away from one another, while the knife will sever the more stubborn connections.

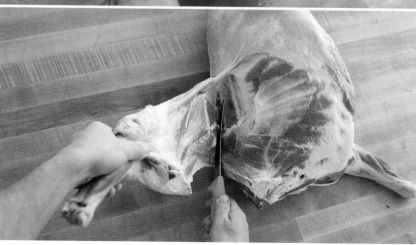

3. **PULL AND SEVER.** If you're working on the table, secure the underlying muscle with one hand while pulling with the other, peeling the muscles away from each other. If you're working on the hook, the underlying muscle will be attached to the hanging carcass; thus all your effort can go into pulling. Continue this back-and-forth action of severing attachments and pulling, keeping the cuts within the natural seam, until the two muscles are completely separated.

Boning

EACH BONE AND JOINT IN AN ANIMAL has its own characteristic shape, formed for specific tasks through generations of evolution. Understanding a bone's purpose can help you understand how that bone plays a role in the overall skeletal structure of an animal. And where there is a bone, there will be a joint. Every joint is formed by the intersection of uniquely shaped ends of two or more bones, often covered in cartilage and held together by ligaments and woven layers of connective tissue called *deep fascia*. Knowing the shape of these bones and the manner in which they fit together to form joints will vastly improve your ability to move quickly and efficiently through the process of boning hams, chicken legs, pork shoulders, lamb loins, and so on.

Basic Boning

Boning is the act of removing one or more bones from a part of the carcass with the goals of maximum muscle removal and clean, separated bones. There is an ideal entry point for the removal of each bone and its associated muscle grouping. Through repetition (and some guidance in this book) you'll learn how to elegantly access a bone and remove it with minimal effort.

REVIEWING THE SKELETAL STRUCTURE

Before you bone an animal for the first time, take a quick look at a diagram of its skeletal structure. I cannot stress this enough. Five minutes with a diagram will save you time, effort, and money in wasted meat. Better yet, keep a copy nearby while you work. Take note of the specific bones in the area you're going to be working on. What is their shape? Where do the bones connect and form joints? What is their orientation? Asking these three simple questions will dramatically improve your results. If you're working on a leg and plan on removing the femur, you will know that it is a long bone with joints on both ends and that it runs from the hip to the knee. If you're boning a shoulder and have to remove the scapula, you'll know that it has a ridge on the side facing outward and that it is flat on the other side; it has only one joint and connects to the humerus in the arm. These are very quick observations that will inform you when deciding where to access bones and how to remove them.

Before removing a scapula, typically one of the hardest bones to follow, it's important to understand its shape and orientation.

Boning

Any boning job can roughly follow the same set of tasks. After some experience, your order of operations may shift here and there, but starting with this approach will help inform you before making adjustments.

In this example we're boning a leg, with the main bone being the largest bone in the carcass: the femur. It's a round bone, so we'll be working around a cylindrical shape rather than against a flat face. This means that cuts will need to be shallow to avoid cutting deeply into the underlying muscles.

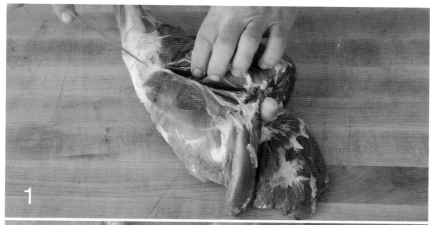

1. IDENTIFY THE BONE AND CHOOSE AN ENTRY POINT.

Familiarize yourself with the shape and location of the bone, along with any joints, then find a muscle seam that leads to the bone. Cut along the seam until you reach bone.

2. CUT ALONG THE BONE.

Upon reaching the bone, cut along the direction that it runs – not perpendicular to it. Move the blade along the bone in long, shallow strokes, releasing small portions of the muscle surrounding it and thus gaining access to the bone. Completely separate the top side of the bone, peeling the muscle away while using long strokes.

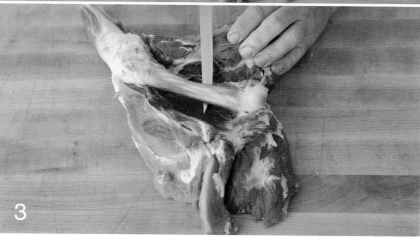

3. CUT UNDER THE BONE.

Find the edges of the bone, follow their shape, and start separating the underside. Stay close to the bone, separating one end before the other. In most cases, separate the joint-end of a bone as the last step.

Joints

Although it's possible to bone an entire carcass and leave the skeleton intact, chances are that won't be your approach. Most likely you'll be breaking down a carcass into primals and then boning, and to do so you will need to cut through some joints. You could do this using a handsaw, or even a bandsaw in some situations, but this would be missing the point. Learning the anatomy of an animal is paramount to butchering, without which the craft and beauty of being able to produce delectable cuts of meat with a simple tool (the knife) would be lost. Modern tools have their place, and can certainly make some tasks faster, but a solid understanding of anatomy is what will make you a great butcher. Mastering the joints is part of this.

A joint is the meeting place of two bones, held together by a combination of muscle, ligaments, tendons, or other connective tissue. Although you can look at a skeletal diagram to see the exact shape of a bone, learning the access points of a joint will take more trial and error — and quite a few sharp knives turned dull. To separate the bones, you'll need to understand the shape of the connecting parts (cartilage-covered bone ends) along with what connective tissue (ligament, tendon, etc.) needs to be severed to release the bond.

SEPARATING THE JOINT

When exploring how to separate a joint, go slowly. Don't try to cut your way into or through the joint with sheer force. Don't insert your knife tip into the joint and try to pry it apart. Not only will these approaches quickly dull or ruin your knife, but you're bound to cut yourself doing it. Instead, take the time to examine the area around the joint and note the obvious connective tissues. Make some exploratory cuts, use the tip of your knife to cut through some ligaments, and then apply force to the joint by moving the bones or trying to bend them over the table edge. Notice which incisions cause more instability than others do. Cutting through large tendons and ligaments is an ideal place to start. Don't be timid — you may at times feel like you're cutting through bone or cartilage because of the force it takes to get through them. After you knife through a joint a few times, the right method of approach will surely stay with you.

Peeling the Bone

Every bone is wrapped in a layer of fascia, which the musculature surrounding the bones attaches to. When boning, sometimes it's more beneficial to "peel" the bone — to get under the fascial layer and remove it with the muscle itself. Examples of such situations are removing the scapula from a shoulder, separating the sirloin tip from a hanging carcass, and frenching the bones of a lamb rack. These examples and others are explained with instructions in their respective chapters.

There are two different approaches to peeling the fascial layer off a bone, and they are easily differentiated.

REMOVING THE MUSCLE. With the first method, your goal is to cleanly remove a muscle, or group of muscles, that are attached to bone; the bone gets left on the carcass.

REMOVING THE BONE. The second method is the inverse: removing a bone while leaving a muscle or group of muscles still attached to the carcass.

Fortunately both methods employ the same skills. As mentioned earlier, butchering is largely a process of identifying and severing key connection points between muscles and bone and allowing parts to fall away according to natural seams. Peeling a bone is no different. Every bone has ideal approaches to releasing its fascial layer, and through practice you will understand how best to approach each unique situation.

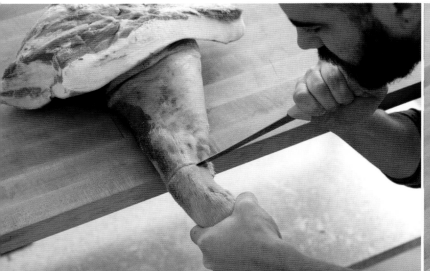
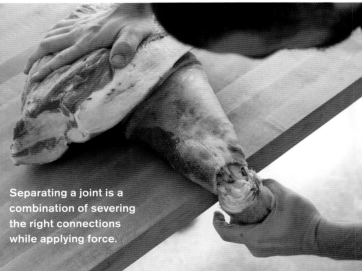

Separating a joint is a combination of severing the right connections while applying force.

Peeling the Bone

Here are the basics to get your explorations in peeling the bone started, using the sirloin tip as an example.

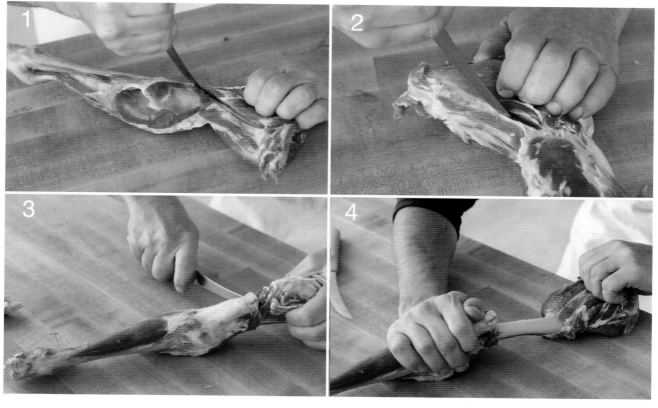

1. Find a muscle seam and make your way to the bone. On one side of the muscle, make a firm incision along the bone using the tip of your knife. (The first time you do this it might feel like you are cutting into bone itself, but you will soon realize that where you think there is only bone there is actually another layer of fascial covering.)

2. Repeat with the other side: find a seam, get to the bone, and then make a firm incision.

3. Sever the upper connection of the muscle, also ensuring that the cut goes deep to the bone and through the fascial covering.

4. Steady the bone with one hand. With a firm motion, pull the muscles toward you, peeling it away from the bone. Sever any remaining connections to finish separation.

MAKING USE OF BONES

Finding a use for bones will maximize your carcass yield. Any of the vertebrae or joints will be excellent stock bones due to heavy amounts of collagen and connective tissue, structures that break down under moist heat into unctuous gelatin. The round bones — femur, tibia, ulna/radius, humerus — can be crosscut for marrow bones, smoked for stocks, or used as soup bones. Ribs, scapulae, and pelvic bones are all great for dogs. By all means, find a use for the bones and get the most out of your animal.

Tenderization

MEAT, IN ITS RAW STATE, is generally unpalatable; our teeth struggle to sever the bundles of muscle fibers and make their way through the tough collagen. Raw meat that is served for consumption has been processed in a manner that overcomes this limitation and allows us to enjoy the experience: sliced exceedingly thin, as with carpaccio, or diced into small bits, as with tartare.

Cooking is the main form of meat tenderization. It unravels proteins; breaks down muscular structures; and, along with increasing the nutritional content, transforms a previously intimidating hunk of flesh into a delectable gastronomic experience. But cooking isn't the only tool we have to increase tenderness. We can also employ techniques to sever the fibers for us (mechanical tenderization) or use acids and enzymes to break down protein structures (chemical tenderization).

Cutting across the Grain

Discussed in the preceding section on trimming, muscle grain direction is an important trait to identify when butchering. Cutting with the grain is important for trimming, but if you employed the same approach when portioning, the results would be subpar. This is because the length of muscle fibers would be a challenge for us to chew through. Instead, we want to shorten the fibers whenever possible, thus increasing the tenderness of resultant portioned cuts. We do this by cutting the meat perpendicular to the grain direction. Kari Underly, a talented butcher and consultant for the meat industry, has named this technique the "90-degree rule," which helps illustrate the idea that you always want to cut exactly perpendicular to the linear direction of the muscle fibers.

An easy example of cutting across the grain is chops and steaks. The main loin muscle, *longissimus dorsi*, has a grain direction that runs roughly parallel to the spine. Hence, steaks and chops, whether bone-in or boneless, are cut perpendicular to the spine. Cutting across the grain becomes more challenging when the grain direction doesn't run so conveniently with the muscle itself, as illustrated in the photos below.

Mechanical Tenderization

Technically speaking, the 90-degree rule is a form of mechanical tenderization, since it employs a tool to do the work, but mechanical tenderization is more frequently applied to techniques that grind, pound, or sever muscle fibers into shorter forms of themselves. You've probably eaten a hamburger or sausage: in this form of mechanical tenderization the meat is ground into small pieces, aiding in mouthfeel while also creating a unique form. Meat grinders use manufactured dies, plates, and knives to break down characteristically fibrous muscles

It can be difficult to cut a steak when the grain doesn't run parallel to the length of the muscle. In these cases, slice parallel with the grain to separate a section of the muscle, equal in thickness to your target portion. Then, slice perpendicular to the grain for final portioned cuts.

into tender morsels. Mallets and meat hammers tenderize meat by separating the fibers, breaking apart the collagen structures that hold them tightly in bundles, and often creating flattened cuts like cutlets. The Jaccard is a tool that utilizes dozens of flat, sharp knives to penetrate meat and sever muscle fibers (see page 51). It serves the purpose of prechewing and can dramatically improve the texture of tough cuts. But be wary of overusing it, which can result in a mealy texture.

Chemical Tenderization

Chemical tenderization is more passive than mechanical tenderization. (Passive for us, though the action of the chemicals is anything but.) With this method, you expose the meat to a solution that aims to break down protein structures. Given adequate time, this will lead to dramatic improvements in tenderness; left to their own devices for too long, the ingredients in the solution will make the meat mushy and overtenderized. Common ingredients that aid in tenderization of muscle cuts fall into one of the following categories.

ACIDIC

The most common form of acidic tenderization is the traditional marinade, which often includes ingredients like citrus juice, wine, vinegar, or anything else with a pH below 5. The action of an acidic tenderizer is highly effective, breaking down both connective tissues and muscle proteins. An additional benefit is the increase in moisture-holding capacity, making the meat juicier as well as more tender.

ALKALINE

Though they are on the opposite end of the pH scale, alkaline marinades react with meat in a very similar manner to acidic mixtures. Frequent ingredients include baking soda, soda lime, lye, or anything else with a pH above 9.

ENZYMES

Fruit juices are a source of enzymes that snip proteins into segments, thus increasing tenderness. The following are the most popular fruits (with their enzyme names in parentheses): kiwi (actinidain), papaya (papain), and pineapple (bromelain). Bromelain is highly effective because it attacks both muscle and connective tissues. These enzymes can be found in liquid form (fruit juice) or in a concentrated powder, the latter being the common

form found for sale in grocery stores. (Enzymes are discussed in more detail on page 19.)

FERMENTATION

The live cultures in fermented products can enact positive changes on meat, allowing the naturally occurring bacteria to attack muscle fibers, breaking them into fragments. Buttermilk and yogurt are two traditional fermented products used for tenderization. They do not break down collagen fibers, like other chemicals used for tenderization do, but they do provide an inhospitable environment for dangerous bacteria, thus retarding spoilage.

Portioning

MEAT MUSCLES CAN BE COOKED in any form, large or small, but there is a frequent need to cut a muscle into portions, be it individual steaks, a two-person roast, or sizes of specific weight. Portioning is a true skill, requiring repetition and time to maximize usable portions from amorphous lumps of meat. Some instances are easier than others — ripping 1-inch-thick pork chops from a loin primal on a bandsaw is more a redundant activity than one of dutiful accuracy — but each action of fabricating portioned cuts should embody the same priority: maximize yield, minimize loss.

Portioning is based on either weight or size. Starting with an entire boneless loin, you can create a set of 6-ounce boneless pork chops by weighing the entire boneless loin and then dividing the total weight by your target portion weight (in this case 6 ounces). You could also portion the loin by size, in this case thickness, by cutting 1-inch chops. When portion weights are not critical, use the easier method of portioning by size.

Portioning muscles of inconsistent and oblong shapes (which, frankly, most muscles are) poses some other challenges to maximizing your yield. Consider this: when portioning for weight, the width for a portion will also affect the thickness. A 6-ounce portion that is 3 inches wide will be much thicker than one that is 5 inches wide. If the muscle is 9 inches wide, choose a width that divides cleanly, like 3 inches or 4.5 inches.

Portioning Based on Weight

Muscle of a consistent and elongated shape, like this trimmed pork loin, is good to use as you start practicing portioning. In this case we are going to cut boneless pork loin chops.

Start by identifying the grain direction and square off one end, cutting perpendicular to the grain. Here, the grain of the loin runs parallel with the muscle. Weigh the whole loin and then determine a portion weight that maximizes your yield. Our loin weighs 2 pounds, 4 ounces (i.e., 36 ounces total). A portion weight of 6 ounces will give us an even 6 chops.

Measure the length of the muscle. Our loin measures 12.5 inches long. Divide that by the number of portions (6) to determine a relative width to cut, in this case around 2 inches.

Measure out the thickness for the first portion, cut, and weigh to determine accuracy. Adjust the width as needed. Cut additional portions, weighing the portion after each cut to ensure accuracy.

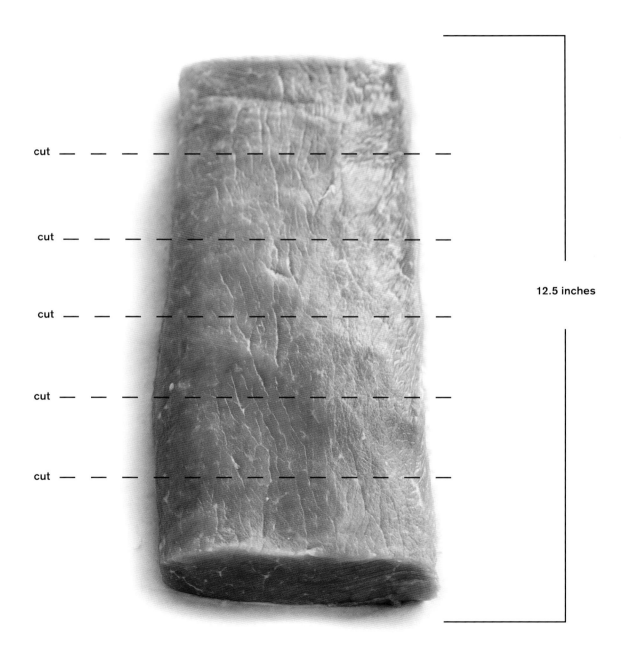

cut

cut

cut

cut

cut

12.5 inches

Portioning Based on Thickness

Here we are going to use the pork loin again to demonstrate portioning based on uniform thickness.

Start by identifying the grain direction and square off one end, cutting perpendicular to the grain. Keep in mind that the grain of the loin runs parallel with the muscle.

Measure the length of the entire boneless loin and determine an appropriate portion thickness that maximizes the yield. In this example the loin is 12.5 inches, so we're going to aim for 1.25-inch-thick portions, which will give us 10 chops. Measure out the thickness for the first portion and cut.

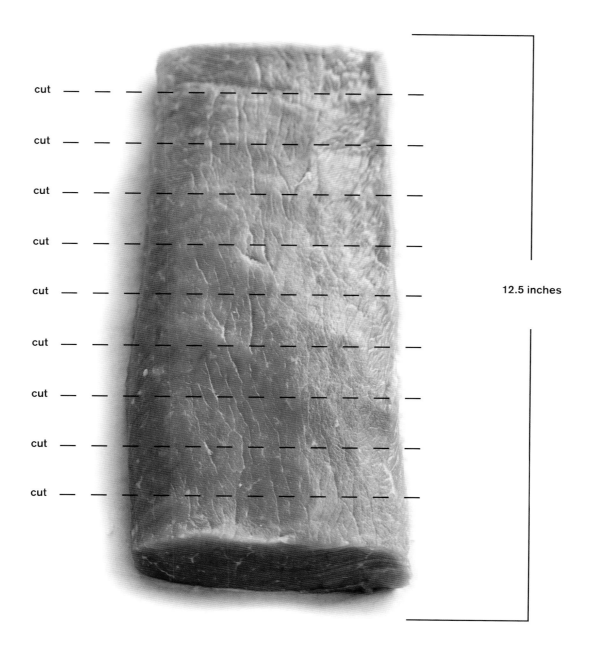

cut

cut

cut

cut

cut

cut

cut

cut

cut

12.5 inches

Deciding What to Cut

Making your way through an entire carcass cut by cut requires a fair amount of planning to ensure that you are maximizing your yield. It is important to have a clear idea of how each primal will be divided so that portioning and proper storage can happen efficiently, because it may take a couple of days to break down and properly package every cut from the carcass, especially if you are new to the process. No matter how large or small the animal is, the best way to organize your efforts is to use a cut sheet. A cut sheet is a simple concept but is the most valuable tool for minimizing mistakes in processing. It is, in essence, a cutting directive: an obvious process of how to break down the carcass will arise when the required products are clearly laid out.

You can create your own cut sheet by simply going through each primal and writing down a list of what you want to produce from it. After you have read through the butchering chapters in this book, you should know what the possibilities are for the appropriate primals and which ones work for you or your customers.

Pork Cut Sheet

Side 1 (Choose one option under each section)

Side 2 (Choose one option under each section)
☐ Use the same preferences as Side 1

HEAD
☐ Boneless ☐ Whole ☐ Jowl ☐ Cheek

BOSTON BUTT
☐ Whole Boneless ☐ Whole Bone-in *Skin on?* _____
☐ Shoulder Chops *Thickness:* _____ *(at least 1" recommended)*
☐ Country-Style Chops *Thickness:* _____ *(at least 1" recommended)*
☐ Boned for Sausage
☐ *Use Scrap for Sausage?* _____ *Ground Pork?* _____

PICNIC
☐ Whole Boneless ☐ Whole Bone-in *Skin on?* _____
☐ Boned for Sausage ☐ Ground

BELLY
☐ Spare Ribs Boneless *Skin on?* _____
☐ Spare Ribs Bone-in *Skin on?* _____

RIB LOIN
☐ Rib Chops *Thickness:* _____ *(at least 1" recommended)*
☐ Boneless Rib Chops ☐ Back Ribs
☐ Rib Chops *Thickness:* _____ *(at least 1" recommended)*
☐ Bone-in Roast Weight (lbs) _____
☐ Boneless Roast Weight (lbs) _____

SHORT LOIN
☐ Porterhouse Chops *Thickness:* _____ *(at least 1" recommended)*
☐ Loin Chops *Thickness:* _____ *(at least 1" recommended)*
☐ Boneless Loin Chops *Thickness:* _____ *(at least 1" recommended)*
☐ Bone-in Roast Weight (lbs) _____
☐ Boneless Roast Weight (lbs) _____
☐ Tenderloin ☐ Riblets

SIRLOIN
☐ Chops *Thickness:* _____ *(at least 1" recommended)*
☐ Bone-in Roast
☐ Boneless Roast

LEG (HAM)
☐ Whole Bone-in *Frenched?* _____ *Skin on?* _____
☐ Whole Boneless ☐ Small Roasts ☐ Kebabs ☐ Cutlets
☐ Ham Steaks *Thickness:* _____ *(at least 1" recommended)*
☐ *Use Scrap for Sausage?* _____ *Ground Pork?* _____

HOCKS
☐ Whole *Skin on?* _____ ☐ Cross-cut *Skin on?* _____

VARIETY MEATS
☐ Brains ☐ Fat Back ☐ Leaf Lard ☐ Stomach
☐ Caul Fat ☐ Heart ☐ Liver ☐ Tongue
☐ Cheeks ☐ Kidney ☐ Spleen

HEAD
☐ Boneless ☐ Whole ☐ Jowl ☐ Cheek

BOSTON BUTT
☐ Whole Boneless ☐ Whole Bone-in *Skin on?* _____
☐ Shoulder Chops *Thickness:* _____ *(at least 1" recommended)*
☐ Country-Style Chops *Thickness:* _____ *(at least 1" recommended)*
☐ Boned for Sausage
☐ *Use Scrap for Sausage?* _____ *Ground Pork?* _____

PICNIC
☐ Whole Boneless ☐ Whole Bone-in *Skin on?* _____
☐ Boned for Sausage ☐ Ground

BELLY
☐ Spare Ribs Boneless *Skin on?* _____
☐ Spare Ribs Bone-in *Skin on?* _____

RIB LOIN
☐ Rib Chops *Thickness:* _____ *(at least 1" recommended)*
☐ Boneless Rib Chops ☐ Back Ribs
☐ Rib Chops *Thickness:* _____ *(at least 1" recommended)*
☐ Bone-in Roast Weight (lbs) _____
☐ Boneless Roast Weight (lbs) _____

SHORT LOIN
☐ Porterhouse Chops *Thickness:* _____ *(at least 1" recommended)*
☐ Loin Chops *Thickness:* _____ *(at least 1" recommended)*
☐ Boneless Loin Chops *Thickness:* _____ *(at least 1" recommended)*
☐ Bone-in Roast Weight (lbs) _____
☐ Boneless Roast Weight (lbs) _____
☐ Tenderloin ☐ Riblets

SIRLOIN
☐ Chops *Thickness:* _____ *(at least 1" recommended)*
☐ Bone-in Roast
☐ Boneless Roast

LEG (HAM)
☐ Whole Bone-in *Frenched?* _____ *Skin on?* _____
☐ Whole Boneless ☐ Small Roasts ☐ Kebabs ☐ Cutlets
☐ Ham Steaks *Thickness:* _____ *(at least 1" recommended)*
☐ *Use Scrap for Sausage?* _____ *Ground Pork?* _____

HOCKS
☐ Whole *Skin on?* _____ ☐ Cross-cut *Skin on?* _____

BONES
☐ Marrow Bones *Thickness* _____
☐ Stock Bones
☐ And All the Rest!

SAUSAGES

		Qty.		Linked	or Loose
Andouille: *A spicy, smoked sausage from the Bayou region*		_____ lbs.		_____	_____
Breakfast: *A classic sausage with notes of sage and a touch of spice*		_____ lbs.		_____	_____
Italian: *The classic, with fennel and garlic flavors*		_____ lbs.		_____	_____
Spicy Italian: *A spicy variant of the aforementioned classic*		_____ lbs.		_____	_____

Create your cut sheet a day or more ahead of when you plan to butcher, to give yourself time to mull over the options and make any changes. Once you cut the meat there is no changing your mind, and what you cut first will determine what else you can cut. One classic example is the saddle or short loin of an animal: it contains the muscles for loin chops as well as the tenderloin; in combination they make porterhouse chops. Thus, if you want to cut porterhouse chops you cannot also have tenderloin and loin chops, and vice versa. It is therefore important to understand the relationship between cuts when determining your cut sheet. (Fortunately, butchering for yourself has custom advantages: you can cut a few porterhouse chops off the sirloin end of the loin and then peel off the remainder of the tenderloin and cut loin chops, giving you some representation of all three categories.)

Lamb Cut Sheet

Side 1 (Choose one option under each section)

Side 2 (Choose one option under each section)

☐   Use the same preferences as Side 1

NECK
☐   Boneless
☐   Bone-in Slices
☐   Stew
☐   Grind

NECK
☐   Boneless
☐   Bone-in Slices
☐   Stew
☐   Grind

SHOULDER
☐   Blade & Arm Chops *Thickness:* _____ *(at least 1" recommended)*
☐   Boneless Arm Roast & Boneless Shoulder Roll
☐   Boneless Shoulder
☐   Stew Meat

SHOULDER
☐   Blade & Arm Chops *Thickness:* _____ *(at least 1" recommended)*
☐   Boneless Arm Roast & Boneless Shoulder Roll
☐   Boneless Shoulder
☐   Stew Meat

RACK
☐   Rib Chops *Thickness:* _____ *(at least 1" recommended)*
☐   Bone-in Rib Roast *Frenched?* _____ *Fat cap?* _____
☐   Boneless Rib Roast *Fat cap?* _ ___

RACK
☐   Rib Chops *Thickness:* _____ *(at least 1" recommended)*
☐   Bone-in Rib Roast *Frenched?* _____ *Fat cap?* _____
☐   Boneless Rib Roast *Fat cap?* _____

LOIN
☐   Loin Chops *Thickness:* _____ *(at least 1" recommended)*
☐   Boneless Loin Roast *Fat cap?* _____

LOIN
☐   Loin Chops | *Thickness:* _____ *(at least 1" recommended)*
☐   Boneless Loin Roast   *Fat cap?* _____

BREAST	**RIBS**	**BREAST**	**RIBS**
☐ Bone-in	☐ Denver Ribs	☐ Bone-in	☐ Denver Ribs
☐ Boneless	☐ Riblets	☐ Boneless	☐ Riblets
☐ Grind	☐ Grind	☐ Grind	☐ Grind

SIRLOIN	**SHANKS**	**SIRLOIN**	**SHANKS**
☐ Sirloin Chops	☐ Bone-in *Crosscut?* _____	☐ Sirloin Chops	☐ Bone-in *Crosscut?* _____
☐ Boneless Roast	☐ Grind	☐ Boneless Roast	☐ Grind

LEG
☐   Bone-in & Whole *With Sirloin?* _____
☐   Leg Steaks *Thickness* _____ *(at least 1" recommended)*
☐   Boneless   *Butterflied?* _____
☐   Small Roasts
☐   Kebabs

LEG
☐   Bone-In & Whole *With Sirloin?* _____
☐   Leg Steaks *Thickness* _____ *(at least 1.25" recommended)*
☐   Boneless   *Butterflied?* _____
☐   Small Roasts
☐   Kebabs

VARIETY MEATS			**BONES**
☐ Brains	☐ Caul Fat	☐ Cheeks	☐ Marrow Bones *Thickness* _____
☐ Heart	☐ Kidney	☐ Liver	☐ Stock Bones
☐ Lungs (Lights)	☐ Spleen	☐ Stomach	☐ And All the Rest!
☐ Sweetbreads	☐ Tail (Oxtail)	☐ Tongue	

SAUSAGES

Lamb Sausage: *A traditional tasting sausage with garlic, fennel, and other spices*	Qty.: _____ lbs.	Linked: _____ or Loose: _____
Merguez: *A spicy sausage with red pepper, garlic, cumin, and other spices*	Qty.: _____ lbs.	Linked: _____ or Loose: _____
Lamb with Parmesan and Mint: *A moderately spicy sausage with bright flavors*	Qty.: _____ lbs.	Linked: _____ or Loose: _____

Making Knots

If you've ever cooked an oblong roast, a pork chop that won't sit flat on the pan, or even a slice of bacon that curls too much, then you've experienced how the shape of meat can dramatically affect the outcome of cooking it. Evenness of cooking is often a priority, and to achieve that we often need to control the shape of meat. This begins with proper cutting and portioning, but even the best-fabricated cuts will sometimes need reinforcement. Tying with butcher's twine is one way to do this.

There are two main knots that I find useful for tying: the butcher's knot and the packer's knot. Both are basic slip knots that rely on a final hitch to secure them. They are virtually interchangeable, so use whichever suits your fancy. In both examples there is a "long" string and a "short" string. The long string will be the one connected to your roll of twine; the short string is the section you are wrapping around the roast. These instructions are written from a right-handed point of view; you'll need to swap hands if you're left-handed.

BUTCHER'S KNOT

Once you get the hang of the butcher's knot, you'll understand its main benefit: speed. You can quickly tie and hitch this knot much faster than with the packer's knot. In simple terms, think of this knot as: loop once, knot, loop twice, hitch.

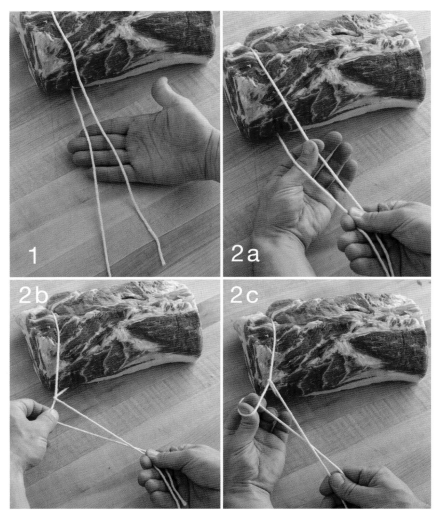

1. **WRAP THE ROAST.** Run the short end of the string under the roast and then over the top. Hold both the short and long strings in your right hand.

2. **FORM THE LOOP.** Continue to hold both strings in your right hand during this entire step.

2a. With your left hand, go under the long string and hold the short string between your thumb and forefinger.

2b. Pull the short string under the long string, and twist your left wrist in the process to form a loop.

2c. Your left fingers should be inside, and form, the loop.

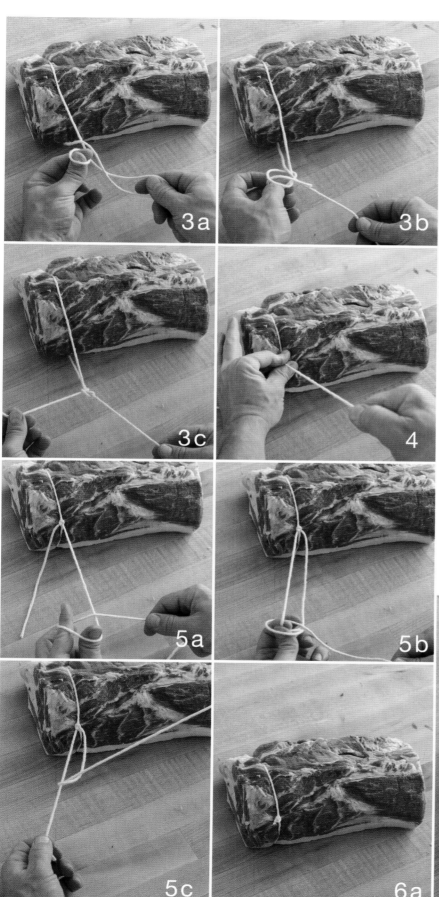

3. FORM THE SLIP KNOT.
Grab the end of the short string in those left fingers and pull it through the loop. This will form your slip knot.

4. CINCH.
Hold the knot with your left hand and pull on the long string with your right hand to cinch the knot tight against the roast.

5. SECURE THE KNOT.
Wrap the long string around the thumb and forefinger of your left hand to form another loop. Again, pull the short string through the loop, in this case forming the hitch that will secure the slip knot. Pull the hitch tight and, if you want, make another hitch and double it up.

6. TRIM.
Tighten the knot and trim off the excess string and repeat knots every 2 inches down the roast.

PACKER'S KNOT

The packer's knot is an attractive knot formed by making a figure-eight. Unlike the butcher's knot, which has to be secured with a hitch, the packer's knot has the ability to hold on its own. In simple terms, think of this knot as: under, under, and through the loop.

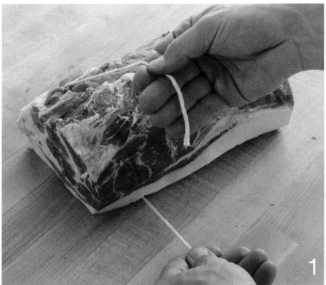

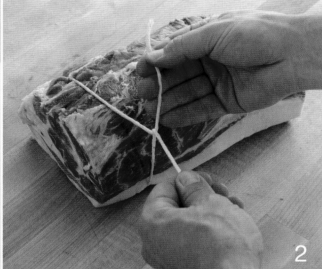

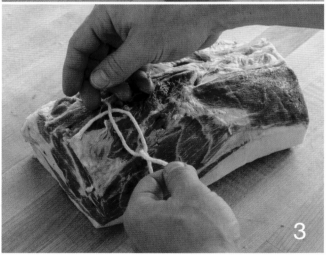

1. **WRAP THE ROAST.** Run the short end of the string under the roast and then over the top. Hold the short string with your left hand and the long string in your right.

2. **UNDER ONCE.** Pass the short string under the long string from your left hand to your right. There is now an intersection of string, with the short running under the long.

3. **UNDER TWICE.** Run the end of the short up and back under the portion of short string that is lying atop the roast.

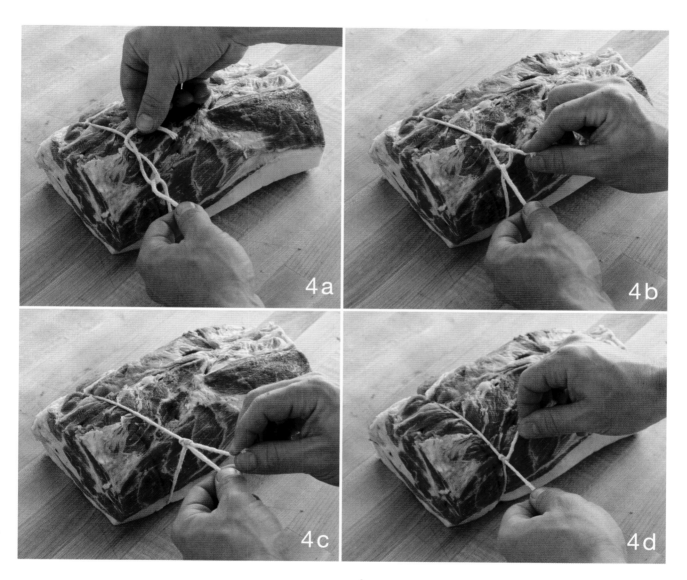

$4a.$ **THROUGH THE LOOP AND HITCH.** There is now a loop formed by the short string.

$4b.$ Run the end of the short string back over itself and then up through the bottom of the loop.

$4c.$ and $d.$ Pull on both ends of the string to tighten, then pull on the long string to cinch it against the roast. Hitch to secure. (See step 5 of Butcher's Knot, page 75.)

PRE-SLAUGHTER CONDITIONS & GENERAL SLAUGHTER TECHNIQUES

THE RESPECTABLE HARVEST of any animal is less a single event than it is the culmination of many considerations and preparatory procedures taken to ensure the highest-quality outcome. You cannot expect ideal results if the animal has not been tended to properly, equipment is absent or inoperable, additional help is not arranged for when needed, or a cavalier approach trumps the humility of inexperience. It is for these and many more reasons that I impress upon anyone harvesting livestock the obligation of making adequate arrangements, for both the well-being of those creatures destined for death and the perishable products they will provide. You simply can't be too prepared.

General statements in this chapter will apply to all animals discussed in this book (except when noted otherwise). Animal-specific topics, though mentioned here, are explored in more detail within their respective chapters – for example, chicken pluckers are covered in the poultry chapter (page 105).

The Day Before

THE HARVESTING of any animal will benefit from some basic day-before measures.

Separate Animals

Separate the animals destined for slaughter from the larger group. Unless the slaughter will take place in the field, the holding location for the animals should be near where the slaughter will happen. It's best to give animals 24 hours or more to acclimate to their new surroundings; this will help reduce their stress and anxiety, making handling easier while also helping to preserve the meat quality. The animals' new environment should include easy access to shade and, for animals that need it, shelter. Herd animals such as sheep, goats, and cattle are more comfortable with at least one other animal around, so consider dispatching them in pairs or groups.

PROVIDE ADEQUATE BEDDING, especially if the separated animals are being held in a trailer. Without bedding, animals tend to resist urination, making for a full bladder during evisceration.

DO A QUICK ANTEMORTEM INSPECTION of the animals you'll be harvesting. Should you find anything questionable, call a veterinarian and then decide whether to postpone the slaughter or choose another animal.

Withhold Feed

Plan on withholding feed for a period of time, but ensure that the animals have plenty of access to water. This will help clear their digestive systems of fodder and waste, which in turn reduces visceral weight (making handling easier), reduces the potential for contamination (less waste in their entrails), and also improves their dressing percentage (if that matters). Hydrated animals also make for easier skinning, especially if you choose to fist the hide (see page 211). A general guideline for feed withholding time is "the larger the animal, the longer the time," ranging from 8 to 24 hours. Longer is not always better, especially for poultry: the limited feed combined with access to liquid can make for an intestinal tract full of runny excrement — watery guts — which carries the risk of leaking and contamination.

Stunning

STUNNING AN ANIMAL is less about setup than it is about method. In a home setting there is only one way to stun the animal — a blow to the brain — but there are numerous tools available to achieve the same result. Make sure that you have a highly effective approach to stunning, which may include primary and secondary methods, before day-of-slaughter activity commences. Test the equipment far ahead to make sure it works, and provide time for repairs if needed. Accurate aiming is critical; if you miss, you will likely just maim the animal, causing incredible distress and pain in a terrified and unpredictable animal.

Insensibility Is the Priority

When harvesting an animal, our first priority is to reduce the discomfort the animal experiences in its last moments of its life, and for most species this will mean stunning. The main intent of stunning an animal is to render it insensible prior to bleeding. In most cases, the cause of death is exsanguination and the lack of oxygen (via blood) to the brain. To quicken the arrival of death, it's helpful to keep the animal's heart pumping while it's insensible. This will also help promote a more thorough bleed, especially when combined with gravity.

Do not, under any conditions, proceed with the bleeding of an animal when its insensibility is questionable. Extensive research has been done into indicators of insensibility. Dr. Temple Grandin is the leading proponent on the subject; her studies focus mainly on conditions within industrialized abattoirs, but the evidence and guidance contained in her work can be applied to any situation in which animal well-being is a priority during slaughter. The following is a list, prepared by Dr. Grandin, of required conditions to confirm that an animal is insensible and, as you may expect, to confirm the death of an animal prior to skinning or scalding:

>> For the correct captive bolt position when stunning pigs, turn to page 331.

>> For the correct captive bolt position when stunning sheep and goats, turn to page 204.

- The legs may kick, but the head and neck must be loose and floppy like a rag. A normal spasm may cause some neck flexing, but the neck should relax and the head should flop within about 20 seconds. Check eye reflexes if flexing continues.

- The eyes should be wide open with a blank stare. There must be no eye movements.

- The animal must NEVER blink or have an eye reflex in response to touch.

- The tongue should hang out and be straight and limp. A stiff curled tongue is a sign of possible return to sensibility. If the tongue goes in and out, this may be a sign of partial sensibility.

- In captive bolt–stunned animals, insensibility may be questionable if the eyes are rolled back or they are vibrating (nystagmus).

- Shortly after being hung [with your method], the tail should relax and hang down.

- No response to a nose pinch or pinprick on the nose. The painful stimulus should be applied only to the nose. Animals entering a scald tub must not make a movement that is in direct response to contact with the hot water. For all types of stunning this is an indicator of possible return to sensibility.

- No vocalization (moo, bellow, or squeal) after the captive bolt stunning.

Know the signs of insensibility, so that you can ensure the animal is feeling no pain before bleeding it out.

- Rhythmic breathing must be absent. Count as rhythmic breathing if the animal's rib cage moves two or more times.

- When the animal is hung [with your method], its head should hang straight down and the back must be straight. It must NOT have an arched back righting reflex. When a partially sensible animal is hung on the rail it will attempt to lift up its head. It will be stiff. Momentary flopping of the head is not a righting reflex.

HUMANE HOUSING FOR THE FINAL DAY

All animals deserve to exist in a safe environment without stress and fear, especially during their final hours. Handling and housing the animals with respect is not only dignified and ethical but also essential to producing high-quality meat. (See page 18 for more on the science of stress before slaughter and its effect on meat.)

Livestock animals are timid creatures, susceptible to being frightened by unfamiliar factors. Avoid housing them in areas where they may be subject to startling noises — driveways, construction sites, buildings with loud plumbing — or to moving objects that may startle them, such as fan blades, flags, and wind chimes. Like people, different animals have different needs, so plan on giving them time to adjust to the new surroundings in their own way.

When moving the animals, refrain from inciting fear. Any contact with blunt or sharp objects, such as fenceposts or prods, may cause bruising, even without evidence on the hide. Bruised areas of the meat will need to be cut out, resulting in a loss of product.

There is no fail-safe method of stunning: at some point you will miss or something else will go awry. Therefore, prepare for an ineffective stun, be it from a misaligned shot, a misfire, or some other aberration. Do not panic if the first attempt does not work. Keep an extra cartridge immediately available — reload, steady the animal, take your aim, and make sure the second attempt is successful. If you are not using a bullet, start from the beginning of the process and quickly deliver a second blow. In either instance, if the second attempt also fails, then proceed immediately with bleeding using the transverse method (page 84) and either restrain the animal while bleeding it or step away to avoid personal injury.

A stunned animal will collapse immediately, falling to the ground with all its weight. On hard surfaces like concrete or asphalt, this can result in bruising, especially with larger animals. Therefore, consider where the animal is going to be when it falls and what potential measures you can take to soften the landing, like adding bedding materials to a hard floor or stunning in the grass.

Captive Bolt Stunning

The ideal tool for close-range stunning is a captive bolt pistol, which comes in two varieties: nonpenetrating and penetrating. Both varieties have similar methods of operation. You aim the device at the proper point on the skull and then pull the trigger; a metal bolt, propelled either pneumatically or by a blank cartridge, creates an impact that renders the animal insensible to pain. The effect is instantaneous though not always permanent. Hence, immediately following a proper stun with bleeding is paramount.

NONPENETRATING captive bolt pistols hit the skull with a flat metal disc, causing temporary unconsciousness in the animal.

PENETRATING captive bolt pistols use a narrow metal bolt to penetrate the skull and disrupt brain operations, also causing temporary unconsciousness.

Because the nonpenetrating models are prohibitively expensive, as well as pneumatically powered, the penetrating captive bolt guns are more frequently used in small-scale operations.

Captive bolt pistols are suitable only for animals that can be restrained without excessive duress or anxiety. This will in most cases include sheep, goats, pigs, and sometimes calves or cattle that have been habituated to close-range interaction. Captive bolt pistols should be operated according to the manufacturer's directions and treated with the same precaution as any other lethal weapon. They must be regularly cleaned and maintained to ensure proper operation and efficacy.

Projectile Stunning

Anyone who has been around livestock, especially cattle, knows that the area in which they remain calm and stationary is limited; this is called the "flight zone." Invading this safe zone will cause the animal to saunter or flee in a safe direction. Commercial slaughterhouses have equipment that restrains the animal (supposedly in a manner that minimizes stress on the animal), enabling the use of close-range stunners like a captive bolt gun. Animals on the farm can be restrained with various types of specifically designed equipment. But retrieving an animal and coercing it into the restraining device is inevitably stressful.

Therefore, at-home stunning for skittish livestock is most often done from afar with a projectile-loaded gun.

Ensuring the efficacy of a captive bolt requires both maintenance of the gun and knowing the correct placement for each species.

Stunning with a bullet will often be the best method for larger animals.

A basic .22-caliber bullet will suffice, but something a bit heftier, like a shotgun slug, will reduce the margin of error. It's important to recognize, however, that gun use entails a separate set of conditions. The person conducting the slaughter must be licensed to operate a firearm. For safety reasons, using a gun for slaughter may also limit your options for where to perform the stunning. Even so, a marginally skilled marksman can get close enough to most animals to ensure a positive result and limit any unintended consequences.

If using a projectile method is not an option, then animal restraint followed by a close-range method of stunning is the best approach.

Pithing (Debraining)

Poultry is the only species that is challenging to stun in a home-processing environment. Commercially raised poultry is stunned using electricity or gas, two options that are too expensive for home processors. Pithing is a method of puncturing the bird's brain, leaving it unconscious and essentially brain-dead but with a pumping heart. It is the only economical method for minimizing distress in a bird before exsanguination. Pithing will also relax the muscles and follicles around the feathers, allowing for the option of dry plucking or scalding at lower temperatures. Turn to page 110 for instructions on pithing.

BEHEADING

For centuries, animals bound for our bellies have been killed via head removal. As you can imagine, this is less a method of stunning than it is a mode of death. Technically this process is called cervical dislocation, or the separation of the skull from the spine and spinal cord. It results in immediate death and, therefore, immediate suspension of the heart. Done properly, it will kill the animal with little suffering. However, the subsequent exsanguination will be less effective than with methods that maintain a beating heart. Rabbits and poultry are the two groups of animals that are typically dispatched with cervical dislocation. Further details for each of these animals can be found in their respective chapters.

Exsanguination

THE VOLUME OF BLOOD that comes from exsanguination is copious. The total percentage of live weight can range from about 4 to 7 percent, depending on the species and age, which for the average market-size pig works out to be approximately 2 gallons of blood (or 16 pounds). A successful exsanguination will drain around 60 percent of the blood, leaving you with about 1 gallon (or 8 pounds) of liquid that will need to be either saved — for cooking, composting, or creating blood meal for fertilizer — or disposed of. The majority of the remaining 40 percent of carcass blood is contained within the viscera; less than 10 percent will be left within the muscles.

Collecting and Disposing of Blood

Depending on the slaughter location, disposal of blood may be simple. When bleeding an animal in the field, where blood can be left where it lands, be aware of the incline and make sure you're uphill, allowing the blood to flow away from the work area. Handling disposal inside a barn or outbuilding will require more consideration. Floors made from cement, tile, or another cleanable material and fitted with integrated drains are ideal, because they allow blood to be diverted into septic or sewage systems. Regional regulations will apply for this method of disposing of wastewater, a catchall term that covers all the wash water and drainage coming from bathrooms, hand wash stations, kill floors, processing, and other areas of slaughter operations. The reason for this is that wastewater is likely to contain a host of things not normally present in sewage — fats, bits of bone, blood, sanitizing solutions, and microorganisms, to name a few — all of which need to be disposed of in a way that is safe for the environment and the community's water system. Slaughtering for personal consumption, or in low volumes, may fall under exceptions to wastewater disposal regulations, but it behooves you to confirm what requirements you need to meet in your area. If you're unsure about disposing of wastewater in such a way, I recommend collecting it in a container that makes hauling it out to a field for dumping as easy as possible.

Methods of blood collection largely depend on the method of exsanguination and the size of the animal.

The larger the animal, the greater risk of injury from death throes, so holding a blood collection bucket next to the throat of an 800-pound boar is probably not a great idea. Hanging large animals, those weighing more than 300 pounds or so, over a blood collection vat is the safest approach. This allows you to step out of harm's way while the animal bleeds into a container set below it. Hanging the animal above a wide collection vat also applies to those bled using the transverse method, explained further below, because it involves a large incision across the neck, making it hard to predict the exit point of the blood. The spray area for blood from any hanging animal can be quite broad, especially considering the ferocity of death throes, so the wider the catch container, the better. You can improve your odds of collection by obtaining a wide plastic funnel. You can buy one manufactured for just this purpose, or you can fashion one from a large piece of bendable plastic sheeting.

Collecting blood from small animals being bled with the thoracic or chest stick methods, both described later, can be done with the animal hanging or on the ground, the latter being certainly easier on a farm. Because the incision is small, the exit point for blood is predictable, allowing someone to hold a container near the incision for collection, during which the legs of small animals can be restrained to prevent any chance of personal injury from death throes.

Collecting Blood for Culinary Purposes

Collection of culinary-destined blood requires a clean food-grade container, a utensil for stirring, and of course clean blood, which can be a challenge when the bleeding animal is covered in manure and flailing above the container. Large plastic lug bins and 5-gallon buckets are both good for collecting blood from hanging animals; you can stretch a piece of cheesecloth across the opening to help keep out solid matter. Large bowls, small buckets, and the like work for collecting blood from small animals on the ground. If you plan to compost the blood or turn it into blood meal, then any kind of collection bin will suffice; still, you should give it a good rinsing before use.

Transverse Method (Poultry, Rabbits, Sheep, and Goats)

The easiest exsanguination method is a clean cut across the neck (i.e., a transverse cut), starting right under the spine, that severs the carotid arteries and jugular veins along with the trachea and esophagus. This method works well and can be used effectively with all animals profiled in this book except for pigs.

There are a couple of disadvantages to the transverse method, however. First, tests on exsanguination efficiency have shown that blood flow from transverse severing of the vessels is slower than that from the thoracic method (described next). Second, the blood becomes contaminated with stomach contents released through the severed esophagus, so if you plan on saving your blood you will want to use the thoracic method.

Thoracic Method (Sheep and Goats)

The second method of exsanguination severs just the carotid arteries and jugular veins, leaving the trachea and esophagus intact. An intact esophagus also pre-

>> See page 205 for a demonstration of the transverse method.

COLLECTING BLOOD FROM PIGS

The blood from pigs is often collected for consumption. You can easily do this with any clean vessel. With a successful stick, the blood will be expelled from the incision site. Hold the collection container near the incision, and catch the blood until the stream becomes inconsistent. Stir constantly, ideally with a nitrile-gloved hand, while the blood collects and until it has cooled down. This prevents coagulation, which will ruin the blood for culinary purposes. The quicker it cools the better, not just in the prevention of coagulation but also to stem the propagation of foodborne pathogens. (Warm blood is an ideal environment for bacterial growth.) You can speed up the cooldown process by placing the blood container on top of a larger container full of water and ice, all the while agitating. Once cold, pass the blood through a fine mesh strainer to remove any clots and foam. Blood can be safely stored at 34°F to 36°F for up to 4 days.

If you plan on saving the blood, have someone to help and a vessel ready before you stun and bleed the animal.

Unless it's constantly stirred, blood will quickly coagulate while it cools.

vents stomach discharge from contaminating any blood that you may collect. The blood vessels are cut nearer to the heart than with the transverse method, producing quicker death and a more efficient bleed. The cut is started with a shallow incision from the anterior point of the sternum to the jaw, exposing the carotid arteries and the external jugular vein, which are severed to start the exsanguinations.

When executing this method, it helps to keep the dewlap — the often draping area of skin and muscle between the jaw and sternum — taut. This will help when you make an incision through the hide. If the animal is hanging, the area will become taut naturally; prepare to stand to the outside of either foreleg to perform the incision. If the animal is on the ground, you will want to pull the jaw away from the forelegs to tighten the skin; prepare to stand behind the head to perform the incision.

Chest Stick Method (Pigs)

Pigs typically have immense amounts of fat in their neck and jowls, and these areas are often used for culinary purposes. Therefore, the preferred method of bleeding a pig avoids making long, deep incisions across the neck (as with the transverse method), or along the neck (as with the thoracic method). Instead, the goal is to make an incision just large enough to allow for severing vessels and an exit for the blood. The ideal location for severing vessels is just anterior to the sternum, close to where they crisscross and connect to the heart.

Sticking a pig is a challenging task to get right. It is best learned firsthand, from someone who has experience sticking pigs. Use a 7- to 8-inch knife with a rigid, straight blade.

>> For a demonstration of the chest stick method, turn to page 332.

Testing for Life (and Death)

Death from exsanguination can take anywhere from 90 seconds to a few minutes. You will gain nothing by jumping the gun, so err on the side of caution and wait out the few minutes. With any animal, the eyes should be open; tap the eye to make sure there is no reflex response. There should be no signs of breathing or movement. The tongue should be limp and hanging out of the mouth. If you prick the nose with a needle or knife tip, there should be no response.

With a chicken, look to see if the cloaca is pulsing or moving at all. A rabbit should not be moving its nose.

With a sheep and goats, test for signs of life by squeezing and pinching the flank fold, a flap of skin connecting the abdomen and the hind leg. Make sure you are standing on the dorsal (spine) side, reaching across the animal and keeping yourself away from a kicking hind leg should the animal hold the remnants of electrical responses. Pour a bit of scalding water on either leg of a chicken or the hind leg of a pig, and look for movement. Proceed only when you have confirmed that the animal is no longer living.

Hoisting

ANIMALS ARE HOISTED during slaughter for a variety of reasons, and one is to take advantage of gravity. An animal may bleed more thoroughly when gravity is aiding the heart in expelling blood than when it is lying on the ground. The more complete the bleed, the more resistant the meat is to spoilage. Moreover, when an animal hangs by its hind legs, gravity works on the viscera, pushing them against the diaphragm. This greatly reduces the chances of puncturing entrails when making initial incisions in the groin area. Hanging the carcass also allows you to use both your hands so that you can work quickly and efficiently. Ideal setups for hanging will provide 360-degree access to the carcass; this makes your job easier and quicker while also preventing any contaminated surfaces — walls, fences, poles — from rubbing against an otherwise clean carcass.

Animals are hung by either their feet or their gambrel cords (the animal equivalent of an Achilles tendon). The gambrel cord is an incredibly strong tendon that can easily support the weight of the carcass as well as any additional downward force applied during slaughtering or butchering. There is a space between the gambrel cord and the calf muscles that allows for the insertion of hooks or rope.

The size of sheep, goats, and market-ready pigs makes for easy hanging; rudimentary setups are often just as effective as those with more equipment. (Information on hanging chickens and rabbits are found in their chapters, on pages 108 and 170, respectively.) The deadweight may be cumbersome for a single person, but with some minor forethought you can set up to go at it alone. Fortunately, most farms already have the basic equipment needed for heavy lifting.

Lifting Options

Bringing the lifting to the animal may be easier than vice versa, which makes a portable method of hanging a huge benefit. A backhoe or a tractor with a loader is an easy and versatile solution. In fact, any mobile machinery that's capable of lifting several hundred pounds can be fashioned to work, as long as the clearance it provides is enough to hang the animal so that it does not touch the ground.

In addition to vehicles, there are dozens of affordable options available for heavy lifting. A block and tackle, a come-along, a winch, and a chain hoist are all good choices. Even a single rope and pulley will work. (Attaching the rope to a vehicle will even save you the effort of having to do the lifting.) These solutions are all relatively stationary, since they need a stable point to elevate from, like a sturdy rafter or a tall A-frame, so you'll need to coordinate where you're going to hang the animal with where you're going to stun it.

With any of these solutions, consider what will be directly above the hanging carcass. A dirty backhoe, a loader used for silage, or maybe a rafter that has never been wiped off? Make sure these areas are relatively clean to prevent contamination from falling debris while processing.

Securing the Animal

Where and how to lift the animal is not the only part of the process to consider — to prevent slippage, a mistake that can result in contamination of the meat or serious personal injury, you must be able to secure the carcass.

GAMBRELS. Any solution for securing the hanging animal should involve two hooks and something that keeps the legs spread to aid in processing. One common method is a gambrel, which is a solid piece of steel with a hook on each end and a centered eyelet for hanging and balancing the weight. You can find gambrels in two formats: a single solid construction or a construction with independently hanging hooks. The former has one major disadvantage: the center eyelet does not allow for imbalanced hanging. This can heavily impact a common scenario: wanting to remove one side of the carcass and leave the other side hanging. For this reason, if you're considering using a gambrel with animals that will be split while hoisted — for example, beef and sometimes pigs — I highly recommend using the kind with independent hooks.

Gambrels are sized for weight and distance of spread. The appropriate-size gambrel for sheep and goats can also be used with most pigs (not including fully mature sows and boars), deer, and other animals of similar weight. These small modern gambrels are often marketed

Most farms won't have a mobile slaughter truck like this one; a bucket loader or simple rope and pulley setup will work just as well.

to hunters and include built-in hoisting methods, making it easy to hang an animal anywhere you want to bleed it. Animals in excess of 300 pounds will need a hefty gambrel to support the weight.

SINGLETREE. A singletree is another option that operates identically to a gambrel and is available in the same two versions. Singletrees are most commonly constructed of wood, which is why the meat industry has replaced them with durable stainless-steel gambrels, but for home use a singletree is fine. And because singletrees have fallen out of use, you may be able to get a good deal on a used model.

PARACORD. The animal's back legs can be secured using a heavy-duty nylon twine or cord, often marketed under the term *paracord*. Make sure it has a weight rating well above the weight of your animal to accommodate any additional force. Legs can be tied independently, which is especially helpful if you plan on splitting the carcass. Know your knots before hanging hundreds of pounds

above your head! A simple double-loop paracord wrapped around the legs or through the gambrels will often suffice.

GRAB HOOKS. If you plan on using a loader or backhoe to hoist the carcass, there is another easy option: a chain with grab hooks on either end and a couple of meat hooks, all of which are easy to obtain and affordable. With the grab hooks secured, you can insert the meat hooks into any one of the chain links to determine how far spread you want the animal's legs to be.

Skinning

THERE MAY BE MORE THAN ONE way to skin a cat, but one is all you need to know to get the job done right. Rabbits, sheep, and goats are always skinned, while pigs and poultry are often processed skin-on (though skinning is an option for those wanting to avoid the extra steps required for cleaning the skin of a carcass). Regardless of which skinning method you choose, there are some tenets to which you should adhere for the sake of sanitation, efficiency, and condition of both the meat and the pelt.

Avoiding Contamination and Damaged Meat

Keep in mind that until you break skin the interior of an animal is completely sanitary. The only contaminants on a final carcass are those that we introduce, from outside sources or from puncturing the digestive tract or other viscera. Thus, anytime you use a knife while skinning, remember to cut from the inside out. This means that making initial cuts through skin is always done with the knife edge facing away from the animal's body, as shown on page 207. This will also save the knife edge: wool, hair, and fur are often loaded with dirt and grit, and cutting through them will dull a knife.

Skinning is as much about removal of hide as it is about producing a clean carcass and avoiding any unnecessary damage from sloppy knife work. Keep your cuts shallow to avoid nicking the muscles, reproductive organs, or abdominal walls. During much of the skinning process, you must keep the hide taut. To avoid bruising, do not pull excessively on areas of the hide that are still attached to the carcass. As you make cuts between the

Make sure your gambrel provides adequate spread and support for the species you are working with.

hide and the carcass, the knife edge should be angled slightly toward the hide rather than toward the muscles.

Wet Carcass, Dry Hide

When the hide is removed properly, the interior face should be devoid of membranes, muscle, and fat. It will appear almost dry when compared to a hide that has been improperly removed. This applies to both fisting and knifing — stay above the membrane when skinning. This leaves the most meat, fat, and surface tissue on the carcass, helping prevent excessive evaporation during aging while protecting the underlying muscles from any surface contamination.

Avoid making short cuts while skinning; they cause more uneven surfaces on the hide and carcass, along with carrying a greater risk of puncturing the hide. Keep your knife strokes long and smooth. This will be quicker, cleaner, and produce a better appearance.

Allow gravity to help keep the carcass free from hide grime: the pelt should be removed in such a manner that it naturally drapes away from the carcass. Hanging carcasses are skinned from the top down so that the exterior of the hide can only touch parts of the carcass that have yet to be skinned.

The exterior of an animal's hide is undoubtedly dirty, requiring you to rinse your hands regularly to minimize contamination. You should also designate a "clean hand" and a "dirty hand": hold your knife or fist the hide with your clean hand; handle the exterior of the hide with your dirty hand. Do not confuse the two; wash the clean hand when it gets dirty, and keep the dirty hand away from the exposed carcass.

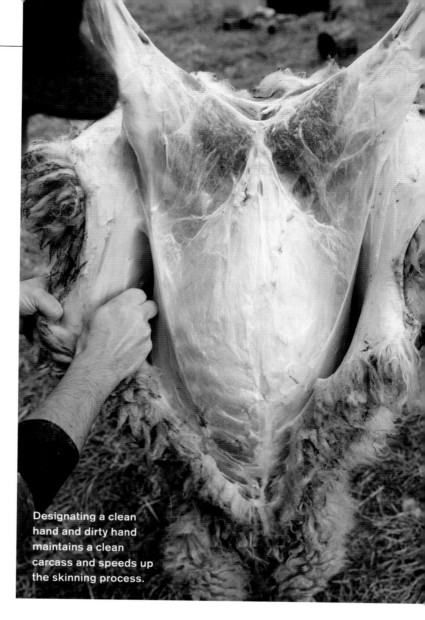

Designating a clean hand and dirty hand maintains a clean carcass and speeds up the skinning process.

HANGING DURING BLEEDING

You may choose to hang the animal for bleeding. This is not required, and there are some who assert that hanging before death damages the hind-leg meat by making the muscles conform to an unnatural position and causing a detrimental shift in pH. With nonlethal methods of stunning, do this only if you have a way of hoisting that can bring the animal off the ground in 30 seconds. With anything longer you take the risk of the animal regaining sensibility before the bleeding.

Hanging will involve a separate way of securing the back legs, since the gambrel cords will not yet be exposed and you don't want to make any incisions to the still-living animal. One easy method is a steel chain with an eye on the end allowing for a loop. Have enough length to quickly and easily wrap the loop around one or, better, both shinbones (also known as *cannon bones*). Remove any slack in the chain as you wrap it around the bones; as you lift the animal, the carcass weight will further tighten the chain and secure it. After bleeding, the animal will be lowered for skinning.

Evisceration

EVISCERATION IS the removal of the digestive tract and other internal organs, collectively known as viscera. The process of removal accounts for a large portion of the work done during a slaughter. It also poses some of the greatest risk of contamination, since you are working among areas containing digestive and fecal matter.

Avoiding Contamination

Preventing the spread of fecal matter is the first concern during evisceration. Avoid pressure to the abdomen, which can in turn put pressure on the intestines and bladder, causing feces and urine to spill out. With small animals, such as poultry and rabbits, the process of evisceration starts with the safe removal of the anus. With large animals the anus must be tied shut, preventing spillage when the rectum is pulled down through the cavity.

Again, remember that the abdominal cavity is airtight and uncontaminated until we pierce the lining. Initial cuts should be made with caution, especially because they often occur around the groin, potentially with a full bladder right inside. The walls of the bladder are the thinnest of any viscera you will encounter and, while urine is typically considered sterile until it passes through the urethra, the spilling of it throughout the inside of the carcass is certainly something best avoided.

Evisceration should happen as quickly as possible while maintaining control and a priority of cleanliness. This is especially true with ruminants, whose stomachs are full of partially digested material that continues to be broken down by enzymes even after death. The stomach begins to fill with gases, a by-product of the enzymatic process, and expand the moment death occurs because the gases become trapped. (One step to composting the carcasses of ruminants is to lance the stomach before burial; if this is not done, the gases become trapped and eventually will explode!) Therefore, the longer you wait, the larger the stomach will get and the more compressed the space inside the abdominal cavity will be.

Remember that total cleanliness is impossible; the last step of slaughtering — trimming — only aims to reduce the surface contaminants by removing areas where fecal matter, digestive waste, and other obvious contaminants have landed.

General Methods (Sheep, Goats, and Pigs)

The process of evisceration is nearly identical across the larger species covered in this book, since the components that make up the different species' viscera are dramatically similar. The following sections provide methods for specific stages of the evisceration process; the complete process of eviscerating any of these species is discussed in their respective chapters. (For chickens and rabbits, their smaller size makes the process a bit different, and simpler; those details are discussed in their chapters.)

TYING OFF THE BUNG

One of the first stages of evisceration is closing the opening of the bung (another name for the rectum), thus preventing more feces from being expelled. The bung runs through the pelvic opening, at the center of the pelvis. The bones around the pelvic opening have changing angles: the sides tend to be around 45 degrees, whereas the front (ventral face) and back (dorsal face) are almost 90 degrees. As you cut around the bung, change the angle of your knife tip accordingly.

Start cutting along the sides of the bung using the tip of your knife. Grab hold of the bung once you are able to,

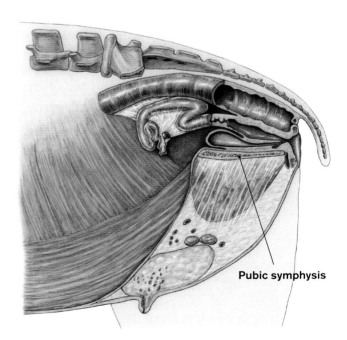

Pubic symphysis

The shape of the pelvis allows for access from either the anal opening or from the front through the pubic symphysis.

then finish by following the shape of the pelvis until you can pull the bung out far enough to tie it off. Use a piece of sturdy twine or cord to tie the bung closed. (Two people are needed for this process: one holds the bung, and one ties the bung.) Alternatively, you can place a plastic bag over the end of the bung and secure it with a rubber band or string to prevent any contamination around the bung opening from getting on the carcass.

OPENING THE ABDOMEN

The initial cuts into the abdomen are the most dangerous, so proceed slowly. Pinch the abdominal skin about 4 to 5 inches below the rectum and pull it outward. Make a small, light cut across the pinched skin into the abdominal cavity. If you chose not to split the pubic symphysis, you may hear a rush of air — the abdomen is essentially a vacuum until you introduce air through this initial cut. With this change, the viscera will fall more toward the diaphragm, giving you a bit more room to work in. Pull the skin away and carefully enlarge the initial incision, cutting downward, until it is large enough to insert your hand. The bladder sits right behind this point, so do not drive your knife into the cavity.

There are two ways you can cut down the midline of the abdomen while avoiding visceral punctures and resultant contamination: the whole-handle method and the knife-tip method. Note that once you start cutting, you should not stop until you reach the sternum. When you stop, the guts want to fall out, especially with larger animals, and restarting the cut will be incredibly difficult.

WHOLE-HANDLE METHOD. To use the whole-handle method, hold the knife handle and insert it completely into the cavity with the blade pointing out and the edge facing down. The abdominal wall should sit right at the heel of the knife, where the handle and the blade meet. With this position, you can drive the knife straight down the middle of the abdomen. If you are right-handed, stand to the right side of the carcass and reach around to do this, or vice versa. This is a good habit: the weight of falling viscera can cause the knife to lunge forward mistakenly, so no one should be standing in harm's way.

KNIFE-TIP METHOD. To use the knife-tip method, insert your non-knife hand into the cavity, fingers pointing down. Insert the knife tip between two fingers. (I use the index and middle fingers.) Move your hands down the abdomen together, using the inside hand to keep the viscera away from the knife tip. This is not appropriate for use with cows or other large animals, because their abdominal viscera are too heavy.

Whole-handle method

Knife-tip method

SPLITTING THE PUBIC SYMPHYSIS

There is one notable variation to tying off the bung that you may choose to incorporate or not. The two sides of the pelvis are joined by a weak strip of cartilage (called the *pubic symphysis*) that runs vertically right in the middle of the hind legs. It can be split, giving you a wider opening to work with when separating and tying off the bung. But proceed with caution: the bladder sits right behind it, and any sudden jerk of the knife can cause a flooding of urine.

Cut through the leg muscles that join at the center of the groin, following the connective tissue seam between them, until you reach bone. Notice the centerline of cartilage. When working with small animals, you can carefully apply pressure to the seam with your knife. Wiggle the knife back and forth, making your way into the cartilage and pry the bones apart.

Larger animals will require a bit more force, and for that you can use the tip of your knife. Place the tip about one-third of the way up from the bottom of the pelvis and, while holding the knife perpendicular to the bone, tap on the back of the handle to set the tip into the cartilage. You do not need to go too far. Once the knife is set, rotate the handle downward until the tip is released. Do this repeatedly as you work your way up the seam. When the structure of the cartilage is compromised enough that you can push the two legs apart, the seam should split open.

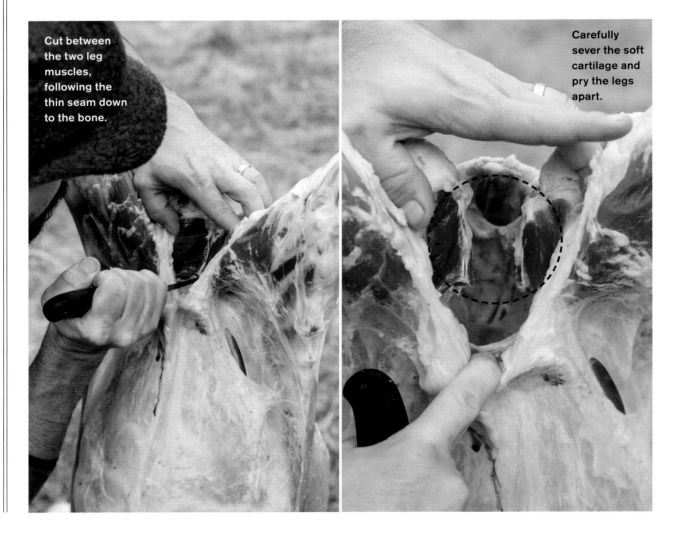

Cut between the two leg muscles, following the thin seam down to the bone.

Carefully sever the soft cartilage and pry the legs apart.

Edible Offal

EXCEPT FOR BONES, most everything within an animal is edible and should be considered before disposal. Cultural tradition is the largest determining factor in what we choose to save for consumption and what we consider inedible or waste. The glands and viscera we choose to eat are commonly referred to as *offal*. The organs you're not interested in eating can be saved and ground into natural dog food, a value-added product for those looking to increase revenue per carcass. Any organs intended for future use should be removed from the carcass and frozen immediately. (For more information on freezing organs, refer to page 422.)

The most commonly consumed offal is heart, liver, and kidneys. In all animals, these organs are routinely saved, even sold individually aside from the carcass or muscle cuts as another revenue source. The tongues of sheep, goat, pigs, and cattle should also be saved. Their intestines and stomachs, once cleaned, offer edible containers for stuffing minced meat and other goodies into, without which we would never have cased sausage. Chambers of the beef rumen, culinarily known as tripe, are edible after a vigorous cleaning. Even bladders and beef bungs are used for stuffing. Spleens from the same four animals can be grilled, braised, or even simply pan seared.

Oddly enough, even cheeks, jowls, and oxtail are considered offal these days, despite their apparent external origin; they are all delicious. Brains of any kind can be extracted from skulls and cooked, though sheep, goat, and pig brains are the most commonly consumed. Younger animals still have their thymus gland, which is largest at adolescence and slowly atrophies as the animal ages; these glands are better known as sweetbreads and are considered a delicacy. In poultry, the gizzards are standard giblet fare. Duck tongues are an Asian delicacy and, after the center bone is removed, can be prepared in many different ways. In short, I encourage you to find a way to waste as little as possible.

Brains

The easiest way to extract a brain without damaging it is to partially split the skull in half using an axe or heavy cleaver, allowing you to pull apart both sides and safely remove the undamaged brain. It's also helpful to have a hammer (with the axe) or a wooden mallet (with the cleaver) to improve the cutting accuracy. This can be easily done anywhere as long as you have the tools, so it's perfect for on-farm extraction.

The brain is exposed after the sides are split. Use a knife to carefully cut around the brain, helping to coax it out of its protective cavity. Follow with a good rinse and the following cleaning instructions.

Brains should be handled carefully, since their delicate structure is easily damaged. Peel away the exterior membrane if it was removed with the intact brain. Soak the brains for a day or more in lightly salted water, which helps pull out blood and other impurities. Change the water frequently, every few hours or so, after which the brains will be ready for cooking.

>> For a demonstration of brain extraction, turn to page 220.

Cheeks/Jowls

Cheeks, also known as the *masseter* muscle, are found on all animals and responsible for mastication. Pigs have cheeks, too, but they are smaller (because their diet doesn't require as much chewing) and are found beneath their jowls, the fatty area under and around the lower jaw. Cheeks from ruminants and pork jowls will both need to have salivary (parotid) glands removed. They exist in clusters and are clearly identified by their beige-to-brown coloration and spongy texture. Use your knife to slice them away while maintaining a relatively flat surface for the cheek or jowl as a whole, avoiding removing any unnecessary meat.

Heart

Sever any remaining tubes attached to the heart as well as the pericardium, the sac that surrounds the heart. Cut open the heart by slicing partially through the top, exposing the chambers and allowing you to expel any blood clots by squeezing. Rinse thoroughly, inside and out.

The hearts of large animals — which exclude poultry, rabbits, and small game — require some additional cleaning prior to cooking. Cut the heart into manageably flat pieces, often corresponding to chambers. Trim all the exterior fat from the pieces, leaving a clean surface while trying to remove as little of the heart as possible. Flip the pieces over and trim away the tough interior membrane that lines the chambers. Cardiac muscle that is ready for cooking should be dark with a dull finish.

Kidneys

Kidneys require only a thorough cleansing prior to storage. With large animals, ensure that the kidneys have been "popped" (a process covered in their respective chapters) and the outer translucent membrane removed. Also, slice them lengthwise along their midline and remove the fat deposits within.

Liver

The liver needs minimal cleaning in order to be palatable; mainly, removal of the gallbladder while working to avoid spilling any bile on the organ. Remove the gallbladder by pinching the duct where it connects to the liver and slowly tear the duct away from the liver in the direction of the gallbladder. Stubborn duct connections can be severed with the tip of a knife. Look for signs of parasites or infection such as dead or alive flukes and white lesions; if you find any, then you must dispose of the liver.

Spleen

After you trim away any membranes or connective tissue, thoroughly wash the spleen and store it frozen until use.

Sweetbreads

The sweetbreads require the same preliminary steps prior to cooking whether they be from the thymus or the pancreas. Rinse the sweetbreads in a bowl of cold water, gently cleaning the surface of any collected residues or blood. Soak them for up to two days in lightly salted water or milk, which cleanses the glands by drawing out stagnant blood and impurities, changing the liquid every few hours. After a good soaking, briefly poach the sweetbreads and then shock them in ice-cold water. Once they are cooled, peel away the exterior membrane, including any veins or gristle, leaving them recipe-ready.

Tongue

Thoroughly rinse the tongue, scrubbing the papillae-covered surface with your hand or a brush. Tongues need to be peeled before they can be consumed. To do so, remove the tough, tastebud-lined surface by poaching the tongue for 2 to 3 hours, depending on the size of the animal. After poaching, remove the tongue and let it cool until it can be handled. The thick outer layer should separate quite easily; if not, poach longer and repeat.

Carcass Cooling

PROMPT COOLING OF ANY CARCASS is imperative for limiting microbial growth, thereby preventing spoilage and a loss of product. You'll need a meat thermometer to test the internal temperature before further processing. The target temperature for the carcass is 36°F to 40°F, and the ideal time it takes to get there is dependent on the species and size. Chickens should take about 1 to 4 hours; larger animals can take up to 36 hours. Faster is not better in this case, because quick cooling carries the risk of cold shortening. (For more on cold shortening, see page 18.)

All carcasses benefit from a period of aging. The specific length of time is determined by species, size, and personal preference. Following the cooldown period, proceed with aging based on the methods described on page 18.

A Quick Rinse

Before heading into the final cooling process, it's beneficial to rinse the carcass. For small carcasses (poultry and rabbits) this could simply be a large, iced water container in which you dunk the carcass for a few seconds before transferring into the final cooling container (which may very well be another iced water bath). Larger carcasses headed for cooling in an environment with good air movement — e.g., a blast cooler or cold space with fans — will benefit from a quick rinse with cold water sprayed from a hose. All carcasses, regardless of cooling environment, benefit from a vinegar rinse, which helps fight any pathogenic growth on the surface. This can simply be a 1-to-1 mixture of vinegar and water spray on the carcass with a quality spray bottle. Go light on carcasses that will cool in environments with little to no air movement — e.g., large chest coolers — as excess moisture can be problematic for the aging.

Cooling Small Animals (Poultry and Rabbits)

There are many industrial solutions for small-animal carcass chilling, but using ice-cold water is the most common method for small processors and on-farm slaughters. Coolers or other insulated containers full of cold water and ice should, in total, be large enough to accommodate the entire day's processing while maintaining a temperature between 32°F and 36°F. Chipped or cubed ice works best, though frozen 2-liter bottles also work well. (A general industry guideline is approximately 1 pound of ice per 2 pounds of meat for proper carcass cooling.) Keep a thermometer on hand or floating in the bath to monitor temperature fluctuations. Keep an additional cooling container, also full of ice water, for any edible organs you plan on removing and saving.

BONE IT WHILE IT'S HOT

It is highly recommended that all carcasses go through the proper cooldown process. There is no shortcut for food safety, and this is one of the most important steps. Yet the carcasses of larger animals take up a considerable amount of cold space, and this can be hard to come by. Hence, quickly breaking down a carcass into primals prior to cooling may be the only viable option.

Breaking down a warm carcass is known as *hot boning*, a process generally not recommended for home butchery, for a couple of reasons. The temperature of a warm carcass falls right in the middle of the "danger zone." (Body temperatures are ideal for pathogenic propagation.) Therefore, the breakdown process into constituent parts must be swift and efficient, preventing any excessive delays before the meat makes it into the cooler; the butcher's handiwork must be deft, and many home butchers are not accustomed to working at such speeds. Furthermore, the ambient temperatures within home processing environments are often higher than ideal (40°F), doubling the risk of delays prior to cooling.

Nonetheless, some home butchers may have cold storage restrictions that limit their options; fitting a whole pig in a stand-up refrigerator is impossible, for example. In these cases the carcass must be promptly processed into primals or subprimals, and then packaged for cooling and aging. The process of primal and subprimal breakdowns is included in each butchering section and should be followed for hot boning as well. Keep in mind that primals and subprimals can be further broken down into portions and cuts after they have reached a safe temperature, 32° to 36°F.

Here are some recommendations if hot boning is your required course of action:

- Sanitize your tools and processing station ahead of time. Sanitation, while always important, is critical when working with a warm carcass.

- Set up your processing environment before you begin the slaughter. This includes having packaging ready and ensuring the final cooling environment is at its target temperature.

- Bring the ambient temperature of your processing environment as close to 40°F as possible.

- Bring in some help and set up workstations. For example, you perform the initial butchering into primals; someone else butchers primals into subprimals; the next person packages them; and the final station weighs, labels, and deposits in the cooler. An assembly-line approach will dramatically speed up the process and get items into the cooler quickly.

- Immediately begin hot-boning after the slaughter is complete.

- Use less force than usual when separating muscles and exposing seams. Warm, unaged muscle is much more delicate than its cooled counterpart. Tearing can easily happen, irreparably damaging muscles.

- Do not immediately freeze hot-boned meat, which runs the risk of cold shortening (page 18); it still needs to age, after which it can be frozen. (Aging times are found on page 20.)

A more advanced, and resourceful, solution for those slaughtering in higher volumes is used dairy equipment — bulk milk coolers with stainless-steel tanks. The equipment prevents the water from going below 32°F, and a submersible pump dropped in the bottom will agitate the water to prevent the bottom carcasses from freezing.

Demarcate the first and last animals into the cooling bath, using butcher twine, colored twist ties, or something similar. After a reasonable period, based on the species and size, use these animals to ensure proper cooling prior to packaging. Test the temperature by inserting a meat thermometer into the thickest part of the thigh.

Cooling Large Animals (Sheep, Goats, and Pigs)

The most difficult stage to accommodate in large-animal processing is the cooling down and aging process, simply because of the size of the animal. The animal needs to come down to below 40°F in less than 36 hours. That requires either a cold day and night (nature-assisted cooling) or a large walk-in cooler (mechanical-assisted cooling).

NATURE-ASSISTED COOLING

Nature-assisted cooling can happen when the temperatures for an outdoor enclosure range from 26°F to 40°F. Anything outside of that range will cause either spoilage from heat or freezing from cold, both of which should be avoided. If nighttime temperatures are expected to fall slightly below 26°F, you can wrap the carcass with cheesecloth or muslin to prevent any considerable freezing. Obviously, this temperature range is hugely dependent on your local climate, but for areas that have cold winters this range usually will occur in the late autumn or late winter/early spring. Choose a clear day without precipitation or high winds to help reduce excess moisture and airborne contaminants.

Nature-assisted cooling also presents challenges for aging. How consistent the temperature will be over the weeks following slaughter is uncertain. Keeping the carcass enclosed to prevent attracting predators or scavengers may be difficult. With this method there is no real control of humidity, which is usually much lower than the ideal range. Do your best to control the environment for aging; if, however, conditions become detrimental to the meat yield, it is time to butcher.

MACHINE-ASSISTED COOLING

Machine-assisted cooling is an ideal option for those wanting to slaughter year-round or in areas where tem-peratures never get as low as needed. Walk-in coolers should maintain a temperature of around 30°F for ideal cooling and 34°F to 38°F for aging; a relative humidity of 70 to 80 percent is ideal in both circumstances. Make sure the distance between the ground and the hooks or other hanging implements provides enough clearance to accommodate the whole, halved, or quartered carcass. You don't want any part of the carcass touching the ground or lying on its side while it's cooling or aging. Allowing for ample air circulation around the whole carcass will decrease cooling times, reduce bacterial growth, and prevent overall spoilage.

No matter which method of cooling you choose, ensure that it's progressing at a satisfactory rate by taking a temperature reading every 12 hours or so. Test by inserting a thermometer into the thickest part of the thigh meat, near the bone.

Cleanup and Disposal

PROPER POSTSLAUGHTER CLEANUP is an important step; you should leave the work area free of blood, entrails, or other material that will attract predators or produce foul odors.

Disposing of Inedibles

First deal with the inedibles: entrails, blood, and feathers. There are myriad options for disposing of slaughter detritus, but the most common are compost, trash, or a rendering company. Do not dispose of animal waste by dumping it and leaving it for scavengers. This is a high-risk solution with regard to disease transmission on your farm and poses risks to the safety of the surrounding water, which in turn endangers not only your livestock but also the surrounding wildlife and any pets in the area. Moreover, attracting scavengers can result in higher predation on your more vulnerable animals.

COMPOSTING

Composting is by far the best option. It is the cheapest and least wasteful, and it results in a nutrient-rich compost. It will require having wood chips, sawdust, and other organic material on hand, but with a bit of forethought you can easily acquire the necessary materials. If you're processing any species on a regular or even semi-regular basis, it will definitely help to maintain a compost pile; such piles are relatively self-perpetuating with the right maintenance.

TRASH

Disposing of inedibles within a municipal trash system is another option, and is often the simplest for those dealing with small animals or minimal amounts of waste. First, double-check with your town or garbage company to ensure that what you intend to dispose of is allowed. If so, then choose thick plastic bags, often called contractor bags, to help prevent punctures and leaks. Also, make sure the area where the bags are left for pickup is secured from scavengers and other animals looking to bury their nose in a fresh or not-so-fresh kill. Or place remnants in metal garbage bins with heavy weights — like cinder blocks — on top to prevent inquisitive creatures from getting inside.

RENDERING

Most likely, there's a rendering company in your area that aids farms in the disposal of sick animals, placental membranes, butchering residue, and other renderable matter. Prices and requirements for handling the waste differ — many companies do not allow rumen manure, for example — so make sure to contact the company well in advance to allow for any preparations, like procuring a 50-gallon drum to hold the waste.

STORING AND MOVING INEDIBLES

Depending on the volume of animals being processed, the by-product of small animals may be stored in 5-gallon buckets and disposed of as you work. Moving the inedibles of larger animals is going to be more difficult. A loader or backhoe will make moving 200 pounds of guts an easy task. A good-size garden cart or pushcart lined with plastic (to avoid staining and other additional cleanup measures) may also work, depending on the terrain you need to cover. As a last resort, a wheelbarrow can do the work, but it may take a few trips. After the animal has been hoisted, lining the ground beneath it with a durable plastic tarp will help keep the area clean and can also help in collecting and moving the waste.

Work Area and Tools

Your work area should be cleared of all blood and other waste, which may attract vermin, especially if it is in an enclosed area or near where you're hanging the carcass for cooling and aging. After cleaning, spread sawdust to absorb any remaining liquids; the sawdust can then be shoveled to a compost pile or discarded.

All the equipment should be washed thoroughly with hot water and soap, and then sanitized before being stored. Anything not made of stainless steel should be hand-dried to prevent corrosion. Knife steels should be completely dried immediately after cleaning to prevent rust buildup. (The capillary action of paper towels works best for drying grooved steels.) Other cutlery can be left to dry naturally.

Tools and Clothes for Slaughtering

THE HAND TOOLS REQUIRED for slaughtering are quite minimal, especially when compared to those required for additional processing.

Knives and Other Tools for Slaughtering

Knives should be recently sharpened and should come with handles constructed from a material that makes sanitization an easy task.

A SMALL, THIN-BLADED KNIFE or scalpel is preferred for bleeding and pithing poultry, whereas a 4- to 5-inch knife will be best for evisceration.

A PINNING KNIFE or dull paring knife can help remove stubborn pinfeathers.

A KNIFE FOR ALL-AROUND TASKS should have a 5- to 6-inch semi-stiff curved blade. I prefer one of these for all-around tasks when harvesting species other than poultry.

A SKINNING KNIFE will help in the skinning of larger animals. It has a wide and more severely curved blade, but it's not required to do the job well.

A STICKING KNIFE, specifically designed for bleeding pigs, can make the job of hitting the right blood vessels easier.

A HONING STEEL is needed wherever there are knives. Your knives will dull quickly, especially when you're trying to cut through ligaments and cartilage or removing the hide.

SLAUGHTERING ESSENTIALS

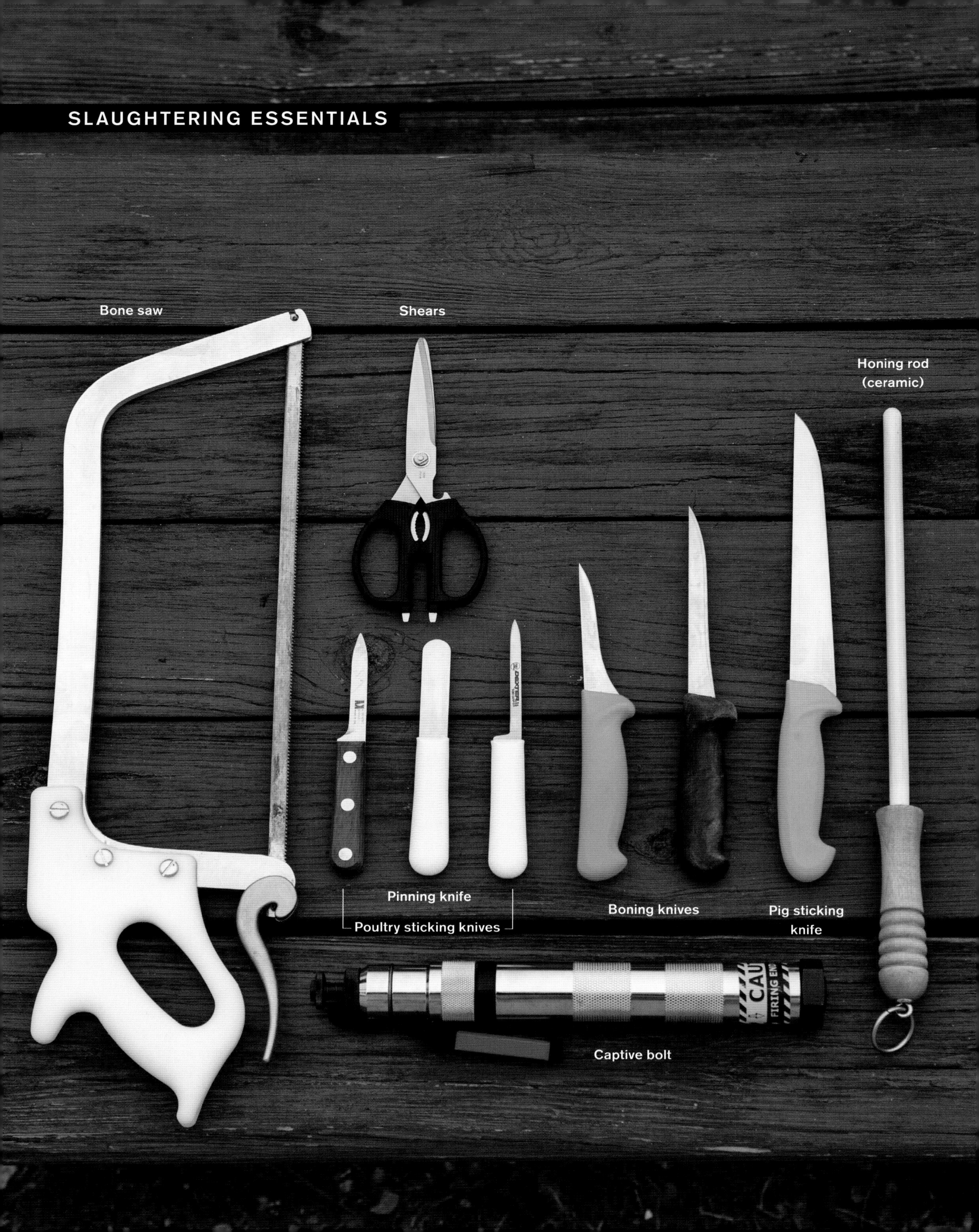
Bone saw
Shears
Honing rod
(ceramic)
Pinning knife
Poultry sticking knives
Boning knives
Pig sticking
knife
Captive bolt

A SAW, ideally intended for cutting through bone, is needed to get through the sternums of older animals. Electric meat saws are excellent but costly. Handsaws are another option that will get the job done. Look for a saw with a blade length of 23 to 25 inches.

TRIM HOOKS can aid the process of cutting away small areas of contamination from a carcass before it heads into the cooler.

A PAIR OF HANDHELD BOLT CUTTERS or heavy-duty shears simplifies foot removal of rabbits and poultry, reducing the jagged bone edges often left by poultry shears, which easily puncture packaging.

STURDY TWINE should be kept on hand. Cut an 18-inch length for tying off the bung.

Clothes

When referring to the accumulated muck from slaughtering and butchering as compared to the typical daily grime, a mentor of mine once said, "There are different kinds of dirty." Expect any clothes you wear to get heavily soiled with blood, feces, fats, and everything else around. Rubber aprons, similar to those worn by restaurant dishwashers, are easy to clean and can be purchased online. Cloth aprons work too, though staining is permanent. Sleeves should be rolled up to the elbows to avoid smearing contaminants anywhere your arm goes.

Footwear should be nonslip if you're working on tile or other smooth surfaces.

Nitrile or latex gloves are helpful to have for a few reasons: you may not always want to work with bare hands; the tackiness will help you grip rabbit pelts and aid in hand-plucking poultry; nicks and wounds happen, and when they do you should cover your hands to prevent contamination.

Required Space

IT TAKES SURPRISINGLY LITTLE SPACE to process even the largest animals. The equipment you'll use may be the main factor in figuring out what your space needs will be. Animals that can be stunned, skinned, and eviscerated all in the same space require very little room. However, if you're going to use a tractor to lift an animal, the tractor — not the process — will define the amount of space you need. This is the case for sheep, goats, veal, and beef: determine what equipment you will use, especially for hoisting, and that will determine how much space you need. Poultry and pigs are the two families of animals that require larger accommodations due to the scalding and plucking. When working in an outdoor space, pick a location where the ground is flat and clean, has good drainage, and isn't prone to excessive dust. (For species-specific space requirements, see the Setting Up for Slaughter sections in the following chapters.)

Access to potable water is a requirement. Running water is helpful for spraying carcasses prior to cooling and cleaning offal. Keeping yourself and the carcass clean during slaughtering will require repeated hand washing, so plan ahead on having a hands-free solution: options include two buckets, one with soapy water and one with clean rinse water; a foot-operated faucet; or even a secured hose with a ball valve. Try to get plastics — buckets, cutting boards, cooling containers, and hoses — that are food-grade.

CHICKEN SLAUGHTERING

IT'S NO WONDER that chickens were among the first animals to be domesticated: their small size, hardy disposition, and omnivorous appetite make them ideal for raising in many conditions where other animals would have difficulty flourishing. For us, their relatively simple anatomy makes poultry a perfect starting place for learning about slaughtering. They are easy to handle (when done correctly) and can be dispatched quickly. The setup tends to require some special equipment, but nothing that can't be fashioned from cheap materials if you are working on a tight budget. And for butcher novices, the carcasses will be a good place to learn the basics of boning and anatomy while the resultant volume of meat won't overwhelm the freezer.

Anatomy

BEFORE ATTEMPTING TO SLAUGHTER or butcher an animal, you must perform enough preliminary research to ensure that the work you do is efficient and humane. A good place to start is with the anatomical facts about the animal you're going to be working with. The basic anatomy of poultry includes many of the same working parts that any other animal's system does, though there are a number of unique components that are important to be familiar with. Visceral anatomy for poultry is covered on the following page. It's important to understand the basic anatomy so that you know what makes for the most humane slaughter, what is edible, what to discard, and what may cause contamination.

The skeletal structure of a chicken is representative of most avian animals you'll be working with. And, since it's the most common bird in the meat world, we'll be using it as the primary example for describing efficient methods of slaughtering and processing. Later in this chapter we'll cover some of the differences for other birds such as turkeys, ducks, and geese.

As we cover processing later in this chapter with step-by-step instructions, you may find it helpful to use the following diagram as a reference while familiarizing yourself with specific parts of the skeleton.

Setting Up for Slaughter

YOUR PLAN FOR SLAUGHTER should be laid out well in advance. Make sure to read chapter 5 in its entirety, since it covers many of the basic methods and tenets for responsible slaughtering. Some equipment for the job is required, and several optional tools and machines can aid in the process. (Additional information about sources for equipment is available in Resources, page 440.) The more birds you decide to process, the more advantages you'll gain through the use of specialized equipment, but the basic premise of slaughtering poultry is the same no matter what your volume is. You can effectively slaughter poultry on your own, but many would say that it's best experienced as a group activity, allowing for the separation of processes into specialized stations that help with efficiency and flow.

Choose a workspace for slaughtering that will have plenty of room for all the equipment and stages. Within that space, it's helpful to have one distinct area reserved for the killing and plucking process (the dirty part) and another defined area for the evisceration, cleaning, and cooling (the clean part). Keeping space between these two phases of slaughter will help you maintain sanitary conditions. Potable water is required throughout any slaughtering process, so make sure easy access is available when considering your setup. All plastics, ideally including even your evisceration water hoses, should be food-grade.

Hanging and Bleeding Setup

Any acceptable plan for hanging the birds during exsanguination should include an easy way to restrain the bird and a method for minimizing struggling. The most common methods, in order of specialization, are killing cones, shackles, and rope or cord. Whatever hanging method you employ should allow for the easy removal of a slaughtered bird and the insertion of a live chicken that may be flapping its wings and struggling a bit, even when handled correctly.

KILLING CONES

Cones are a convenient and highly effective option. Inserting the bird is simple, and the shape of the cone restrains the wings from flapping, thus minimizing potential bruising. (A friend of mine also attaches a short bungee cord around the legs of the inverted bird to reduce the kicking, flapping, and sometimes escape of a bird, while further minimizing wing bruising and fractures.) The cone also keeps the bird inverted during bleeding. Manufactured killing cones are widely available, but with relatively little effort you can make your own killing cones using modified construction cones or by securing sheets of any sanitary material — galvanized steel, aluminum flashing, foldable plastic cutting boards — into a cone shape.

WEIGHTS

If you're not using killing cones, attaching a weight to the bird's beak once it is hung will help minimize movement during the bleeding process. This can be done with a medium-size fishing hook attached to fishing line that has a several-ounce weight attached to it.

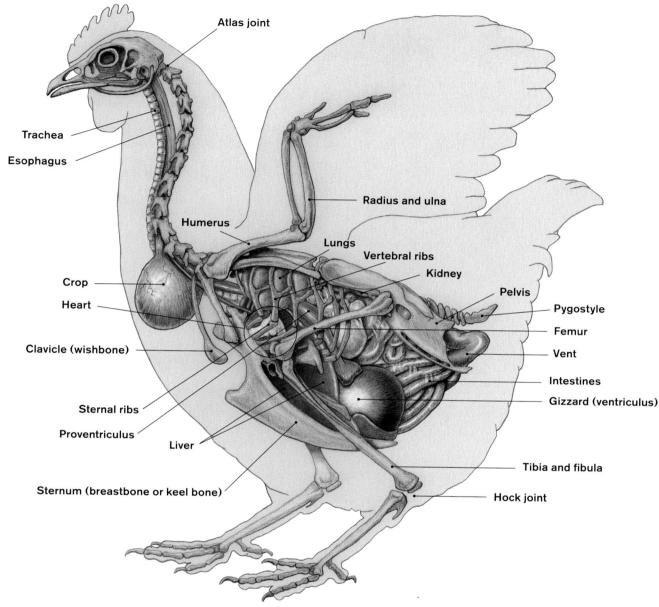

Atlas joint

Trachea

Esophagus

Radius and ulna

Humerus

Lungs

Vertebral ribs

Kidney

Crop

Heart

Pelvis

Pygostyle

Femur

Vent

Clavicle (wishbone)

Intestines

Gizzard (ventriculus)

Sternal ribs

Proventriculus

Liver

Tibia and fibula

Sternum (breastbone or keel bone)

Hock joint

THE ATLAS JOINT connects the base of the skull to the first cervical vertebra (C1). It is this connection that is disjointed in order to remove the head, either during cervical dislocation or butchering.

THE ESOPHAGUS AND TRACHEA leave the skull and run together until they enter the thoracic chamber. The blood vessels that need to be severed run on both sides of these tubes. To avoid severing the trachea and esophagus, one must sever the vessels with an angled cut, like a V.

THE CROP can be challenging to locate when empty. Follow the esophagus toward the body cavity, and you are sure to find this fleshy sack when needing to separate it from the neck skin.

THE PERINEAL GLAND, which is most often removed, sits atop the caudal vertebrae, at the rear just before the pygostyle, or pope's nose. Cut wide and shallow to remove this gland, erring on the side of more removal rather than leaving remnants.

THE LOWEST JOINT of the leg, just below the drumstick, needs to be disjointed to remove the feet. With a bit of pressure on the joint, this is a simple task. Save the feet for stock.

HEART, LIVER, AND GIZZARD are the most common offal saved from poultry for consumption. The liver is best when frozen immediately after slaughter, though all offal benefits from this approach.

HANGING STRUCTURE

The hanging area is part of your slaughter workspace and should have enough room for you to easily unload chickens, hang and bleed them, and move them to the next step (scalding). An uncomplicated and stable wooden structure for hanging birds from cones or other implements can be constructed with a few 2×4s and is useful if you are going to be processing birds with any regularity. Prefabricated setups for kill cones are also available in numerous formations.

BLOOD CATCHMENT

During bleeding you will need something to catch the blood. If you are hanging multiple birds at once, a large or metal plastic trough — something easy to clean afterward — placed underneath will catch all the blood. For individual birds a 5-gallon bucket placed just below the head (or hanging weight) of the bird will suffice. Blood collected from bleeding can be easily composted and turned into blood meal, a mixture rich in nitrogen and other soil nutrients.

Scalding Setup

Scalding poultry is the process in which you repeatedly dunk a bird in heated water to help relax the feather follicles and break down the protein holding the feathers in place. Follicles anchor the feathers into the skin; when they are relaxed, you can easily pluck the feathers. Because all the birds in a slaughtering session are scalded in a common water bath, one concern is the propagation and transference of bacteria and other contaminants between birds. Even so, scalding is generally recommended, especially if you are choosing not to pith (debrain) the bird during the bleeding process. Otherwise, you may find that dry plucking suits your needs. (For more info about the effects of pithing on plucking, see page 83.)

A simple scalding setup, with a pot and a burner, is enough for the occasional poultry processor.

Deceptively simple in appearance, tub pluckers quickly turn a feathered bird into a clean carcass.

Any scalding setup should achieve two main goals: to maintain a relatively consistent water temperature and to allow for repeated dunking of the bird(s) being scalded over a period of about 30 to 45 seconds. Rudimentary set-ups can involve a simple heat source and a large stock pot or other metal container. You can maintain the temperature through intermittent monitoring with a thermometer and do the dunking by hand. More advanced setups, some of which can be built at home, involve thermostatic controls and mechanized dunking. A modest investment in a propane-heated turkey fryer may be useful for a small-scale operation.

Scalding vessels need to be large enough to accommodate the number of birds you plan to process simultaneously as well as enough water to cover them adequately while maintaining a high temperature. (You may plan to slaughter 100 chickens in a day, but the container only needs to be large enough to accommodate the number you plan on killing simultaneously.) A container may fit the birds, but if it doesn't have enough space for a considerable volume of water, the scald will never be effective. Plan on a minimum of several gallons of water per bird. You can also keep a separate container of boiling water as backup if the temperature in the scald bath drops too severely or if water needs to be replaced during scalding.

Plucking Setup

Birds can be plucked using one of three methods: dry, wet, or wax. Dry plucking is exactly what it sounds like: plucking the feathers out of a dry bird. Wet plucking is the process of plucking a bird after it has been scalded. Wax plucking involves dipping a scalded or dry bird into a vat of heated paraffin wax. The wax covers the carcass and adheres to the feathers; it is then hardened in cold water and, when removed, takes with it the feathers, pinfeathers, and (in the case of waterfowl) down.

Kill cones

Scalder

Wax setup

Tub plucker

EQUIPMENT OVERVIEW

The basic equipment requirements are simple, though if you're routinely processing greater numbers of birds, you will most likely find that investing in some time- and effort-saving equipment will be invaluable. Listed items are in order from bare-bones to ideal.

BLEEDING: small (about 3-inch) sharp knife

HANGING: homemade or industrially fabricated kill cones

SCALDING: large pot, turkey fryer, scalding tank, or combination dunker/scalder

PLUCKING: by hand or with a tabletop plucker or tub plucker

EVISCERATION SURFACE: cutting board, butcher block table, or stainless steel water table

CHILLING: ice-filled cooler or temperature-regulated stainless steel tubs

AGING: refrigerator, blast chiller, or walk-in cooler

WET PLUCKING

Wet plucking may be done by hand or by machine. The most common machines are wet-tub pluckers; a good tub-style plucker can clean the feathers off several chickens at once in 15 to 30 seconds.

Although they speed up the process, tub-style pluckers do have some drawbacks. You'll certainly encounter more broken bones in the wings and legs of machine-plucked birds, though breakages may occur in as little as 5 percent (or even fewer). Birds with broken bones are still safe to consume, though the breaks affect the carcass appearance and reduce the value of the bird as a salable product. Another drawback is that, according to a University of Arkansas study, the meat from mechanically plucked chickens is 2.5 times tougher than meat from hand-plucked birds. Even so, the vast majority of chickens in the United States are wet-plucked in a tub plucker for the sake of efficiency and effectiveness.

While hand plucking (wet or dry) is effective, it can be done only one bird at a time. This works if you're handling small batches of birds or have other folks on hand to help; otherwise you may consider building your own plucking machine or purchasing a prebuilt plucker.

DRY PLUCKING

Dry plucking can be done either by hand or with one of the many cylinder-based dry pluckers, primarily marketed for game birds. If you plan on using one of the hand-plucking methods — dry, wet, or wax — then having a means of securing the bird will be helpful. This can be something as simple as a modified coat hanger or wire that is bent to accommodate insertion of the bird's feet, or a table that has a way to secure the feet. Securing the bird allows you to work with both hands and provides resistance while you're pulling out feathers. Having a tarp or bucket underneath or near the bird to catch the feathers is helpful, as is a tub of water, so that you can periodically dunk your hands to clean them of feathers. For wet or dry plucking, having a propane torch on hand will allow you to singe any carcass hair that may remain after plucking.

The Double-Dunk

After plucking, the carcass can benefit from a quick double-dunk rinse before proceeding into the clean part of the process, which includes evisceration and final cooling. Between the dirty and clean areas of your slaughter setup, place two large containers of iced water. (Ten-gallon buckets work well for small processing; for processing large flocks food-grade containers of 20 gallons

or more are ideal.) Label one container "step 1" and the other "step 2." As birds are plucked, toss them in the "step 1" container and leave them for a minute or so. Transfer them to the "step 2" container for another minute, then continue on to evisceration. This will dramatically help in maintaining the sanitary separation of two sides of the process. In setups where both the dirty and clean sides of the process are working simultaneously, the "step 2" container can also act as a holding space for birds, should the dirty side produce carcasses faster than the clean side can process them.

Evisceration Setup

Keeping clean during evisceration is imperative, and to do so you'll need an easily cleanable work surface, cold potable water, and a few containers. The work surface should be stable and large enough to safely accommodate the number of people who will be working simultaneously. A plastic folding table or an enamel farmhouse sink across sawhorses can work nicely, though a stainless steel table is ideal. Tables designed for poultry evisceration often have pass-through holes built into the tabletop, allowing you to dispose of inedibles into a container below. Having a hose or faucet with a spray tip simplifies the task of cleaning the bird outside and inside.

Packaging Setup

Chapter 14 covers packaging materials and methods, so be sure to read that prior to bundling the birds. A drying rack allows you to invert multiple birds for drip-drying before packaging. Simple racks can be built from PVC pipe or other materials that can be sanitized.

For the packaging process, choose an easily sanitized, stable surface that can get wet. Make sure it's large enough to accommodate all your tools and containers without having to resort to setting items on the ground and risking contamination. Packaging the birds close to where you'll be storing them, especially with large numbers, will make things easier and quicker, along with preventing carcass temperatures from rising.

Slaughtering

CHICKENS AND OTHER POULTRY are some of the easiest animals to slaughter. In fact, gathering the equipment and setting it up will often be more work than the slaughter itself. Though small compared to other animals, chickens still require the same forethought and respect during slaughter. Furthermore, the carcasses are often left skin-on, which requires special attention since they carry a higher risk of foodborne illness (see page 24).

The Right Age

To determine whether an animal is the right age for slaughter, you must consider breed, desired characteristics, and cooking method. When choosing the right age, keep in mind that the carcass yield — the weight after slaughter and evisceration divided by the weight of the bird while it was alive — will be about 70 percent. So, for example, a 6-pound live-weight chicken will yield a carcass that's a little over 4 pounds.

Preparation

Prior to planning the slaughter day, take the time to read through all of chapter 5, as it outlines pre-slaughter necessities, day-before preparations, and techniques used during the process. Planning to collect the birds before dawn will help reduce excitement and, in turn, the chance of injuries like broken wingtips and carcass bruising. If the birds will be in crates for extended periods, avoid placing them in direct sunlight or overly warm environments.

Hang and Bleed

Chickens will naturally relax when properly restrained and can then be moved into a hanging position with little resistance. Before you pick up the chicken, hold on to it for several seconds and make sure its wings are kept close to its body. If this is done properly, you'll feel the bird relax. The calmer and more assertive your actions are, the more effective this is. Once the bird is relaxed, slide one hand under it and use that hand to pick it up by both legs, with the chicken's back facing you.

You do not want to sever the spinal cord by cutting off the head, since this will cause the heart to stop pumping immediately. Evidence also suggests that severing the spine causes the follicles of the feathers to tighten, making the plucking process more difficult. Taking the time to properly bleed your bird will result in a superior product and efficient processing.

The chicken may exhibit death throes after bleeding. This is normal and is not an indication of trauma or pain. A proper bleed will render the bird unconscious in less than 30 seconds, despite the dramatic display of physical movements. The movements will subside after a minute or so, and the bleeding should be finished in less than 4 minutes.

You must wait until your bird has stopped moving and all bodily functions have come to rest before moving on to scalding. Along with being traumatic to the person conducting the slaughter, a remaining reflex in the bird to inhale while in the scalding tank may result in internal contamination to the bird through the circulatory system. Refer to page 80 and familiarize yourself with indications of insensibility and death. Proceed with scalding only after you have confirmed that the animal is no longer living.

POULTRY CLASSES

As with most livestock, chickens are categorized based on size and age. Chances are, you're going to be slaughtering the equivalent of a broiler, roaster, or stewing bird, and fortunately the process is basically the same, no matter the age and size of the bird. Keep in mind that the age and size specified in these classifications is based on the commercial breeds; pasture-raised and heritage breeds will most likely differ.

CLASS	CARCASS SIZE	AGE
Poussin	1–1.5 lbs.	Under 5 weeks
Cornish Game Hen	1–2 lbs.	5–6 weeks
Broiler/Fryer	2.5–4.5 lbs.	Under 13 weeks
Roaster	5–9 lbs.	3–5 months
Capon (castrated male)	6–9 lbs.	5–8 months
Fowl/Stewing Hen	5–8 lbs.	Over 10 months

THE SLAUGHTER PROCESS FOR POULTRY

There are very few options for slaughtering poultry, so the steps to follow are quite straightforward:

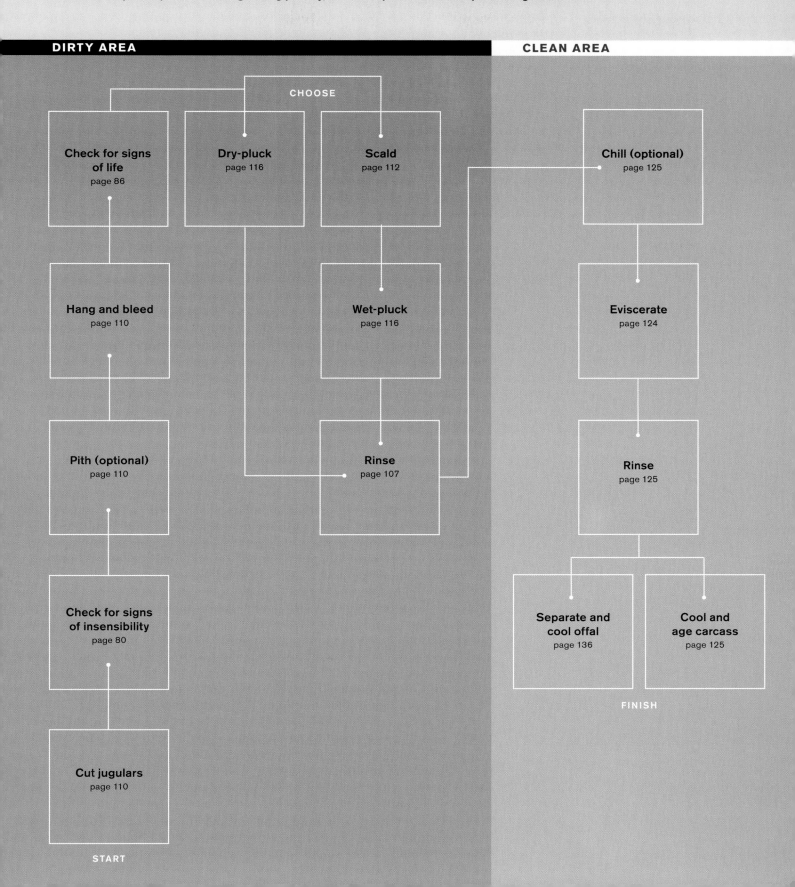

DIRTY AREA

CLEAN AREA

CHOOSE

Check for signs of life
page 86

Dry-pluck
page 116

Scald
page 112

Chill (optional)
page 125

Hang and bleed
page 110

Wet-pluck
page 116

Eviscerate
page 124

Pith (optional)
page 110

Rinse
page 107

Rinse
page 125

Check for signs of insensibility
page 80

Separate and cool offal
page 136

Cool and age carcass
page 125

Cut jugulars
page 110

FINISH

START

Bleed and Pith the Bird

1. INVERT AND CALM THE BIRD.

As the bird's blood starts to rush to its head, it will flap its wings for a few moments and then settle into a surprisingly copacetic state, making it easier to start exsanguination.

2. HANG AND SECURE THE BIRD.

Make sure the breast side is facing toward you. With your knife in one hand, take the other hand and hold the bird's head by its comb, arching the head back a bit to help tighten the skin around the neck. (With kill cones, you may need to reach into the cone and pull the head through the base.) With the neck skin taut, proceed with cutting the jugular veins.

3. CUT THE JUGULARS.

Place the knife at the base of the skull and jaw and ensure that it is touching skin, not feathers. (It's impossible to cut through feathers and get an effective bleed, not to mention that it will quickly dull your knife, making subsequent bleeds less effective.) The goal is to sever the two jugular veins on either side of the neck without severing the trachea or esophagus. To do this, you will be cutting in a V shape, the first stroke being toward you and the second away from you. You'll know when you've made an effective cut — blood will come gushing out. A trickle of blood is an indication that you haven't cut deep enough or hit the right vein.

4. PITH THE BIRD.

Once you've cut the jugulars and the chicken is bleeding, you can pith (debrain) the bird. Pithing will relax the follicles around the feathers, allowing for the option of dry plucking the bird or scalding it at lower temperatures. Hold the head steady with your non-knife hand. Using the tip of your knife, find the slit, or cleft, in the roof of the mouth. Keep the blade at a slight angle — almost parallel to the beak — and push it into the skull. (Be careful not to plunge too deep and hit the hand that holds the head.) A squawk indicates success. Alternatively, you can pith the bird first, rendering it immediately unconscious, and then bleed it out.

ALTERNATE METHOD: STARTING THE BLEED WHILE HOLDING THE BIRD

Beginning exsanguination while holding the bird is another approach, especially useful for those not working with kill cones. Start by inverting the bird, holding its legs with your knife hand. Once the bird is calm, hold the head with your non-knife hand and cradle the bird close to your body, securing the wings between your arm and body similar to how you might hold a football. Pinch the skin on the back of neck opposite of where you will cut, making it taut, and then proceed with cutting the jugulars, immediately followed by hanging to bleed out the bird.

Scald to Loosen the Feathers

1. SUBMERGE THE BIRDS.

Once your water bath is holding at the target temperature, take your birds — up to two birds per hand — and fully submerge them in the water, making sure you cover all the feathers up to the base of the feet. Jostle the birds up and down in the water for 3 seconds or so, then pull them out and repeat.

2. DO A TEST PLUCK.

After five or six cycles, remove the birds from the water, grab one of the larger wing feathers, and try to pull it out. When the feather comes out with no resistance, scalding is complete. Repeat the cycle of dunking and test plucking until the birds are ready.

ALTERNATE METHOD: ONE BIRD, ONE POT

If you're only occasionally scalding a few birds, all you need is a large pot of hot water. The approach is the same: dunk, hold, repeat, and test-pluck. A tool, like the honing rod used here, will help submerge the bird within a tight-fitting pot.

Scald the Bird

After your bird is sufficiently bled out — its body is listless, there is no more expulsion of blood, and there is no cloaca movement — you're ready to scald. Just to make sure the animal is dead, consider pouring a bit of the scalding water on its backside; if there is any reaction, then wait another minute or so for the bleed to finish.

There are different levels of scalding, which depend on two factors: water temperature and time of submersion. It will take some trial and error for you to arrive at the right combination based on your priorities — for example, easy plucking, carcass appearance, or skin condition — but I recommend starting with either a soft or hard scald.

If you plan on dry-plucking your birds, you can skip ahead to page 116.

SOFT SCALD (125°–140°F)

Sometimes called a semi-scald or slack scald, the soft scald will help keep the epidermal layers of the skin intact and will help prevent any tearing from plucking. This will also help if you're concerned with carcass appearance or plan to air-chill your birds.

HARD SCALD (140°–160°F)

The hard scald is hotter and quicker than the soft scald, and can make plucking easier, but it will remove the upper epidermal layer. Even so, there should be little risk in the skin tearing from plucking, unless the bird is overscalded (kept in the water too long). With chickens, scalding at temperatures or exposure times greater than a hard scald will likely damage the skin and can even cause some of the external muscles to become slightly cooked.

The process is relatively the same no matter what combination of water temperature and submersion time you go with. Knowing the "doneness" of the scald is where the trial and error will most likely come into play. Do not be discouraged if, in the beginning, birds show signs of underscalding (difficulty when plucking) or overscalding (torn skin, slightly cooked meat). If your temperature hovers in one of the recommended ranges, start experimenting with adjustments to the submersion time. If results do not improve after a couple of rounds, start making adjustments to the temperature range. Also try adding liquid dishwashing soap to the water to help the water penetrate the feathers. If you're planning on saving the feet (they make great stock), make sure the feet are fully submerged during the scalding process, enabling you to peel off the skin later.

Plucking

PLUCKING IS THE MAIN TASK that determines the final appearance of the poultry carcass. For those with customers, proper plucking is profitable because it results in a good-looking carcass that will garner a high price. There are many ways to pluck a chicken; in the end each one provides a clean carcass, so choose one that best fits your situation. Your choice will be determined by the space you have, the number of birds you will be processing, and your budget. Follow plucking, the last step in the dirty process, with the double-dunk method (described on page 107) to quickly rinse off the carcass and help bring the body temperature down.

Mechanical Plucking

If you're using a mechanical plucker, follow the manufacturer's directions for how to proceed through this step. The key to any successful mechanical plucking is a quality scald, so if you're using the machine correctly and the results are unsuccessful, adjust the scalding method before you decide that the machine is underperforming.

Hand Plucking

Plucking should begin immediately after you've completed scalding or pithing. When hand-plucking, don't grab large swaths of feathers and try to pull them out. This will undoubtedly result in torn skin and an unsightly carcass. Rather, grab large flight feathers individually or in pairs, keeping them taut, and pull them with an assertive but not jerking motion in the direction they grow. Grab smaller feathers by their base, between your thumb and forefinger, and use the same motion; they should come out with little resistance. Sometimes it is helpful to use one hand to hold pressure on the skin around the feathers and the other hand to pluck. If the scald or pithing was successful, plucking will be a process not of force but of finesse and speed.

The rumble tumble process of a tub plucker enables the soft, rubber fingers within to remove the feathers, resulting in a near-clean carcass after only 25 to 45 seconds.

FINISHING: REMOVING PINFEATHERS AND SINGEING HAIR

Pinfeathers are undeveloped feathers in the bird's skin. They are stubborn and require an entirely separate method of removal.

Start by taking a dull blade — a butter knife or the equivalent — and move it across the skin in strokes, starting near the neck and traveling toward the tail. Apply enough pressure to pop the pinfeathers out of place but not so much that the skin gets caught under the blade, potentially tearing.

The pinfeathers that persist may need to be disposed of individually. Grasp the pinfeather between a dull blade and your thumb and pull them out, or use your thumbs to pinch the skin — similar to popping a pustule — and the pinfeather should come out. Truly pesky pinfeathers may need to be removed using needle-nose pliers or tweezers (at which point you may question the necessity of a purely clean carcass).

Unless you're plucking mechanically or with wax, some hair tends to remain on the carcass, most often on the wings. You can singe off the hairs during the slaughtering process (unless you're planning on roasting the chickens whole, in which case it's easier to just let the hairs singe later in the oven). If you choose to singe them off now, you can do so using a handheld propane torch or even the open flame of a cooktop burner. In either application, make sure the flame moves across the chicken skin very quickly and exposure is brief; otherwise you risk burning the skin or even slightly cooking the chicken, which will cause spoilage during storage.

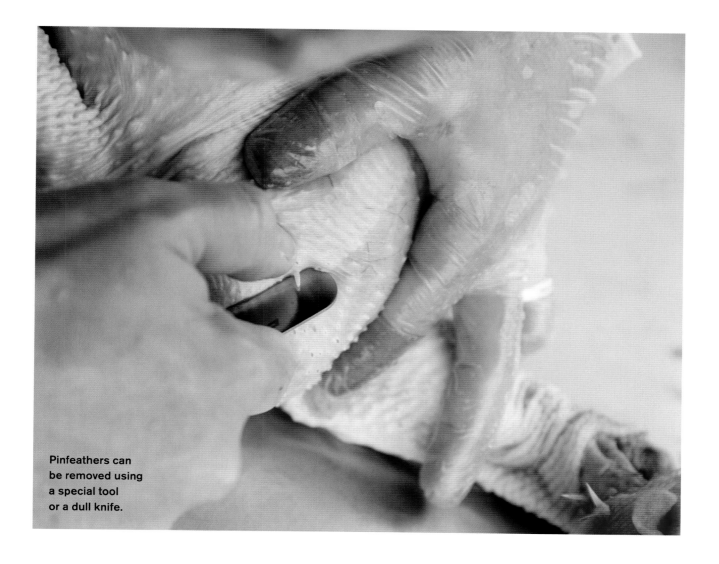

Pinfeathers can be removed using a special tool or a dull knife.

Dry Plucking by Hand

As soon as you have confirmed that the bird is lifeless, start dry-plucking it; you have about three minutes until the muscles begin to cool and the follicles that were relaxed from pithing begin to tighten. You won't pluck the whole bird in three minutes, but it's good to start as soon as possible. Start with the wings and legs; these areas will cool the quickest, causing the follicles to tighten. Move to the tail and work your way down the carcass, working in order from what bled out first.

1. REMOVE THE LARGEST FEATHERS.

Begin with the wings, removing the large flight feathers first, one or two at a time, and then moving on to the smaller ones. Move on to the legs and, once finished, continue with the tail (pope's nose).

2. FINISH PLUCKING.

Move down the front of the carcass until you reach the breasts. Grab a few feathers at a time, between your thumb and forefinger, and pluck against the direction of the feathers. Finish with the back and neck.

ALTERNATE METHOD: WET PLUCKING BY HAND

With your bird suspended or on a table, start plucking the extremities first, then move on to the tail, the breast, the back, and the neck. Dip the bird back in the scalding tank a couple of times if you find that the feathers are becoming too difficult to pull out. Check to see if the feathers have released with each motion. Once the follicles have relaxed again, continue plucking and repeat until you are finished.

Cleaning and Evisceration

THE GOAL OF EVISCERATION is to remove the viscera (interior anatomy) without any contamination, resulting in a clean, consumable carcass. One gram of gut content can carry a billion bacteria, which can be transferred or propagate either on your work surface or within the water bath that you may be using to cool down the carcass. Responsible slaughtering means honoring the animal; discarding any meat due to avoidable contamination is a contradiction. Therefore, take proper precautions and act with intention while eviscerating birds. Finally, sanitize all your equipment before continuing.

Removal of head, neck, and feet is completely optional. If you prefer to leave them on — in what is sometimes called the Buddhist style — then skip over those areas described below and focus on the internal viscera.

Ensure that edible offal is clearly separated from inedibles. In this case, the feet could have been saved for stock.

A stainless steel table is the ideal sanitary surface for eviscerating poultry.

LEG TENDON REMOVAL (OLDER BIRDS)

You may prefer to remove the inedible tendons from the drumstick when working with older birds. If so, this can be done with the bird's breast either up or down.

Cut through just the skin on the back of the shank, proceeding carefully to avoid severing the tendons. The tendons (white bands of connective tissue) should be exposed. Using a small hook — a tie-wire twister or a game-bird gut hook works well — or pair of needle-nose pliers, grab the tendons, twist them, and yank them away from the knee joint. The tendons should release from the leg muscles while remaining attached to the shank.

Remove the Feet

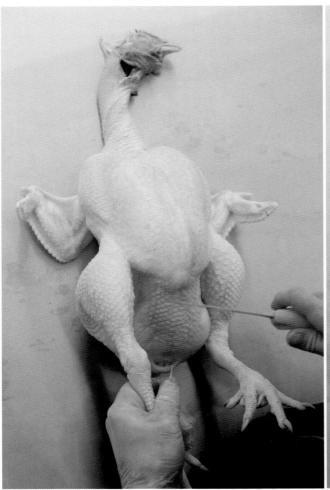 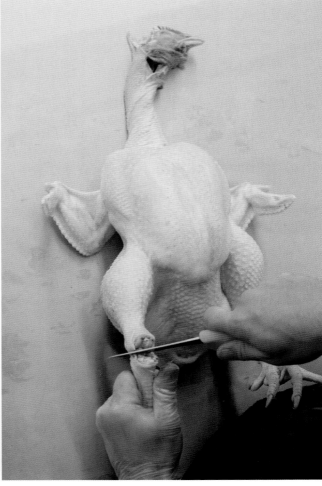

1. STRAIGHTEN THE LEG.

Place the bird breast-side up. Grab a foot and place downward pressure on it, straightening the leg.

2. CUT ACROSS THE HOCK JOINT.

Maintaining the pressure, take your knife and cut across the hock joint. When the cut is done correctly, the joint will pop open and allow you to remove the leg completely by cutting through the connective tissue and skin on the back. Repeat with the other foot. You can choose to save the feet (they're great for stock), feed them to your dog, or toss them in the inedibles container.

Remove the Preen Gland

In the United States, the preen (uropygial) gland, an oil gland at the base of the tail, is commonly removed, though its removal is optional.

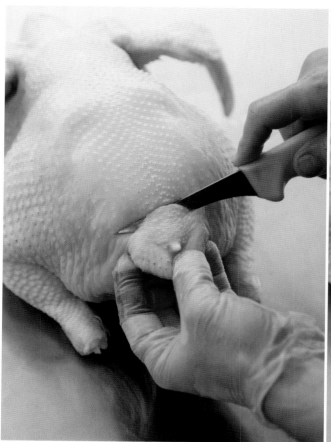 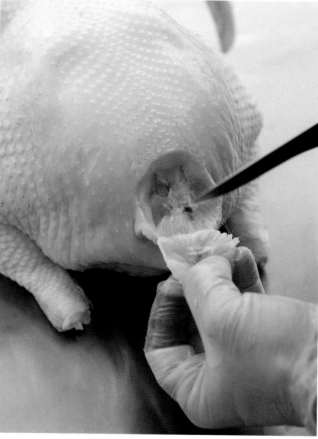

1. FIND THE GLAND AND START THE CUT.

Place the chicken breast-side down and find the gland, a deep yellow nipple-like papilla anterior to the tail. Start cutting about an inch in front of (anterior to) the gland and head toward the tail at a downward angle — 70 degrees or so — until you reach the spine.

2. REMOVE THE GLAND.

Continue cutting along the spine and scoop upward and to the right before you reach the tail. If you're not able to remove the entire gland with one fell swoop, make sure to remove the remaining gland pieces — they'll be the bright, orangey-yellow bits — which may be excreting small amounts of oil. If you should get oil on your knife, clean it off before continuing.

Remove the Head and Neck

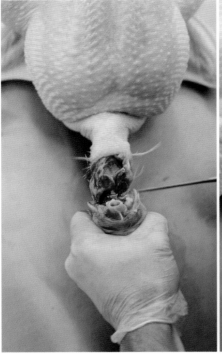 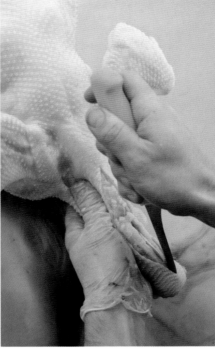 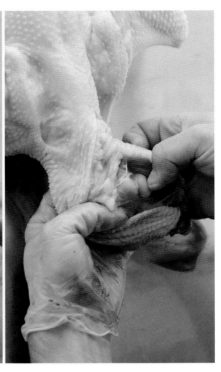

1. REMOVE THE HEAD.

Cut through the muscles between the base of the skull and the atlas joint to remove the head. Do not try to saw through it with your knife (in knife against bone, knife loses every time), but rather slice around the neck just until you reach bone. Once the muscles are severed, a short twist of the skull should easily separate it from the neck.

2. CUT ALONG THE NECK.

On the back of the neck, and with your blade edge facing up, use the tip of your knife to make a shallow incision along the entire neck, starting at the connection between the shoulders and ending where the head used to be.

3. PEEL THE SKIN AWAY FROM THE NECK.

Keep the esophagus and trachea (located on the breast side of the neck) attached to the skin.

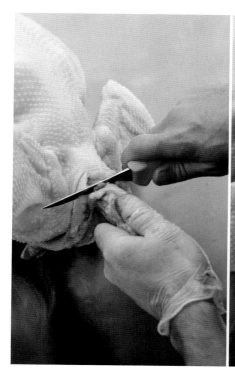
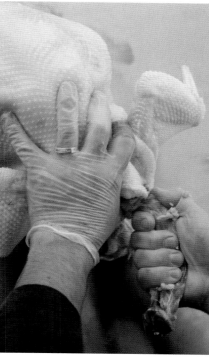
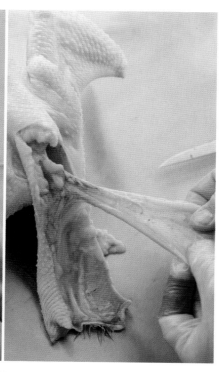

4. SEVER THE NECK MUSCLES.

Find the base of the neck and cut through the muscles — again not sawing, but severing muscle and barely hitting bone.

5. TWIST OFF THE NECK.

Hold the carcass with one hand, grab the neck with the other, and give it one full twist to snap it off. Save the neck for stock.

6. SEPARATE THE SKIN.

Separate the esophagus and trachea from the flap of skin that was covering the neck. If you plan to keep your bird whole, leave this skin intact; otherwise remove and discard it.

Separate the Trachea, Esophagus, and Crop

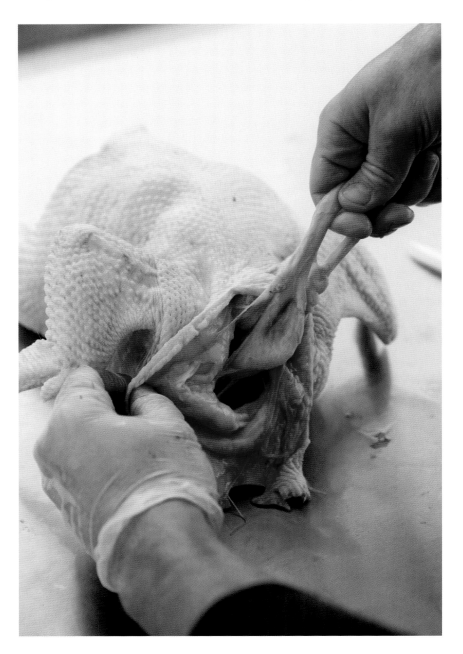

The crop should be relatively empty, assuming food was withheld from the bird, and may even be difficult to identify.

Follow the trachea and continue using your hand to separate the food pathway from the skin until you are barely entering the body cavity. Locate the crop. It will be the enlarged, baglike skin formation that holds food before it enters the interior cavity of the bird.

Leave the esophagus, trachea, and crop loose and separated. The empty and dislodged crop will allow these parts to be removed with the rest of the viscera through the back end in a later step.

A full crop is easy to locate, but pulling it through the back end may cause it to rupture, spilling its contents inside your bird. If the crop is full, use your hand to dislodge it from the skin, cut the trachea and esophagus after the crop, and discard the whole bunch.

If you happen to break open a full crop while separating it from the skin, don't panic — it's just undigested food. Remove the trachea, esophagus, and crop. Follow with a good rinse, making sure not to blast water into the body cavity.

Remove the Vent and Large Intestines

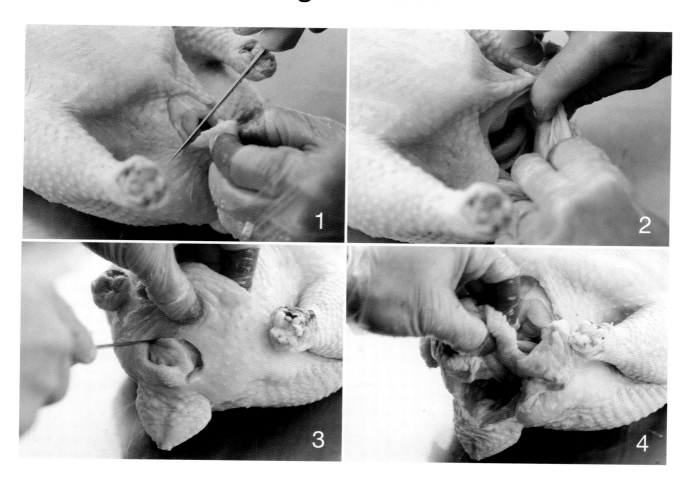

1. OPEN THE CAVITY.

Place the bird on the table breast-side up. Note the location of the cloaca (vent or anus). (If you did not withhold food from the bird prior to slaughter, take extra precaution to avoid fecal contamination, because the fragile intestines will most likely be full of it.) Pinch a small section of skin about 2 inches above the cloaca, keeping the skin below it taut to make it easier to cut. An inch above the cloaca, make a shallow 2-inch-wide horizontal incision. This should expose the underlying fat layer and provide a small opening into the inner cavity.

3. SEPARATE THE VENT.

Hold the skin around the vent, and make a shallow cut all the way around it until fully separated. (Be careful not to cut too deeply.) Connect the vent incision to the cavity opening.

2. ENLARGE THE CAVITY.

Carefully insert a few fingers and pry the opening to make it larger. (With a full intestine, this action may cause some fecal matter to expel from the cloaca. If that happens, stop immediately. Rinse the exterior of the bird with cold potable water, keeping water from entering the cavity. Rinse your work surface before continuing.)

4. PULL OUT THE VENT AND INTESTINES.

Carefully pull the vent and the attached large intestines out of the cavity and away from the carcass. With the vent out of the way, make sure the opening is large enough to accommodate one hand in the cavity; the necessary size may be much smaller than you initially think. Err on the side of keeping the cavity small, because you can always continue tearing a larger opening if needed.

Remove the Entrails

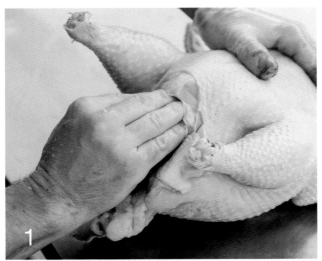

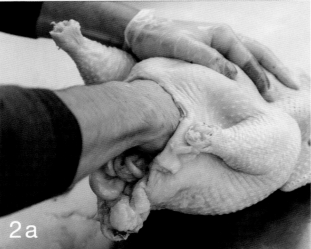

1. **HOLD THE BIRD IN PLACE.**
Position one hand near the shoulders, making sure not to put pressure on the cavity. Your other hand should try to emulate the curvature of the chicken's interior cavity.

2. **INSERT YOUR HAND INTO THE CAVITY.**
Keep the top side of your fingers against the keel bone while breaking any attachments you encounter. Proceed slowly, and once you reach the front, curve your fingers down and pull back to begin removing the guts. Don't squeeze anything in the process, which can cause contamination from a broken gallbladder or intestine; instead, create a clawlike barrier with your hand and ease the guts out, breaking cavity connections as you expel them. It might take a couple of trips into the body to get the whole digestive system and other viscera.

You'll also need to remove the lungs. There are four lungs, each nestled in between the vertebral ribs. It can take a few tries to get them all. Although you can use a tool to remove them, your deft fingers will be effective at extracting them once you get the hang of it.

Take one last exploratory trip into the cavity to see if anything else remains – gonads, ovaries, and kidneys are elements that get easily overlooked.

Cool the Carcass

1. RINSE THE BIRD.

After eviscerating, give the bird a thorough rinse inside and out. It is good to flush the inner cavity from the neck end since the water can drain much more easily out of the tail end, but however you do it, make sure the outside is clean and the interior bits have been rinsed out.

2. CHILL THE BIRD.

Transfer the bird to a cooling container of ice-cold water. When it goes into the cooling container, the bird's internal temperature will be between 104°F and 107°F, depending on how long processing took. Make sure your containers have enough room so that all the carcasses are completely submerged and you've followed the recommendation of ice-to-meat ratio mentioned on page 95. If you crowd your container with hot birds, you may be surprised at how quickly the water temperature rises, thus slowing down the cooling process of the carcasses and causing increased water uptake.

Clean the Gizzard

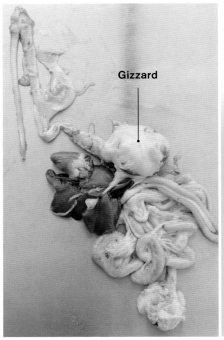

Gizzard

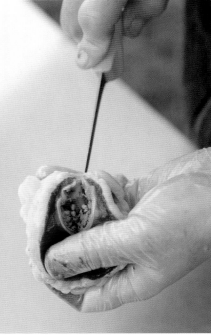

1. SEPARATE THE GIZZARD.

The gizzard shouldn't be difficult to identify, considering that it'll be the largest and hardest item in the pile. Separate it from the digestive tract, and discard the rest of the tract (assuming that you've removed the other edible offal you want to save).

2. SLICE OPEN THE GIZZARD.

This will expose the potentially surprising array of items your bird considered food.

3. PEEL OFF THE LINING.

Rinse out the gizzard and then take hold of the dark yellow coarse lining (made from a material called koilin) and peel it away, discarding it afterward. Grabbing an edge of the lining may prove difficult. If so, try gripping it between the edge of your knife and your thumb or grab a towel and use that to help with holding. Give it one final rinse and then store the gizzard for later use.

Other Poultry

PROCESSING THE OTHER common types of fowl — turkeys, ducks, and geese — is very similar to processing chickens, so most of the information covered in the discussion of slaughtering and butchering chickens applies here as well, with just a few distinctions.

Turkeys

After chickens, turkeys are the most common domesticated fowl. (An alarming 99 percent are a breed named broad-breasted whites, the product of commercialized cross-breeding with a focus on large breast muscles, fast growth, and efficient feed conversion.) Despite its popularity being overwhelmingly concentrated around Thanksgiving, properly prepared turkey can be a marketable product year-round, though probably not in whole-bird form.

SLAUGHTERING

Turkeys are generally much larger animals than chickens, with toms potentially reaching 50 or more pounds. Despite the anatomical similarities of the two species, you will need extra pieces of single-purpose equipment for processing turkeys in addition to those you use for chickens.

Turkeys definitely need larger kill cones than chickens do, and chickens will fall through or will not be properly restrained in cones sized for turkeys. You can choose to hang turkeys outside a cone — like chickens, they also become quite docile when hanging upside down — but during death throes their heavier mass and stronger muscles make bruising, and the resultant loss of product, an even higher risk than with chickens. Cones are definitely preferred.

Turkeys are bled in the same way chickens are, including the option of pithing. One clear advantage when bleeding turkeys is that their unfeathered necks provide easy access to the veins and carotid arteries necessary for exsanguination. (This also keeps your knife blade sharp.) Bleed times are similar, though the volume of blood will be much greater with turkeys, so take that into account when designing the collection system. Beak weights will need to be heavier than those for chickens to overcome the increased neck strength of turkeys. Keep in mind, though, that the proper mass for a turkey-beak weight will most likely be too heavy for a chicken and may cause the chicken's head to get pulled off.

Turkey processing follows many of the same steps as chicken, with a few notable exceptions that mostly address the variance in size.

Scalding and wax-plucking vessels can be used for both chickens and turkeys but must be sized for your largest bird. The only inefficiency with using a large scalder (meant for the largest toms) might be the energy it takes to keep a much larger volume of water at the proper temperature when being used for smaller batches of chickens.

Mechanical pluckers are just as common with turkeys as they are with chickens. Pluckers are most effective if you've taken a few moments to remove the swaths of large wing and tail feathers, and any others that come off easily, before transferring the scalded turkey. Because they're so large, though, you'll want to limit plucking to one or two turkeys at the same time, though the length of time for plucking should be more or less the same.

It will take longer for rigor mortis to set in after you slaughter a turkey than after you slaughter a chicken, averaging 2 hours for the onset and up to 24 hours for the resolution. Cooling turkeys will also take longer, up to 8 or more hours longer depending on the weight of the carcass. Furthermore, with turkeys versus chickens, you will need much more space for chilling for an equal number of birds.

Packaging turkeys can be done in all the same ways as chickens, but clearly the bags will need to be much larger. Vacuum-sealing turkey meat might be unachievable for a small operation, mainly because the bags, not to mention a unit that will allow for such large bags, are prohibitively expensive. Therefore, it's understandable that most small producers opt for using shrink-wrap or a simple plastic bag when packaging whole turkeys. For turkey parts, as with chicken parts, vacuum sealing is preferred.

BUTCHERING

There are two main differences between butchering turkeys and butchering chickens; both have to do with the anatomical structure and physical needs of a larger bird.

BONES. A turkey's substantial weight requires a skeleton that is both stronger and denser than that of a chicken, making bones more challenging to cleave and cartilaginous areas more resistant. When boning a whole breast by splitting the keel bone, not only will your scoring require a bit more knife pressure but the force to cause the actual fracture will also be greater. Furthermore, when disjointing extremities, the structures around the joints will be composed of thicker connective tissue and cartilage and, once cut through, additional force will need to be applied to disconnect the bones. Cutting out the backbone will be less forgiving on a dull knife; you may find poultry shears to be ideal for this task.

TENDONS. The increased body mass of turkeys versus chickens also means that leg tendons connecting the tibia and shank are much thicker, making them entirely inedible and more of a challenge to remove. The technique mentioned on page 117 for removing the tendons of older chickens will work with turkeys.

With older, larger turkeys you can also try the following method: Carefully cut through the skin around the hock joint; don't sever the connective tissue, thus leaving the connection points of the tendons that run into the leg muscles. Take the leg over the edge of a table and break the hock joint by pressing down on the shank. The shank will be separated from the drumstick, but the tendons will still be attached. Hold the end of the drumstick and, with your other hand, quickly pull the shank off with the tendons, which will release from the internal meat of the drumstick.

One last approach is to wait until the meat is cooked and then remove the tendons with a pair of needle-nose pliers. After the drumstick (or full leg or full carcass) is cooked, look for the ends of the tendons near the hock joint. Using a small piece of paper towel around the tendon to improve the grip, take the pliers, grab the tendons one at a time, and pull them out from the drumstick. With the meat cooked, the tendons should give out relatively easily. As with so many other areas of butchering, test the methods and choose what works best for you.

Ducks and Geese

Ducks are the most popular domesticated waterfowl we consume, with geese following that but in substantially lower numbers. Differences in processing, mostly in how to pluck, are attributed mainly to the biological needs of these migratory birds — a dense layer of down feathers and increased fat deposits.

SLAUGHTERING

Compared with chickens, waterfowl have thousands more feathers, primarily in the form of down, making the process of scalding and plucking more of an endeavor. The feathers of ducks and geese are water-resistant to boot. Thus, dry plucking the swaths of large feathers, followed by wax plucking the remaining feathers, is recommended for waterfowl. For small batches of birds, dry plucking is the easiest and quickest way to remove the majority of feathers even if you are planning on wax plucking afterward. If you're scalding the bird, adding a small amount of dish soap (Dawn works best) will help the hot water penetrate the inherently water-resistant feathers.

Because ducks and geese have an insulating fat layer, they will take longer to cool after being slaughtered than chickens will. Their larger carcass sizes will further delay the effects of cooling. Other effects from the larger size that should be taken into consideration are mentioned above in the section on turkeys.

WAX PLUCKING

Wax plucking is most commonly done with waterfowl and birds with large amounts of down feathers — the small insulating feathers that sit close to the skin. These feathers are challenging to remove by dry plucking, and even tub-style pluckers tend to leave residuals that need to be removed by hand. Wax is added to a vat of hot water, in which it melts and settles on the surface. Carcasses are submerged, covered in the liquid wax, and then immediately plunged into ice water to harden the wax. The wax adheres to the feathers so that when it is removed the feathers all come with it. The result is a clean carcass with little or no damage, making it a great approach for those working with small volumes of chickens, ducks, or other poultry.

SLAUGHTERING A DUCK

- **CALM THE DUCK.** Gently take hold of the duck, hold it by its legs, and allow it to hang inverted. It will probably flap its wings a bit, but after several seconds the duck will relax. Not only does this improve carcass quality, but it also allows for easier handling without the risk of hurting the bird.

- **SECURE FOR BLEEDING.** Insert the calm bird into the killing cone, carefully pulling its head through the bottom. The shape of the cone will squeeze the duck, helping to keep it calm.

- **CUT SKIN, NOT FEATHERS.** The thick layer of feathers present on ducks can make cutting for bleeding a challenge; a knife won't cut through feathers. On one side of the neck, use the edge of your blade to push back some of the feathers until you are against the skin. Then proceed with cutting at an angle that severs the blood vessels but keeps the esophagus and trachea intact. Repeat with the opposite side of the neck.

Wax plucking requires two large containers big enough to dip the entire bird into, and one smaller container for the used wax. One container should be kept hot and large enough to submerge your bird in the mixture of water and wax without overflowing. To pluck birds one at a time, a turkey fryer will work fine, as will any large stock pot above a portable burner. Parafin often comes in 4-ounce blocks (four blocks per 1 pound box). For chickens and ducks, you'll need roughly one block per bird, depending on its size. That gives you about three to four birds per pound of paraffin. Keep in mind that the wax is reusable (though not indefinitely). Your vat of hot water and wax should settle into a stable temperature of about 150°F. Also, make sure that your water is warm before adding the paraffin. Paraffin added to cold water can produce a film on top of the water that runs the risk of explosion.

Your second container will hold cold water, preferably with ice, and is used to quickly set (harden) the wax. The size of the container will depend on how many birds you plan to wax simultaneously. It can be as simple as a large cooler with ice and water or a large stainless steel tub.

BUTCHERING

Although the ducks and geese raised on farms are domesticated, they still share common traits with their wild-game relatives, such as fat distribution and the types of dominant muscle fibers. (Thus, the advice given here can be applied to other kinds of waterfowl.) Waterfowl anatomy has evolved to support a life in flight and in the water — very different from the sedentary lives of chickens and turkeys.

FAT. Abundant fat stores will be one of the first noticeable differences when butchering waterfowl versus butchering chickens. Up to a third of the carcass weight may be intramuscular and subcutaneous fat stores, providing both fuel and insulation. Breaking down a whole bird will provide ample amounts of skin and fat trim, so as you butcher waterfowl freeze the trim until you have enough to render into a usable amount of fat.

Unlike with chicken, you'll need to strip a waterfowl carcass of skin and fat before using it for stock. The reason for this is twofold: a better use for the fat is rendering, and bones roast cleaner without fat to give you a purer stock.

Proceed with butchering a whole bird as you would with a chicken. With the wings, breasts, and legs removed, place the carcass breast-side down. Remove the sizable preen (uropygial) gland. Grab hold of the neck skin (again, a fabric or paper towel can help) and begin to peel the skin away from the carcass, working toward the tail.

Use your knife to release stubborn areas and ensure that you are keeping most of the fat with the skin. Pull out any intercavity fat deposits, and inspect the carcass for any remaining fat stores. Check the neck area for glands, and remove them. The vast deposits of fat can be rendered, saved, and used for many culinary purposes, including the common duck-leg confit.

Because of the large underskin fat stores, the skin of ducks and geese is often scored, especially with breasts, to give the rendering fat a place to escape during cooking. This is customarily done immediately before cooking using a cross-hatch diamond pattern that goes no deeper than the skin.

GLANDS. Glands play a vital role in the lives of waterfowl and are therefore larger in size and number. The preen gland, located at the base of the tail, secretes a complex combination of waxes, fatty acids, and fat, and is spread over a bird's feathers through the act of preening. One function of this coating is to make the bird's feathers water repellent, which in turn enables the bird to float for long periods. Given the higher demands, the preen gland in a duck or goose is larger than that in a chicken, but its larger size also means that it will be easier to find. Follow the instructions on page 119, keeping in mind that recommended measurements will need to be adapted to the results of your own carcass inspection.

There are also a host of glands in the neck area of the bird, and these must be removed to prevent their vile taste from infiltrating your dishes. Their color in a healthy bird will be beige to dark tan. Take the time to look over the neck area and remove anything that looks suspect; when in doubt, cut it out.

The preen gland on a duck, or other larger poultry, will be easier to identify for removal.

Wax Plucking

1. MELT THE WAX.

Heat the water and, once warm, add the wax.

2. DRY-PLUCK THE MAJOR FLIGHT FEATHERS.

Start by dry-plucking the major flight feathers on the wings and the tail feathers, as well as any other large feathers that are removed with little resistance.

3. ROUGH-PLUCK THE BODY.

The wax will need to penetrate beneath the outer layer of feathers. To give it access, rough pluck the body and wings. Grab a few feathers at a time between your thumb and forefinger. Swiftly pluck them out, working against the direction that they grow.

4. REMOVE THE WINGTIPS.

The wingtips of waterfowl are difficult to clean and their boniness doesn't offer much reward for the effort. It's better to clip them off with a sharp pair of shears, preferably a pair that has a notch in the blades for cutting through bone.

5. DIP THE BIRD IN THE WAX-WATER MIXTURE.

Take the bird by its feet and slowly dip it into the warm mixture of water and wax, all the way down to the shanks. Slowly move the bird up and down, making sure the layer of wax on top of the water is adequately adhering to the bird.

6. REMOVE THE BIRD.

After 10 or 15 seconds, slowly lift the bird and then hold it above the vat while the wax spreads across the carcass and the excess slinks off, back into the warm mixture.

7. DUNK THE BIRD IN ICE WATER.

Immediately transfer the bird to the cold-water bath and keep it submerged for 2 to 3 minutes to let the wax set.

8. REMOVE THE BIRD.

When the wax is set and hard, remove the bird (hang or secure it, if you're set up for that) and begin peeling the wax off.

9. PEEL THE WAX.

A properly thick coating of wax — about ⅛ inch — should require you to crack it to start peeling. Start with the breasts. Rather than pulling the wax off the bird, pull the skin away from the wax: inch your fingers along the edge of the wax, coaxing the skin cleanly away.

REUSING WAX

Keep in mind that the smaller the surface area of your hot vat, the more frequently you'll need to reload wax during the process. For example, if you take a 20-quart stock pot and fill it with the proper ratio of water and wax, leaving room to submerge a bird, the amount of wax in the container will be sufficient for a single bird. Therefore, you'll need to add more wax after each bird and wait for it to melt to the right temperature. This isn't a problem if you're working with small quantities, but if you plan to wax dozens of birds, you should invest in a large vat that allows for minimal reloading during your process. Plan on reusing the wax and integrate a method of straining feathers into your hot vat setup. A screen cut to the size of your vat or even a pasta colander — again, size depending on your volume — suspended in your hot vat allows you to easily strain feathers out of melted wax before boiling the wax for sterilization. For people slaughtering on a larger scale, there are larger machines that constantly monitor the water temperature, strain the used wax, and add it back to the vat.

Aging, Packaging, and Cleanup

TEST THE TEMPERATURE of the last bird in the cooling tank; if the reading is 40°F or lower, you're ready to start aging (a process that will increase the quality of meat) before any further processing or freezing. For more about the science behind aging, refer to page 18. Follow aging by packaging the birds whole or processing and packaging them as parts. Detailed instructions on breaking down chickens into constituent parts begins on page 140. Although you can age birds in their final packaging, it will benefit the product to age it in an open-air environment if possible.

Drying the Birds

The drier your birds are when aging and packaging, the better. Less water in the bags means less humidity, less microbial growth, and a drier skin (which is better for roasting). There are a number of ways to encourage drying. Remove the birds from the cooling water bath and allow the excess water to drip off. Transfer the birds onto a drying rack or large baking sheets lined with metal racks and allow them to drip-dry for several minutes. A strong fan will help speed the process.

If you have the cold-storage space to fit all the processed birds on racks, then move the birds, uncovered, into the cold storage. Keep them there for 12 to 24 hours and let them "set up," a process that will partially dry and tighten the skin around the bird. This is the ideal way to age a bird that is intended to be sold or roasted whole.

If you have the cold-storage space for your birds but not with racks, then packaging them in open bags is an alternate method. After you've let the excess water drip away, pat the birds with paper or fabric towels, stuff the carcass cavity with a dry paper towel, and, to save on material expenditures, put them in the plastic bag you'll use for sale or storage. Place them in cold storage with the bags left open for 12 to 24 hours. The open bags will allow some of the moisture to evaporate while the muscles continue to relax. Remove the paper towel in the cavity before finalizing packaging.

If you don't have any cold-storage space to hold the birds for 24 hours before butchering or freezing, leave them in the cooling tub for at least 4 hours (6 to 8 hours is even better) and make sure the temperature of the water is kept below 40°F. You run the risk of increased water uptake, but it's a beneficial exchange for more tender meat and an overall better carcass.

Packaging

How you package the birds depends directly on how they're going to be used and stored — fresh or frozen — and the instructions for both methods are available in chapter 14. When packaging whole birds, you may find that trussing helps maintain the shape of the bird.

Offal

The edible parts of the gut pile are the feet, heart, kidneys, liver, and gizzard. Chicken livers are a common ingredient in paté; gizzards are a common ingredient in ethnic foods from around the world; hearts are delicious

Any rack that can be sanitized and allows the birds to drain will be suitable for drying.

marinated and grilled; and necks (and bones if you plan on boning any chickens) can be used for chicken stock. All offal will benefit from being frozen soon after slaughter. If you aren't planning on using it yourself, you may be able to find someone who will take it off your hands and not let it go to waste.

If you saved the chicken feet, rinse them thoroughly while scrubbing with a brush to remove all the dirt. There is no need to peel them now. Instead, freeze them in 1- to 2-pound bags and process them further when you're going to use them. Clean and rinse the gizzard (see page 126).

OTHER OFFAL

Follow the directions on page 93 for further processing of heart, kidneys, and liver. After each step, rinse the organ thoroughly and put it in a container full of ice water; it's best to have separate containers for each type of offal. The kidney, liver, gizzard, and heart are often referred to as giblets (pronounced *JIB-lets*) and are commonly wrapped in wax paper or a plastic bag and included inside the cavity before packaging the chicken. Though the neck is not considered part of the giblets, it's often packaged with them. Otherwise, consider packaging offal of the same kind in ½- to 1-pound packages.

Cleanup

Base your cleanup on the guidelines provided on page 96, ensuring that the result is a clean workspace, sanitized tools, and properly disposed-of inedibles.

A selection of the edible offal from poultry.

PEELING THE FEET

Chicken feet are an often overlooked but beneficial ingredient in earthy chicken stocks. But, before using them, you must remove the exterior, protective layer of scaly tissue. To do so, heat a large pot of water to just under boiling. Carefully drop the feet in and briefly blanch them for 20 to 30 seconds. Remove the feet and place them in a bowl of cold water until they are cool enough to handle. With some minor coaxing, the exterior layer should easily slip off. If it doesn't, scald the feet again for 15 seconds and retry. Be wary of scalding too much, as it can actually fuse the scaly exterior to the feet. Once peeled, you can use a heavy knife to remove the talons at the first knuckle.

CHICKEN BUTCHERING

CHICKENS MAY VERY WELL BE the most common, and easiest, of all animals to raise in a homesteading environment. Chicken meat is also one of the most versatile ingredients in cooking, offering a flavor base that can underscore regional flavors. It is no wonder that today's local food movement has reintroduced the backyard chicken to many families interested in producing a viable animal protein — eggs and meat — on small plots. Butchering a chicken is easy work and, for those raising chickens or buying them whole, processing chickens at home offers unique options for preparation, stock ingredients, and offal.

Butchering Setup, Equipment, and Packaging

PROCESSING POULTRY WILL require some separate equipment due to the risk of bacterial cross-contamination (see page 30). Purchase cutting boards exclusively for poultry processing; choose a size large enough to provide ample working space and made from a material that is easily sanitized, such as wood or plastic (see page 47). I don't generally recommend cutting surfaces with juice grooves (the moat perimeter on some boards); they tend to get full of processing debris and can be challenging to clean. The exception happens to be with poultry that is being processed on cutting boards that will sit atop a separate cutting surface. In this case the grooves can aid in preventing cross-contamination from poultry juices spilling onto other surfaces. Stabilize the board by slipping a moist towel or shelf liner underneath.

Equipment

The basic butchering of a chicken requires one main tool: a sharp 5- or 6-inch boning knife. From there, any additional equipment is based on personal preference or intended preparations. Shears can help you cut through ribs and separate the back; a cleaver can make easy work of breaking down carcasses for stock or separating wings and legs at the joint. The following list describes what is required and what is optional. Don't forget to have a healthy supply of towels around, as well as a sprayable sanitizer. Additional information about any of the equipment listed can be found in chapter 3.

REQUIRED EQUIPMENT	OPTIONAL EQUIPMENT
▪ Bins or lugs	▪ Chef knife
▪ Boning knife	▪ Cleaver
▪ Honing rod/knife steel	▪ Mallet (rubber or wooden)
	▪ Shears
	▪ Trussing needle
	▪ Twine

Packaging

Chickens are often packaged whole, but you should package your birds according to how they will be used. If you generally cook breasts and legs separately, then package them separately. Never make chicken wings? Bag them for stock. Detest skin? Remove it from everything. I tend not to know ahead of time how I will cook things; therefore, I package chickens whole, defrosting what I need and finding a use for the excess. Any method of packaging should adhere to the tenets described in chapter 14: skintight and without air. Vacuum sealing or shrink-wrap bags are your best bets.

Butchering

WE USE A WHOLE CHICKEN as the basis for poultry butchering, because it is the most common and easiest entry point. The techniques covered here are applicable to any other breed of bird you may choose to consume, though changes in size and species may necessitate different methods of handling. Should your chicken still have its head and feet, refer to pages 118 and 120 for instructions on the proper removal techniques.

Stock Parts

Stock is the greatest by-product of butchering chicken. You are left with flavorful, collagen-rich carcass components that allow you to easily make stock in bulk. Freeze what you won't use in the next week, and never look back on buying the tasteless product found on grocery-store shelves. Every time I butcher a chicken, I add the stock parts to a bag in my freezer; when the bag is full I make stock. (I even save the carcasses of whole roasted chickens.)

The best stock parts are feet, backs, and necks — areas rich in collagen, the holy ingredient to getting a gelatin-enriched stock. But just because other bones and body parts aren't so well endowed with collagen doesn't mean they won't add body and benefit to your stock. So toss in everything you have. (Naturally, this doesn't include anything in the inedible category — head, feathers, viscera, and the like.)

THE STANDARD: AN EIGHT-PIECE CUT

Drummette (humerus)

Pectoral muscles (sternum and ribs)

Drummette (humerus)

Flat (radius and ulna)

Flat (radius and ulna)

WING

WING

Tip (phalanges)

Tip (phalanges)

BREAST

Thigh (femur)

Thigh (femur)

Drumstick (tibia and fibula)

LEG

LEG

Drumstick (tibia and fibula)

BECAUSE CHICKENS are so ubiquitous in the marketplace, they're often the only animal for which people have actually heard of every primal. Unlike larger livestock, poultry butchering is generally the same from one region of the United States to the next. The guidelines we follow for primals are those adopted by national poultry organizations and widely recognized as de facto standards.

WINGS: Wings are typically turned into finger food or used to embolden the flavor of stocks.

BREAST: Evolutionarily, most birds are meant for flight, an action necessitating large pectoral muscles to work the wings. Today's chickens are infrequent fliers, making their breasts lean and largely composed of fast-twitch muscle fibers (see page 12), which provide

the characteristic light, or white, meat distinction. It is a tender primal, though the lack of fat and connective tissue can easily lead to overcooking with a dry result. The inner pectoral muscle, known as the tenderloin, sits against the sternum, or keel bone, and can be removed for individual preparations.

LEGS: The legs of a chicken spend all their time supporting the full weight of the animal. This makes for a hearty primal. The slow-twitch fibers, required for constant support, add to the darker coloration of the meat, while the fat and connective tissue enrich the flavor. The leg muscles require slower cooking methods than breasts do, and with patience they provide more nutrients, flavor, and meaty goodness.

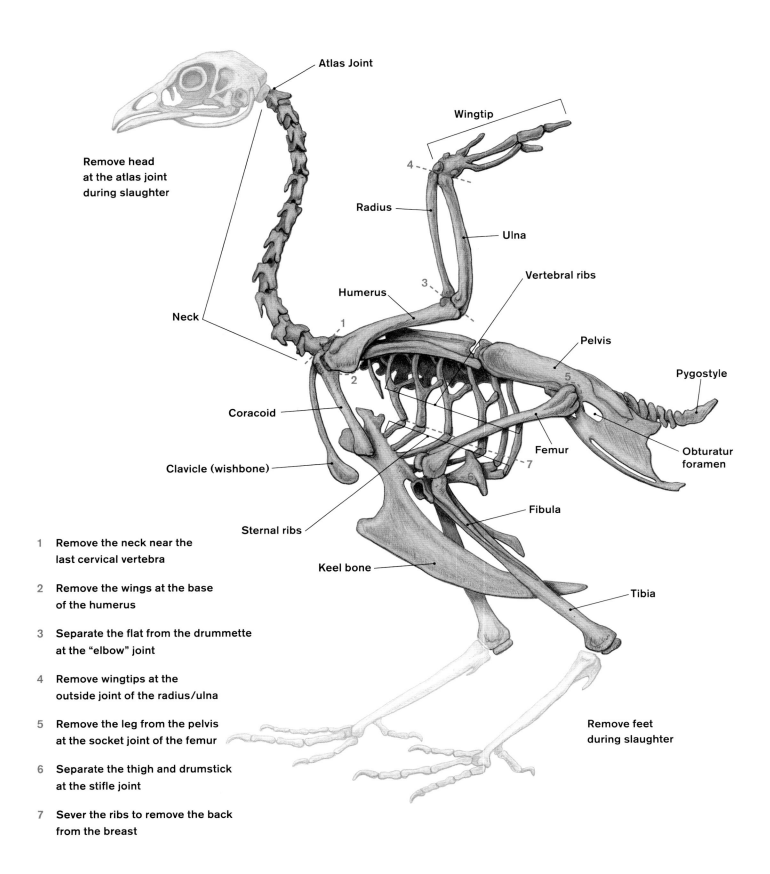

Atlas Joint

**Remove head
at the atlas joint
during slaughter**

Wingtip

4

Radius

Ulna

Neck

3

Vertebral ribs

Humerus

Pelvis

1

Pygostyle

2

Coracoid

5

Femur

**Obturatur
foramen**

Clavicle (wishbone)

7

Sternal ribs

6

Fibula

Keel bone

Tibia

1   Remove the neck near the
    last cervical vertebra

2   Remove the wings at the base
    of the humerus

3   Separate the flat from the drummette
    at the "elbow" joint

4   Remove wingtips at the
    outside joint of the radius/ulna

5   Remove the leg from the pelvis
    at the socket joint of the femur

**Remove feet
during slaughter**

6   Separate the thigh and drumstick
    at the stifle joint

7   Sever the ribs to remove the back
    from the breast

Remove the Wings

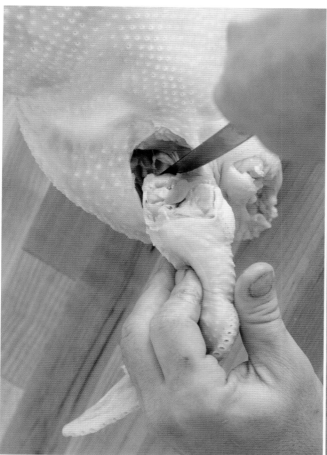 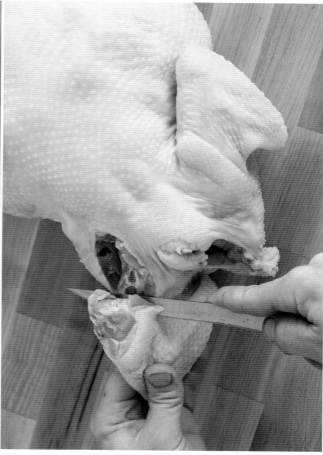

1. CUT THROUGH THE JOINT.

Hold the carcass by the wing. Cut around, and into, the joint, starting at the back edge of the wing. Lift the bird a bit while holding the wing; the weight of the body should help separate the joint with minimal cutting, though you may choose to bend the wing backward to help with the cause.

2. SEPARATE THE WING FROM THE BODY.

Once the wing is out of its socket, cut it from the body with a scooping motion, staying close to the bone and away from the breast meat. Repeat with the other wing. If you like, remove the tips at the joint with a cleaver or knife, and save them for stock.

Remove the Legs

Remove the legs by cutting through the connections between the femur (thigh bone) and pelvis, followed by peeling the surrounding meat off the girdle bones. If you're concerned with keeping a maximum amount of skin on the breast, keep one hand on the breast, holding the skin in place, while making your initial cuts.

 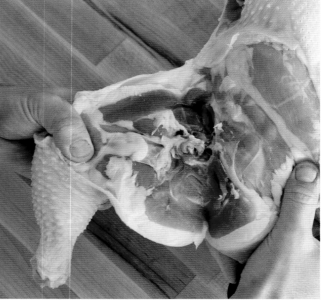

1. SEPARATE THE LEG FROM THE BODY.

Cut between the breast and the thigh, down to the ball-and-socket joint of the femur and pelvis, staying closer to the pelvis. Make sure you've cut the connections atop the joint.

2. POP THE SOCKET.

Grab the leg and, with your thumb on the femur and your other fingers on the underside near the joint, pop the femur out of its socket. Proceed with caution, because too much force can break the pelvis or backbone. If the femur doesn't come loose with moderate effort, grab your knife and cut away more of the connective tissue and try again.

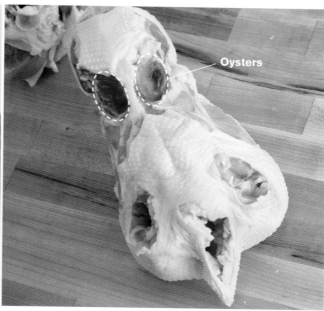

Oysters

3. SEPARATE THE LEG.

Flip the bird over and look for the oyster meat — a solid nugget of meat in a depression in the pelvis. (Technically, it's located in the ilium and is called the *ilioschiatic fenestra*.) You can cut this oyster out of its shell, but you're bound to get a more complete piece if you pry it out. First, with the leg in one hand, cut along the pelvis starting at the posterior end.

When you reach the ball-and-socket joint where the femur used to be attached, stop. Pull the leg away from the carcass, pressing the flat of your blade against the pelvis as a force of opposite pressure. The leg meat should peel off the pelvis with the oyster being the final piece. Repeat steps 1–3 with the second leg.

4. THE RESULTING CARCASS.

Note the depressions in the pelvis where the oysters were.

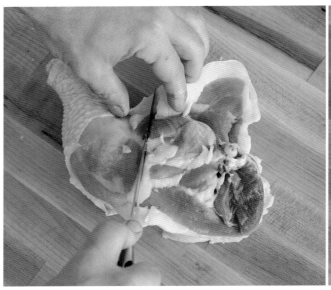

5. LOCATE WHERE TO CUT FOR THIGH-DRUMSTICK SEPARATION.

To further separate the whole leg into thigh and drumstick, bend the leg and feel around the stifle joint for where the bones come together. You should be able to feel a small gap between the two bulbous ends of the femur and tibia, representing the exact location where you want to cut.

If you flip the leg over, you'll see a line of fat that corresponds to where you should cut. Cut from the skin side if you plan to leave the skin on; otherwise cut from either side using the points of reference for the stifle joint.

6. SEPARATE THE THIGH AND DRUMSTICK.

Take your whole unbent leg, and separate the drumstick and thigh by cutting through the stifle joint and the leg meat.

Remove the Back

1. CUT ALONG THE EDGE OF THE PELVIC GIRDLE.

Sit the bird on its neck end. The ilium and the ischium, two of the three bones that make up parts of the pelvic girdle, are located on the outer sides of the pelvis and provide a flat edge for you to follow with your knife as you cut away the back. Keep your knife at a downward angle of about 45 degrees and, following that flat edge, begin to separate the backbone starting at the tail. (If you find that your knife is not sharp enough to cut through the bone connections, use a pair of shears.)

2. CUT DOWN THE LENGTH OF THE SPINE.

Cut all the way down one side of the spine. Make sure your knife stays between the spine and the scapula to avoid running into the coracoid. Repeat on the opposite side. Be careful not to pierce the keel bone or breast meat with the tip of your knife as you plunge downward.

Separate the Breasts

Now you have a whole bone-in breast of chicken, also called the *crown*. You could just as easily leave it whole, roast it, and slice it off the bone when you are ready to serve it, but to get the eight-piece selection of cuts, separate the two sides of the whole breast.

Option 1: Two Bone-In Breasts

1. SCORE THE KEEL BONE.

Place the breast skin-side down on the table. Note the keel bone, a flat skeletal component composed largely of cartilage and best removed by hand. Sever the cartilage at the front (anterior) end of the keel bone and then score the bone by running your knife along its center. Don't bother using a lot of force — the goal here is to disrupt the continuity of the surface so that the connective tissue will pull away when we separate the keel.

2. SEPARATE THE KEEL BONE.

Hold the breast underneath and crack the bones along either edge of the keel bone to separate it. Get underneath the keel bone, using your fingers or thumbs to release the membrane that keeps it attached to the breast muscles. With the membranous connections severed, you can easily "peel the keel," as they say, completely removing the keel bone using your hand. To help prevent tearing the delicate muscular tissue, pin the breast meat against the table with one hand while peeling the keel bone away with the other.

Find the famous wishbone and, using the tip of your knife, free it from the meat, following its natural shape without plunging too deep with your knife. This will allow you to cleanly cut the breasts in two. You can trim away any skin and fat to help shape the breasts, or just leave them (like I do). If you want a 10-piece selection, take your two breasts and cut them in half, width-wise, using a sharp knife to make it cleanly through the ribs.

Option 2: Boneless Breasts

Often the preparation of a dish will require the breasts to be boneless. It is easiest to remove the breasts after you've removed the wings and legs but with the spine still attached, giving you a more stable carcass to work with. The grain of the breast meat runs in from neck to tail; make your cuts in the same direction if you want the underside to be presentable.

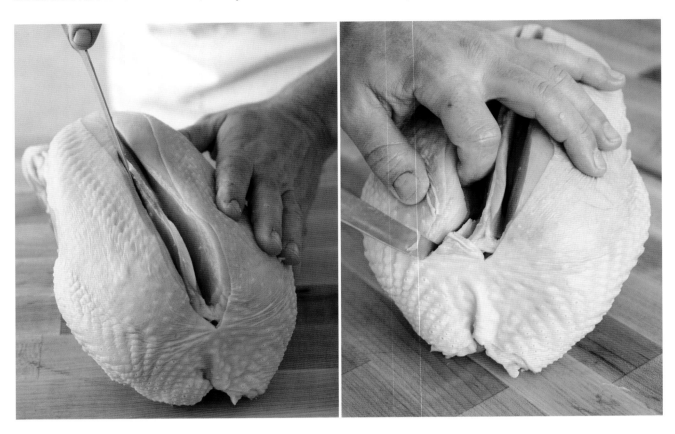

1. CUT ALONG THE KEEL BONE.
Feel around the top of the breast to find the middle ridge of the keel bone. Make two cuts the full length of the keel bone on either side of this ridge. (You can find the wishbone with the tip of your knife and use that as a starting point.)

2. TRACE THE WISHBONE.
Follow your initial cuts back toward the wishbone until you run into it. Trace the shape of the wishbone on one side, staying close to the bone to maximize the breast size. Repeat on the opposite side.

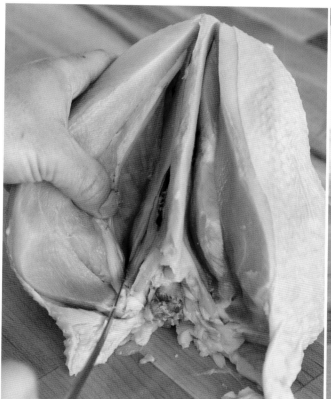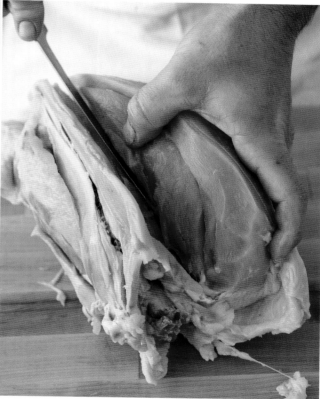

3. SEPARATE THE BREAST MEAT.

Use your thumb or fingers to help separate the meat from the keel bone, giving yourself enough room to insert your knife flat against the bone. The poultry rib cage curves outward from the keel bone and then inward toward the spine, forming a half-moon shape. The initial cut starts at the wishbone. Cut along the outward curve of the ribs, staying close to the bone, while peeling the breast away with the opposite hand. Continue following the rib cage as it curves inward until the breast is fully released from the bones.

4. REPEAT WITH THE OPPOSITE BREAST.

For me, and other right-handers, approaching the opposite breast will be easiest by turning the whole crown around. Finish by trimming the breasts to a desired shape. You're now left with a perfect stock carcass and two boneless skin-on breasts.

Boneless Breasts: Tenderloin Removal

Fabricating boneless breasts gives you the option of removing the tenderloin, also referred to as the chicken tender. (I'm sure at some point in your existence you've been exposed to the American delicacy of breaded and fried chicken tenders. Technically, "chicken tenders" can be strips of breast meat and/or the tenderloin; breast strips are just an attempt to emulate the actual tenderloin.)

The chicken tender sits underneath the breast muscle. After you've removed the breast muscle from the bone, you'll notice the tenderloin — a thin, flaccid strip of meat that,

when attached to the carcass, runs next to and parallel with the center ridge of the keel bone. Because of the fragile connection between the tenderloin and breast muscles, the tenderloin can be easily separated, or damaged, if the two are not properly removed from the keel bone.

Removing the tenderloin is easy. Place the boneless breast upside down on your work surface. Peel the tenderloin away from the breast and use your knife to cut away the minor connective tissue that holds them together to prevent tearing the tenderloin or breast.

TIPS FOR SKINNING

- **The easiest method is to remove the skin from the breast or leg before you cut the meat away from the bone.**

- **Chicken skin is a slimy substance and proves challenging to grip. Do yourself a favor: grab a towel (fabric or paper), and use that to hold on to the skin.**

- **Skinning can usually be done in one motion; if you encounter resistance, it's helpful to apply opposite pressure with your opposite hand.**

Wings: Drummette, Flat, and Tip

Wings are often the ignored stepchild of a pieced chicken. Maybe you've left some breast meat on the drummette to help give it purpose, in which case leaving it whole will suit your needs. Otherwise, wings are bound to become stock or finger food and will require some simple fabrication. Separating them is easy, since the joints are composed largely of forgiving cartilage. Nonetheless, accuracy and the lack of resultant bone fragments will benefit the eater.

1. SEPARATE THE WINGTIP AND FLAT.

Find the joint between the wingtip and the flat by bending the joint and feeling for the small gap; cut through that small gap.

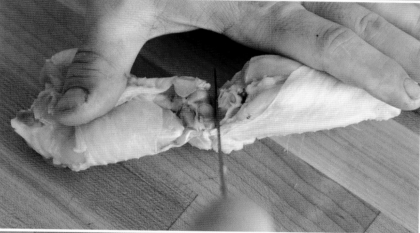

2. SEPARATE THE FLAT AND THE DRUMMETTE.

Next, find the joint between the ulna and the humerus, again by bending the flat and drummette and feeling for the small gap; cut through that gap.

3. VOILÀ: DRUMMETTE, FLAT, AND TIP.

Tips can be saved for stock, while drummettes and flats can be collected and kept frozen until you have amassed enough for a smoked-wings-and-beer session.

Boneless Legs

Take a quick look at the diagram on page 103 and notice the three main bones that you'll be dealing with — femur, tibia, and the tiny fused fibula — plus the patella, which is a cartilaginous kneecap that gets removed in one of the final steps.

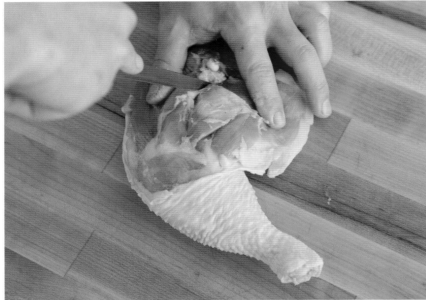

1. CUT ALONG THE INSIDE EDGE OF THE FEMUR.

Place the whole bone-in leg skin-side down on the table. (If you have removed the skin, this would be the side of the leg that faced away from the carcass.) The first cut runs along the entire inside of the femur and tibia. Starting at the ball-and-socket end of the femur, cut along the inside edge of the bone, heading toward the stifle joint.

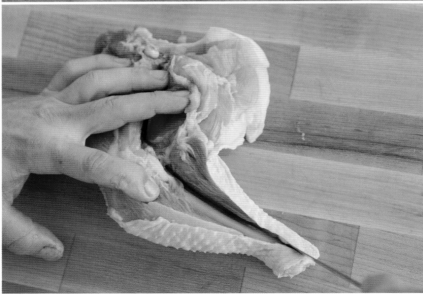

2. CUT ALONG THE STIFLE JOINT.

Stay close to the bone; as you pass on the inside of the stifle joint, cut through the connective tissue by angling your knife tip slightly inward. Continue cutting along the inside edge of the tibia.

3. RELEASE THE MEAT AROUND THE FEMUR.

Trace along the opposite side of the femur, this time starting at the stifle joint and cutting toward the ball-and-socket joint. With both sides released, use your fingers to get underneath the bone, releasing the meat up until the ball-and-socket joint. Insert your blade below the bone and cut around the ball end.

4. RELEASE THE MEAT AROUND THE TIBIA.

Again, begin the next cut at the stifle joint: cut along the outside edge of the tibia, heading toward the hock joint. (Because the fibula starts at the stifle joint, but is unconnected at the opposite end, starting at the stifle also prevents you from damaging the fibula and leaving bone shards in the meat.) Slide your knife between the end of the bone and the meat.

Boneless Legs CONTINUED

5. CUT THROUGH THE TENDONS.

With meat on both sides of the tibia released, you should be able to insert your knife underneath the bone where the meat terminates at the hock joint. There is a grouping of tendons here; cut through them, following the bone.

6. PEEL THE MUSCLES.

Grab the released end of the muscles and peel them away from the tibia toward the stifle joint.

7. CUT AWAY THE PATELLA AND REMOVE THE LEG BONES.

Hold the femur and tibia in one hand. This will contract the knee and pull the patella toward the knee, making it easier to remove. The patella is in direct contact with skin, so if you're planning on stuffing or rolling this boneless leg, take extra precaution when cutting out the patella so that you don't puncture the skin. With the knee contracted, cut away the connective tissues (collagen, ligaments, and tendons), starting at the inside of the joint and around and outward. Locate the patella with your fingers and carefully remove it by cutting between it and the skin.

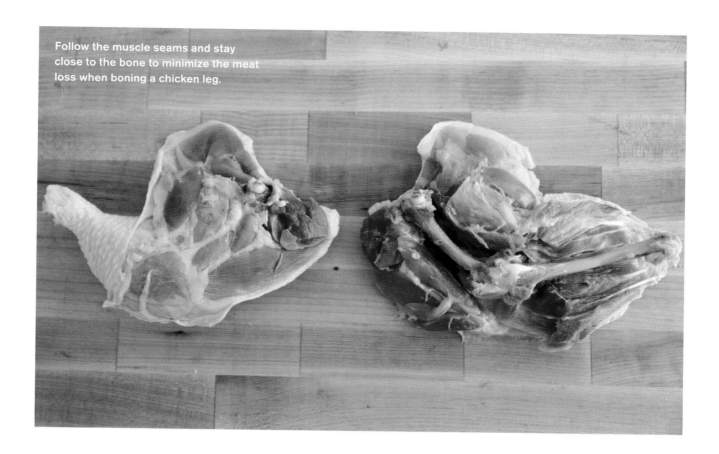

Follow the muscle seams and stay close to the bone to minimize the meat loss when boning a chicken leg.

OPTIONAL: REMOVE THICKER TENDONS.

This last step is optional but helpful when you consider the inedible toughness of tendons. Notice the group of tendons at the end of the drumstick meat. Remove the thicker tendons by running your knife underneath them toward the stifle joint; cut through them once they transition to muscle. The thinner tendons generally terminate about half an inch from the hock joint. Find that point and cut through the muscles and tendons; take precaution to avoid cutting through the underlying skin if you plan on stuffing or rolling the leg.

Take one last moment to look over the meat. Use your fingers to feel for any cartilaginous remnants or bones that may need to be removed; blood clots, blood vessels, and other areas of connective tissue should be cut away.

RABBIT SLAUGHTERING

RABBITS ARE A GREAT SPECIES to raise for consumption because they propagate quickly and provide highly nutritious meat. Raising them can be quite easy, and they require less space than many other livestock, making them good candidates for backyard farmers and urban homesteaders. The process of slaughtering rabbits is also swift and relatively simple. You need to perform the task with caution the first several times through, but soon thereafter you may realize that processing a rabbit can happen in just a few minutes.

Anatomy

THE RABBIT ANATOMY is a good primer for working with larger quadrupeds. Its skeletal structure is basically a miniaturized version of what you'll find in lambs, pigs, and even cows, with some minor exceptions. Familiarizing yourself with the ins and outs will help you build a foundation of knowledge that can easily be extrapolated for application to larger animals. The visceral anatomy for rabbits and other animals is covered on page 161.

The rabbit skeleton is surprisingly light: only 8 percent of its total weight, which for a mammal is very low. (A cat's is 13 percent, and a human's is 30 to 40 percent.) The skeleton is also fragile, making the rabbit the most easily damaged of all the animals covered in this book. For butchering, this offers cautious benefits: the bones will be easier to cut through, with either a sharp chef's knife or a cleaver. But you must handle rabbit carcasses carefully, especially when disjointing, because the bones will often break under minimal pressure.

For the purposes of butchering, a rabbit skeleton is commonly differentiated into the sections indicated on page 182. Within each of these sections are multiple bones, and it's helpful to study the provided diagram to familiarize yourself with the inner workings of each section. The main meat sections are the foreleg, saddle, belly, and hind leg, with other areas being primarily used for stock.

Setting Up for Slaughter

RABBITS ARE THE SIMPLEST ANIMALS to slaughter. Despite the quick process and relatively little equipment needed, honorable harvesting of rabbits deserves the same level of forethought and preparation as for any other animal. Start by reading through chapter 5, which covers the basics for the slaughter setup and techniques used throughout.

The workspace needed to process rabbits can be relatively small, because of their size and the generally simple procedure. If you plan on breaking down the rabbits into pieces, you'll need two separate spaces: one area for the slaughter and evisceration, and another for the butchering. The first stage of the process is quite clean, especially when compared with that for other animals, so those two stations can be quite close. As with any slaughter, potable water is necessary, especially tubs of ice water to chill the carcasses and water to clean the butchering station. All plastics involved should be food-grade — cutting boards, cooling tanks, packaging, and so on.

The Stunning Setup

The setup for rabbit slaughtering needs to include an area for the stun and then a method for hanging, after which you'll proceed with bleeding, skinning, and evisceration — all without moving the animal. No matter what your method of stunning, this entire process can happen at a single small station.

EQUIPMENT OVERVIEW

The first step of a rabbit slaughter — the stun — holds the biggest variability for how you may approach it. If you choose to stun with your hand, your setup and equipment needs may be quite simple. For those slaughtering frequently or in large numbers, investing in some more specialized equipment will be beneficial. The lists here are in order from bare-bones to ideal.

STUNNING: blunt object, Rabbit Wringer, hand, Rabbit Zinger

HANGING: rope, hooks, Rabbit Wringer Butcher Station

BLEEDING, SKINNING, EVISCERATION: 3–5" knife (one with a gut-hook can be helpful)

CHILLING: ice-filled cooler, temperature-regulated stainless steel tubs

AGING: refrigerator, blast chiller, or walk-in cooler

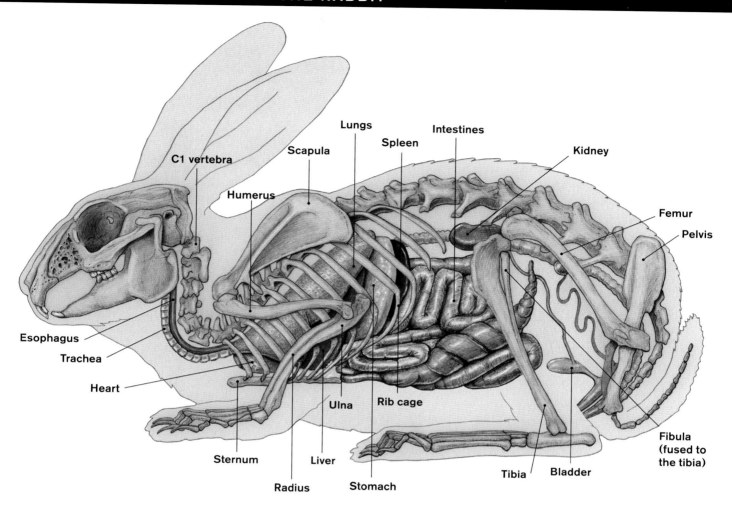

C1 vertebra

Humerus

Scapula

Lungs

Spleen

Intestines

Kidney

Femur

Pelvis

Esophagus

Trachea

Heart

Sternum

Radius

Ulna

Liver

Stomach

Rib cage

Tibia

Bladder

Fibula (fused to the tibia)

THE ATLAS JOINT connects the base of the skull to the first cervical vertebra (C1). It is this connection that is disjointed in order to remove the head, either during cervical dislocation or butchering.

THE PELVIS is the bone surrounding the bung, or rectum. Splitting the pubic symphysis, the cartilaginous midline on the bottom (ventral) edge of the pelvis, is the easiest method to access the bung for separation.

THE ESOPHAGUS AND TRACHEA leave the skull and run side-by-side until they enter the thoracic chamber. The blood vessels that need to be severed run above on both sides of these tubes. Severing just the blood vessels at the base of the skull is far too difficult; thus, the esophagus and trachea are severed along with the blood vessels when using the transverse method of bleeding.

THE LOWEST JOINT OF THE LEG, the animal's equivalent to our wrist, needs to be disjointed or severed to remove the feet. If severing with shears, be careful not to cut too far above the joint, which can result in splintering of the larger leg bones.

THE RABBIT STERNUM can be optionally split when removing the thoracic viscera. Any sharp, marginally stiff knife should be able to split it.

HEART, LIVER, AND KIDNEYS are the most common offal saved for consumption. The liver is best when frozen immediately after slaughter, though all offal benefits from this approach.

STUNNING

METHOD	PRO	CON
STUN GUN (Rabbit Zinger)	Quick, easy, and effective; keeps head connected; great for repeated use	Expensive
HAND STUNNING	Cheap as can be and effective when done right	Takes some practice; may require repeated blows; stubborn rabbits may thrash or try to claw you when you're holding them
STUNNING WITH A BLUNT OBJECT (pipe, hammer, etc.)	Cheap and accessible	High risk of damaging shoulder meat; often feels like bludgeoning to death
CERVICAL DISLOCATION (using a broom handle or similar device)	Cheap and accessible; relatively easy to perform	Lessens the bleed; awkward for one person; doesn't allow for a head-on carcass
CERVICAL DISLOCATION (Rabbit Wringer)	Fast and easy to perform with one person; great for stunning quantities of rabbits	Lessens the bleed; doesn't allow for a head-on carcass; requires installation

There are many ways to stun a rabbit; the most common methods tend to involve dislocation of the head from the spinal cord, called cervical dislocation. This is not an ideal method, however. Although it's easy to implement, it also kills the animal rather than stunning it, stopping the heart thus lessening the bleed. It also prevents you from being able to keep the head attached to the body, a benefit for salability in many markets. Instead, it is better to use a forceful blow to the back of the head, either by using a stunning gun for small game, the side of your hand, or any easily wielded blunt object.

USING A STUN GUN

The designer of the Rabbit Wringer (see page 169) also produces a product called the Rabbit Zinger (see page 169), a rubber-band-powered, nonpenetrating captive bolt pistol that is effective for stunning rabbits or any other similarly sized animal. It operates simply and consistently, and is an effective option for those planning on processing large quantities of rabbits. Like all products from this manufacturer, the Zinger is pricey, which may prevent widespread adoption.

HAND STUNNING

Many folks familiar with rabbit slaughtering would argue that specialized tools aren't necessary for stunning: just use your hand. In what would be the farmstead equivalent to a karate chop, stunning with your hand is surprisingly effective and, like the Zinger, does not carry the risk of bruising shoulder meat, as stunning with a blunt object may. Hold the animal by its feet or back skin, and with the edge of your hand deliver a brisk, powerful blow to the base of the skull right behind the ears (see page 167). When done properly, the animal will elongate and stiffen. You will have less time to hang and bleed the animal than you would if using the Zinger, but the 30 to 60 seconds before the animal regains consciousness should be ample time with the proper setup.

STUNNING WITH A BLUNT OBJECT

The third method of impact stunning is to use a blunt object such as a hammer, a club, a pipe, or an ax handle (see page 167). This method requires a rudimentary amount of hand-eye coordination: missing will inevitably mean bruising useful meat (the shoulders) and an ineffective, and inhumane, stun. But, like all other impact stunning, it works when applied correctly and enables you to bleed your rabbit while keeping the head on. If you intend to use this method, plan to either stun the hanging animal while you hold it with the other hand (similar to the aforementioned karate chop) or place the animal on a hard, stable surface that is easily sanitized. A large cutting board or a few cutting boards pushed together on a table can be sufficient. A soft surface will absorb some of the impact and lessen the efficacy. Stunning with a blunt object is identical to using your hand — just replace your hand with the object, and follow the directions above.

CERVICAL DISLOCATION

Cervical dislocation is still a popular option, despite the drawbacks. For this you will need something that will effectively restrain the head of the rabbit while allowing you to pull the rabbit's body away from its head, all of which happens in a matter of seconds. This may sound complicated, but there's a very simple solution: a broomstick (see page 168). Placed across the neck of a rabbit that has been laid on the ground, a broomstick held in place firmly with your feet can be a very effective method of restraint, as is any easily manageable small-circumference pole, such as a short length of rebar, a rake handle, or a PVC pipe. Bruising often happens as a result of the pressure applied to secure the rabbit's head, and the damaged neck meat may have to be trimmed later.

To perform a cervical dislocation, place the rabbit on the ground or floor. Lay the pole across the neck at the base of the skull, right behind the ears. Step on one side of the pole and ensure that the rabbit is in position. Grab both feet by one hand and then step on the other side of the pole with the other foot. Use both hands on the hind legs to pull straight up on the rabbit until you feel its neck dislocate.

The Rabbit Wringer (shown on page 169) enables quick and easy spinal dislocation. It is well built and, used properly, eliminates the risk of meat bruising. Operation is quick and easy, but the Rabbit Wringer does have a few drawbacks. First, it requires a semipermanent installation — it's small, but it needs to be securely screwed to a surface or bolted around a post. Second, it's expensive. If you're processing rabbits frequently, though, it may be worth installing one in your setup.

Hanging Setup

The hanging area is where all the work happens — bleeding, skinning, and evisceration — so make sure it's at the center of your workspace, where you have easy access to the few essentials of the process. Hanging a rabbit, compared to working on a horizontal surface, has several advantages. Hanging is better for bleeding, since gravity aids the heart's pumping to expel as much blood as possible. This produces cleaner meat that will not spoil as quickly. Moreover, gravity works on the viscera, pulling them down to where they rest on the diaphragm. This greatly reduces the chance of puncturing the entrails when opening the posterior end of the abdominal cavity. Hanging also means not having to transfer the animal after bleeding, which reduces handling and contamination of your area. And, because the rabbit is secured when hung, you're able to work with both hands.

THE RIGHT AGE

Rabbits raised for meat in the United States are usually one of two breeds: New Zealand white or Californian. For these breeds the most common age range for slaughtering is 10 to 16 weeks. Rabbits under 12 weeks are referred to as fryers, those between 3 and 6 months as roasters, and those older than 6 months as stewing rabbits. Younger animals produce more tender meat, but flavor develops with age; the right age for slaughtering will be something based on your preferences or those of your customers. You can assume, no matter what age, that your carcass yield will be around 55 percent of the live weight, give or take, based on diet and breed. So, for example, your average 4.5-pound live rabbit will yield a carcass of about 2.5 pounds.

HANGING BY THE BACK FEET

The best way to hang a rabbit is by its back feet. Make sure that the hanging setup suspends the rabbit in such a way that the carcass doesn't rest against anything that will soil it. Remember, after it's been skinned, this will be a bare carcass; you don't want a post, fence, or wall rubbing against it. Free suspension is one option. Otherwise plan on putting up a sanitary material such as a sheet of Plexiglas or a stainless steel tray where there will be contact.

Hanging a rabbit by its feet limits contact with contaminants from the hanging apparatus to a part of the carcass that's discarded. It also allows you to hang the rabbit immediately, without having to do any preliminary skinning or to cut incisions into the legs to access the gambrel cords (see following section), which helps efficiency and further decreases contamination.

In the simplest form, all you need to hang a rabbit by its feet is a thin nylon rope, at least 3 feet long, secured at the center with two slipknots on either end. The knots allow you to easily insert and remove the feet, while the rope provides enough resistance for skinning and also gives you the versatility of spinning the rabbit around while processing it. Securing the center of the rope can be as easy as tying it to a fence or tacking it up on a garage post. Alternatively you can secure two individual lengths of rope ending with a slipknot, one for each foot. The manufacturer of the Rabbit Wringer also produces an excellent processing station that allows for easy hanging by the feet. It costs more than some other solutions, requires installation, and does not allow for you to rotate the rabbit, but if you plan on processing rabbits with any regularity it is the ideal hanging option.

HANGING BY THE GAMBRELS

Ignoring the advantages of hanging by the feet, using the gambrels to hang a rabbit can be easy. All you really need is one or two sturdy hooks screwed into a post or crossbar, or any hanging hook. Hanging hooks, even a small gambrel bar, will give you the advantage of being able to rotate the animal if you need to.

The Rabbit Wringer's processing station is well-built and a good choice for hanging rabbits.

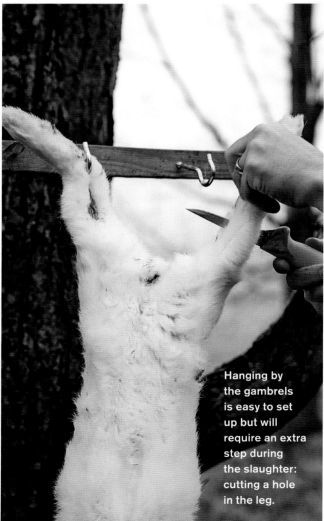

Hanging by the gambrels is easy to set up but will require an extra step during the slaughter: cutting a hole in the leg.

Lift the rabbit carefully by the scruff of the neck. Cradling it like a football (right) will help calm it before stunning.

Evisceration Setup

Underneath the hanging setup, you'll need a vessel to collect the blood and inedibles. The size of the vessel will depend on how many rabbits you're slaughtering. (The combined volume of inedibles from a single head-on rabbit, including the pelt, is about 2 quarts.) The opening of the vessel should be wide enough to easily catch blood and allow you to drop entrails into it without a need for precision. For most at-home situations, a multigallon bucket will work just fine, and if you anticipate having problems with attracting insects, fill the container halfway with water. If you're interested in turning the blood into blood meal, use separate buckets for blood collection and inedibles; if you're planning on saving the pelt, reserve a separate container for that.

Slaughtering

IT'S IMPORTANT TO TAKE A FEW simple steps the day before your slaughter to prepare the animals; those steps are covered in chapter 5, so take the time to read it well before the day arrives. Furthermore, that chapter provides the basic information on how to perform the stunning and evisceration properly. Withholding feed is one of the preliminary steps, since it helps prevent contamination, but be aware that because rabbits practice cecotrophy (the ingestion of soft fecal pellets produced in the cecum), their digestive systems may not be empty when you eviscerate them.

Stunning, Hanging, and Bleeding

It's important to handle the rabbits correctly to prevent unnecessary agitation or fear, which will quickly degrade the quality of meat. Pick the rabbit up by the skin on its back and, if needed, carry it similar to cradling a football, with your arm and hand supporting the animal from underneath, as shown in the image above. Handled correctly, the rabbit should show no signs of stress, but it's up to you to determine how relaxed the animal is. If the rabbit is showing signs of stress, it's better to put it down. Pick up the animal again when it's calm, or work with another animal in the meantime.

If you are using a commercial method of stunning, like the Rabbit Wringer, refer to the manufacturer's instructions about proper usage.

THE SLAUGHTER PROCESS FOR RABBITS

Slaughtering rabbits can be quick and easy, as you'll notice in this flowchart overview. Your method of stunning may be the only variable.

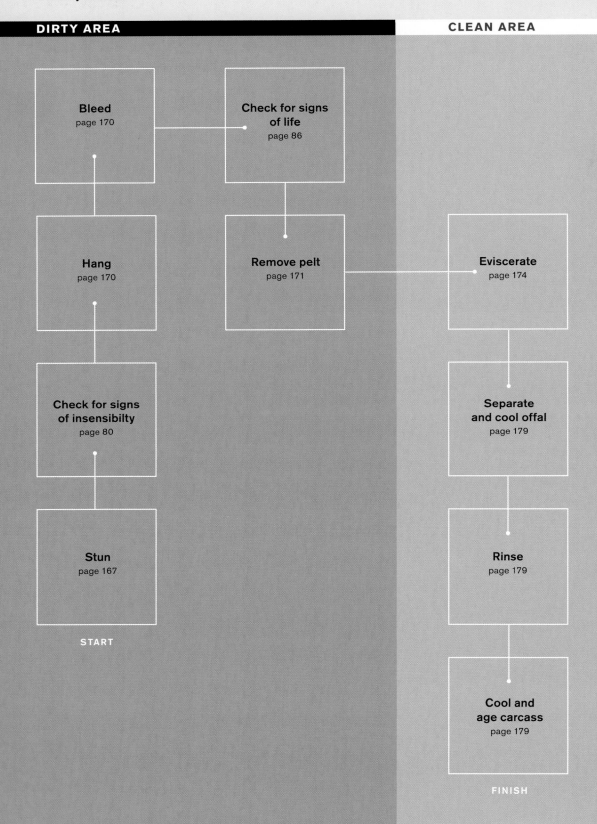

DIRTY AREA

CLEAN AREA

Bleed
page 170

Check for signs of life
page 86

Hang
page 170

Remove pelt
page 171

Eviscerate
page 174

Check for signs of insensibilty
page 80

Separate and cool offal
page 179

Stun
page 167

START

Rinse
page 179

Cool and age carcass
page 179

FINISH

Option 1: Stunning by Hand or Blunt Force

The use of blunt force will stun a rabbit for only a short period; the blow needs to be immediately followed by hanging and a successful bleeding to ensure a humane death. The process of hitting the rabbit on the head is straightforward, but you need to get the location and force right. Hitting the wrong spot will traumatize the rabbit while potentially bruising, and ruining, valuable meat in the shoulder.

1. POSITION THE RABBIT.
Hold the rabbit by its rear legs, wait several seconds, and allow it to relax into the position.

2. DELIVER THE BLOW.
Once the rabbit is calm, deliver a swift blow behind the ears at the base of the skull, near where the skull connects with the neck. You can use either your stiffened hand or a blunt object such as the backside of a cleaver. A successful stun will cause the rabbit to straighten and become stiff. Proceed to bleeding only after you have confirmed a successful stun.

ALTERNATE OPTION:
You can use either your stiffened hand or a blunt object such as the backside of a cleaver.

Option 2: Stunning by Cervical Dislocation

It's important to think through this entire process, step by step, before attempting it for the first time. Committing the sequence of actions to memory will ensure that the process will happen as quickly as possible, thus reducing the animal's stress. This whole sequence should take no more than several seconds.

1. READY YOUR RESTRAINT TOOL, be it broomstick or rake handle, rebar, or other rigid rod. Place the (calm) rabbit on the ground between your legs with the head facing away from you.

2. SECURE THE RABBIT. Hold the animal with one hand, and place the restraint tool with the other. Secure one end of the rod with one foot and place the rod across the rabbit's neck, just behind the ears.

3. PULL UP ON THE HIND LEGS. In one swift motion, step on the other side of the rod and pull straight up on the rabbit's hind legs. The pull should be less a jerk than a smooth, firm, sustained motion. The head should dangle freely, being completely disconnected from the neck.

VARIATION: CERVICAL DISLOCATION, USING A RABBIT WRINGER

1. INSERT THE HEAD.

It's important to proceed swiftly with cervical dislocation to avoid stress on the animal. Slide the neck of the rabbit all the way inside the tapered space of the Rabbit Wringer, with the head on top.

2. PULL ON THE HIND LEGS.

Take the hind legs and firmly pull them straight out from the Wringer until you feel the skull separate from the neck.

Option 3: Stunning with a Stun Gun

The Rabbit Zinger is an effective and easy-to-use stunning method — just follow its directions for operation. Make sure you have a hard surface beneath the rabbit, as soft materials (like the earth) absorb some of the gun's impact, lessening its efficacy.

Hanging and Bleeding

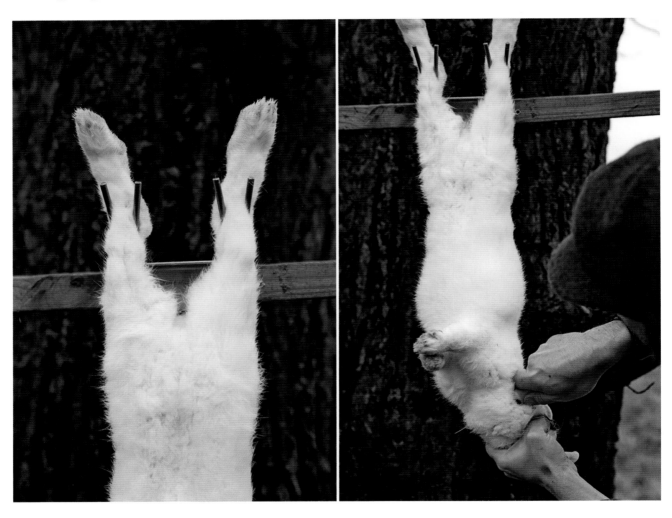

1. HANG THE RABBIT.

Hang the rabbit head-down with its abdomen facing you. If you're hanging it by the gambrels, cut an opening large enough for your hooks, just below the feet and between the tendon and leg bone (see page 164).

2. CUT THE JUGULARS.

For a head-on bleed, hold the head in one hand and follow the instructions for the transverse method, found on page 84. Be sure to get the knife close to the spine, but take care to avoid nicking the hand that holds the head. If you've dislocated the rabbit's neck during stunning, remove the head by cutting just above the jawbone and through the space between the skull and neck.

Allow the rabbit to bleed (into a separate bucket, if you're isolating the blood) for 1 to 2 minutes, or until the flow has dissipated. If you've used blunt force to stun the rabbit and left the head on, make sure that the rabbit is in fact dead (page 80) before you proceed with skinning.

Skinning

Rabbit skinning is a surprisingly easy and quick endeavor. The objective is to sever the pelt where it connects to the posterior end of the carcass and then to pull the entire pelt down and off. Done right, the process resembles removing a T-shirt from the animal. This is also the best method if you're planning on saving the pelt. Review the information about skinning on page 88 before proceeding.

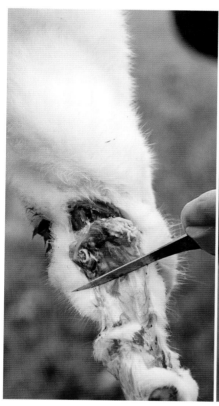

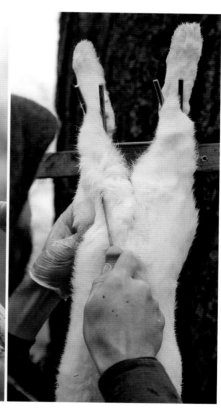

1. START SKINNING THE HEAD.

Take hold of the ears and begin skinning behind them, heading toward the nose.

2. SKIN TOWARD THE NOSE.

Follow the shape of the skull around both sides, connecting with the incision you made during the bleeding until the entire skull is skinned.

3. CUT UP THE INTERIOR OF THE LEG.

Make an incision up the interior side of a back leg, starting at its base and ending at the hock joint. (You can also cut in the opposite manner: start at the hock and work down the leg.)

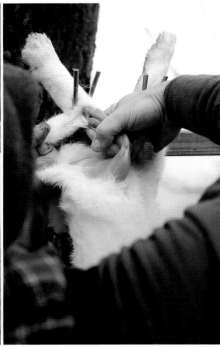

4. FREE THE PELT FROM THE LEG.

Slide one or two (clean) fingers between the pelt and the muscle, all the way around the leg until you can touch your thumb.

5. PULL THE PELT TO RELEASE IT.

Grab the freed area of pelt with your opposite hand, and pull your hands away from each other, releasing the pelt up to the hock joint. Either cut the pelt at the hock joint or simply tear it by hand. Repeat on the opposite leg.

6. PULL THE SKIN AWAY FROM THE GROIN.

Connect the two leg incisions by pulling the skin away from the groin area and inserting one or two fingers, creating enough room for you to insert your knife and cut up and out. This can help prevent accidental puncturing of the abdominal cavity, especially the bladder, which sits right below the skin where you're cutting. Pull down on the front and rear sides of the hide using the areas released from the legs. Stop when you reach the tail.

7. CUT ACROSS THE TAIL.

Take the hide above the tail and pull it down to keep it taut. Cut across the rectum and through the base of the tail.

8. PULL DOWN ON THE HIDE.

The top areas where the hide secures are all released. Grab the two sides of the hide from both hind legs, and pull straight down on the hide with a firm and sustained motion. Proceed carefully over the belly to avoid any tears in the thin belly muscle. Should a tear begin, use your knife to help release the skin around the site. Make your way past the tear with your knife, and then continue pulling the hide off.

9. PULL THE HIDE OFF.

Place your hand on the hide between the legs, and use it as leverage to finish pulling the hide off. After the hide is removed, lop off the front feet at the joint, using shears or a knife, and discard them. You now have a skinned rabbit and an intact pelt. Freeze the pelt if you are not preparing it for tanning on the same day you slaughter.

Evisceration

Eviscerating a rabbit is a great way to begin understanding how to properly disembowel an animal while avoiding contaminants. The solid pellets of rabbit feces are less likely to contaminate the carcass (as compared to the liquid fecal matter in other species) should you accidentally make a wrong cut. Decide what offal you are going to keep. Options for edible offal are heart, liver, and kidneys, all of which can be left attached to the carcass or removed, depending on your preference. When processing multiple rabbits, collecting the offal from all the animals may make sense, since it's likely to be sold in volume. What you don't want, including the lungs, can be fed cooked or raw to pets.

At this point, you can probably see through the abdominal walls of the inverted rabbit and observe how the abdominal viscera have collected atop the diaphragm. This will enable you to make an incision near the top without puncturing the entrails.

1. MAKE THE INITIAL CUT.

Pinch a small piece of belly muscle a few inches below the rectum. Make sure you're just pinching muscle, then make a small horizontal cut right above your fingers.

2. CUT DOWN THE CENTER OF THE ABDOMEN.

Insert two fingers into the abdominal opening. (Widen the opening by tearing if your initial cut was too small.) Hold your knife vertically, handle on top, and insert the very tip of your knife between the two fingers and just under the muscle. Move down the center of the abdomen while holding this position. Use your two fingers to push back the viscera as you move, keeping enough space between the viscera and the abdominal muscles for your knife tip to safely slide down unobstructed.

3. CUT UNTIL YOU REACH THE RIB CAGE.

At this point, the viscera will spill out of the abdominal cavity.

4. SPLIT THE PUBIC SYMPHYSIS.

The pelvis, which is composed of multiple fused bones, has a line of cartilage, called the pubic symphysis, running vertically between the two hind legs. (More about splitting the pubic symphysis on page 92.) The rectum runs through the center of the pelvis; by splitting the pelvis at this cartilaginous point we can safely, and easily, remove the entire colon.

I prefer to have one of the rabbit legs freed to help split the pelvis. Remove one leg from its hanging implement and hold on to it with your non-knife hand, pulling on it slightly to apply outward pressure. Partially insert the flat edge of your knife into the pelvic cartilage, and then pry it apart; to avoid damaging the rectum, do not cut too deep. The seam should separate easily.

5. FREE THE RECTUM.

Reach through the pelvic opening and pinch the top of the rectum. Pull down through the pelvis, tearing connections as you go. Make sure to get hold of the bladder when you feel it. (If you encounter resistance take your knife, hold it vertically, and cut around the rectum, all while keeping against the pelvic bone.)

6. REMOVE THE ENTRAILS.

Continue removing the entrails by pulling them out of the cavity and using your hand or knife to sever any connections. Use your knife to separate the liver from the entrails, leaving the liver attached to the carcass and being careful not to puncture the bile sac. Leave the kidneys and surrounding fat attached to the cavity.

Make sure the esophagus and trachea are severed at the neck if they weren't during bleeding. Use your hand to break any connections around the stomach, but leave the esophagus attached. Finish removing the entrails by pulling the stomach up and away, pulling the esophagus through the diaphragm and away from the neck in the process. Drop all the entrails into the inedibles bucket.

7. A CLEANED RABBIT WITH INTACT PLUCK.

Now you have a cleaned rabbit with the pluck (heart and lungs) intact, liver and kidneys left attached. You can stop here and skip ahead to cleaning and cooling, or continue for offal and pluck removal.

8. REMOVE THE KIDNEYS.

Pull down on the kidneys and cut through the attachments, including the ureters.

9. REMOVE THE LIVER.

Grab the liver, being careful not to burst the gallbladder, and sever any connections that are keeping it attached to the carcass. Pulling gently on the gallbladder, peel it away from the liver without causing it to burst.

11. KNIFE THROUGH THE DIAPHRAGM.
Use the tip of your knife to cut through the diaphragm.

13. REMOVE THE LUNGS AND TRACHEA.
Pry the rib cage apart and identify the lungs, which sit beneath the heart. Grab the heart and lungs together and remove them, pulling out and downward to remove the attached trachea. The connective tissues and veins holding the thoracic viscera should not provide much resistance; even so, cut through any that do. Separate the heart from the lungs and save; discard the lungs.

12. CUT THROUGH THE RIB CAGE.
Hold your knife at a downward angle of about 60 degrees and cut down through the rib cage, just on either side of the sternum.

14. CLIP OFF THE BACK FEET.
Remove the rabbit from the hanging setup, clip off the back feet with shears at the hock joint.

Post-slaughter

AFTER EVISCERATING, GIVE the carcass a good rinsing inside and out, ideally with a hose. Look it over thoroughly and remove any unwanted bits like facial fur, remnants of entrails, or interior glands. Transfer the rabbit to the ice-cold water container you've prepared, and follow the directions on page 94 for proper cooling procedures.

Separated offal can be immediately packaged according to your needs. You can either package single-carcass offal together (e.g., one heart, one liver, two kidneys) and place them inside the rabbit for final packaging; package offal types into volume-specific containers (e.g., 1 pound of hearts); package total volumes of offal into separate containers (e.g., all hearts in one bag). Or, if you're planning on feeding the offal to pets, package them based on meal volume.

Aging

The onset of rigor mortis begins approximately one hour after the death of the rabbit. (For more on the effects of rigor mortis and why aging meat is imperative for quality, refer to page 18.) The length of time before rigor mortis resolves for rabbits is debatable, with estimates ranging from a few hours to seven days. Domesticated rabbits are certainly in the lower range. Studies investigating the pH levels in muscles at various time increments after death have shown that the pH for a healthy, humanely slaughtered rabbit stabilizes at 5.7 (the ideal post–rigor mortis level) in just over 4.5 hours; this suggests that it's best to wait at least 5 hours before freezing or consuming a rabbit. But the rate of rigor mortis activity is heavily influenced by temperature, so while you're keeping your rabbits cool to prevent microbial growth, you're also slowing the rate of rigor mortis resolution. Therefore, it will benefit you to wait at least 12 to 24 hours before butchering or consuming the meat.

If you don't have the cold-storage space to hold the rabbits for aging before butchering or freezing them, leave them in the cooling water bath, changing the water and ice to ensure that the temperature of the bath is kept below 40°F.

Packaging

Shrink-wrap bags are the ideal solution for storing whole rabbits, whose bones break easily and may puncture vacuum-sealed bags. Vacuum sealing is ideal for butch-

ered parts, however. Instructions for both types of packaging start on page 426.

Whole rabbits are also easier to package when they're arranged in the fetal position. This position offers a few benefits: it minimizes empty space inside the carcass, thus preventing interior freezer burn; it protects any offal you've chosen to include, either attached or bagged; and it will help limit any potential damage to the rib cage if you're vacuum-sealing.

If you're including bagged offal, first place the offal package inside the cavity. Take the rabbit's back legs and tuck them into the rib cage, being careful not to damage any attached offal. Insert the carcass headfirst into the bag and arrange the front legs so that they curl back to complete the pose.

Cleanup

Cleanup from slaughtering rabbits is relatively simple. Follow the directions on page 96. If you've taken proper precautions with your setup, the breakdown will be very quick. It's best to tackle the cleanup immediately after the last rabbit is processed, and while it is cooling, rather than waiting until blood dries and entrails potentially attract flies or other vermin.

Carcasses should be immediately cooled in ice water.

RABBIT BUTCHERING

RABBIT MEAT IS HIGHLY NUTRITIOUS – its concentration of protein is higher than in any other livestock species, while its cholesterol and calories are lower – and its leanness appeals to a broad range of palates. Because of their small size, rabbits require deft knife work during the butchering process, providing a good exercise in control. You can also use the process as an anatomical primer, gathering information about features and skeletal contours that can be applied to butchering other species.

Butchering Setup, Equipment, and Packaging

AS YOU MIGHT EXPECT, the setup required to butcher rabbits can be quite small. Any cutting board of greater length than the carcass itself will suffice. Cutting surfaces should be easily cleaned and sanitized; opt for wood or plastic (see page 47). It's helpful to place a nonslip shelf liner or a moist towel underneath individual cutting boards to help prevent movement while cutting.

Equipment

Fortunately, only minimal equipment is required to properly butcher a rabbit; all you really need is a sharp knife. My preference is for a 5- to 6-inch semi-flexible boning knife, but you can successfully perform most methods covered in this chapter with a sturdy, well-sharpened pocketknife. When tracing small bones with my knife, I use a pistol grip that is choked up on the blade (page 54). Notwithstanding, all butchering activities benefit from some specialized equipment that may be helpful to have on hand — especially if you plan on butchering rabbits with any frequency. A cleaver or chef's knife will make cutting through bones an easy exercise. Shears work well for trimming ribs. Having paper or fabric towels and plastic wrap around will also be helpful. Further details about the equipment listed below can be found in chapter 3.

Packaging

Rabbits are most commonly packaged whole, though you may find it beneficial to butcher multiple rabbits and package parts — for example, multiple saddles or groups of hind legs. Ideal packaging is skintight with as much air removed as possible. This is best done with vacuum

sealing, though you will need to be wary of the sharp ribs and leg bones. Detailed recommendations for packaging materials are provided in chapter 14.

Butchering

RABBIT BUTCHERING METHODS can be used for many small game animals. Thus, while the instructions here focus solely on rabbits, use this information as a foundation for butchering anything from raccoon to hares to nutria to squirrels. While proceeding through some of the instructions, you may benefit from referencing the skeletal diagram on page 161.

It is always helpful to understand the breakdown of a carcass based on primals, no matter how small the animal is. The primals are largely differentiated by anatomical features, grouping similar muscle structures together. The primals of a rabbit are not standardized, so you may see information here that differs from other sources. The following breakdown of a rabbit carcass, into three primals and a few subprimals, is based on most butchering methods and common uses. Scientific muscle names, where applicable, are provided throughout the chapter in italics.

FORELEG. Easily separated from the rest of the forequarter, the foreleg provides little meat. Because of the connective tissue (mostly in the lower portion of the foreleg), it is most often stewed, either bone-in or boneless. Boning a rabbit's foreleg is a tedious exercise, though the process can be beneficial for understanding scapular removal.

LOIN. The loin runs the length of the carcass, corresponding to the origin and termination of the loin muscle (*longissimus dorsi*). The rack and saddle are often separated, allowing for individual preparations, or the entire loin can be boned (see page 196). The tenderloin, a very small piece of meat on a rabbit, is present on the saddle. The loin and tenderloin are both very tender; the belly contains more connective tissue and is often removed for moist-heat preparations.

HIND LEG. A rabbit's hind legs are the largest source of meat on the entire carcass. The leg can be separated into a thigh and a drumstick, though it is more often stewed whole or boned because of the small amount of meat and large concentration of connective tissue in the drumstick.

REQUIRED EQUIPMENT	OPTIONAL EQUIPMENT
▪ Bins or lugs	▪ Chef knife
▪ Boning knife	▪ Cleaver
▪ Honing rod/knife steel	▪ Mallet (rubber or wooden)
	▪ Shears

RABBIT PRIMALS

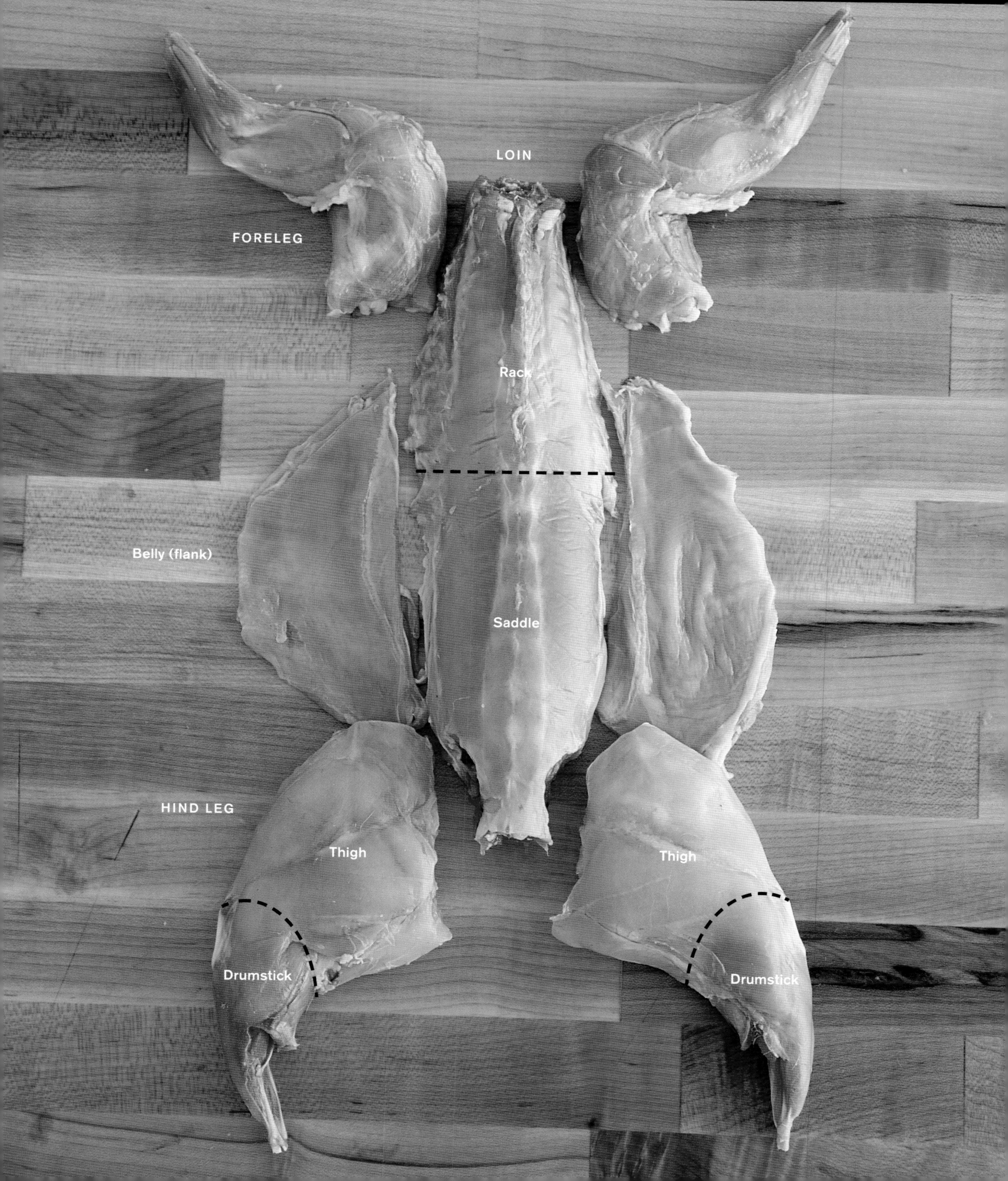
LOIN
FORELEG
Rack
Belly (flank)
Saddle
HIND LEG
Thigh
Thigh
Drumstick
Drumstick

VARIATIONS ON THE STANDARD CUTS

THE STANDARD METHOD of breaking down a rabbit separates the whole carcass into 7 pieces: 2 forelegs, 2 hind legs, 2 bellies, and 1 loin (see page 183). From here, there are a few variations that produce more pieces and separate subprimals, some of which are presented after these foundational instructions. The order of operations is largely based on personal preference, so after you've become comfortable with the basics, reorder tasks to suit your needs. The following are variations on the 7-piece jointed rabbit:

8-PIECE JOINTED RABBIT Follow instructions for the 7-piece jointed rabbit, and then split the loin into the rack and saddle subprimals by cleaving through the spine behind the last rib.

10-PIECE JOINTED RABBIT Follow instructions for the 8-piece jointed rabbit, then split the hind legs into 2 pieces (see page 191) – thigh and drumstick – at the stifle joint.

12-PIECE JOINTED RABBIT Follow the instructions for a 10-piece jointed rabbit, then split the rack and saddle each in half, making for 4 pieces of loin.

8-Piece Jointed Rabbit

10-Piece Jointed Rabbit

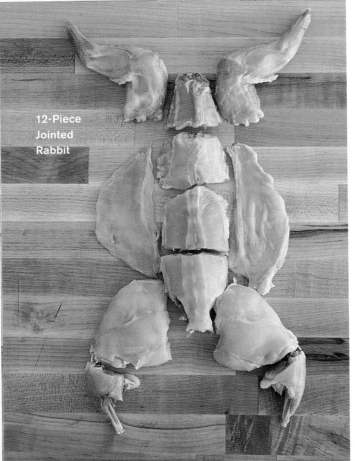

12-Piece Jointed Rabbit

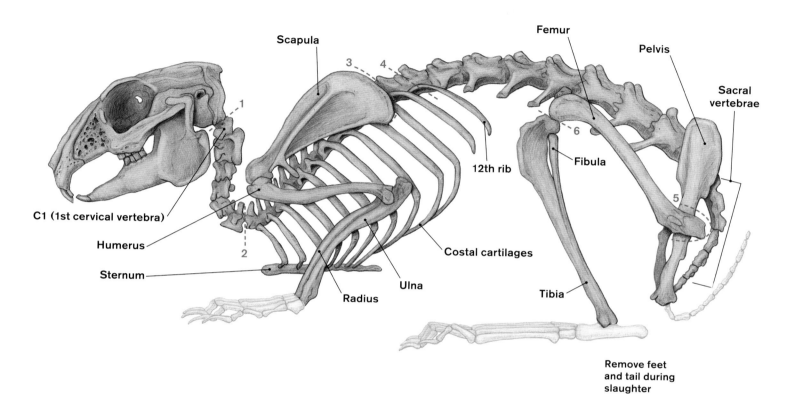

Scapula

Femur

Pelvis

Sacral
vertebrae

3   4

1

12th rib

6

Fibula

C1 (1st cervical vertebra)

Humerus

2

Sternum

Costal cartilages

Ulna

Tibia

Radius

Remove feet
and tail during
slaughter

1   Remove the head at the atlas joint

2   Remove the neck at its base (the last cervical vertebra)

3   Separate the foreleg using the underlying seam

4   Separate the rack and loin between the last two ribs

5   Remove the leg from the pelvis at the ball joint of the femur

6   Separate the thigh and drumstick at the stifle joint

Remove the Head

TWIST OR CLEAVE.
Start by removing the head if it is still connected. The easiest method is to simply twist the skull while keeping the carcass stationary, severing the connective tissue between the base of the skull and the first vertebra (known as the atlas bone). After twisting, simply pull the skull to break any remaining attachments. Another option is cleaving through the spine at the base of skull. Bones as fragile as a rabbit's don't require much force to shear; rather than swinging away, you'll find it more accurate to place your cleaver or chef's knife at the right location and apply the force with a mallet or the palm of your hand.

Remove the Forelegs

The forelegs are the easiest to remove, because they're held against the carcass by muscular and connective tissue; there is no connection between the foreleg bones and the rest of the bones in the body, a fact that applies to every four-legged animal found on a farm. It is along that natural seam that we will want to separate the foreleg from the rib section.

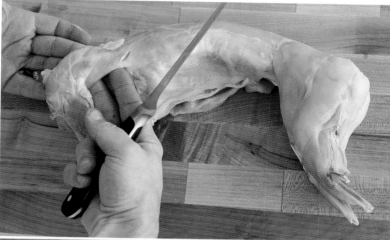

1. IDENTIFY THE SEAM.

Place the rabbit so that the foreleg faces up. Pull the foreleg away from the carcass to help identify the natural seam that you'll use to separate them. Feel for the back edge of the scapula and insert a couple of fingers to help widen the natural seam between the leg and the rib cage.

2. CUT THROUGH THE NATURAL SEAM.

As you cut through the seam, keep your blade against the ribs, and use your opposite hand to pull the foreleg away from the carcass.

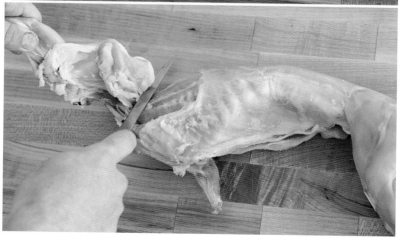

3. REMOVE THE LEG.

Before you reach the spine, the seam will pass on top of the loin meat; be careful not to cut into that by pulling your knife away from the ribs and then finishing the foreleg removal. Turn the rabbit over and repeat with the other side. Look for any inedible or unappetizing areas of the foreleg — clotted blood or connective tissue, for example — and trim them away.

Remove the Flank

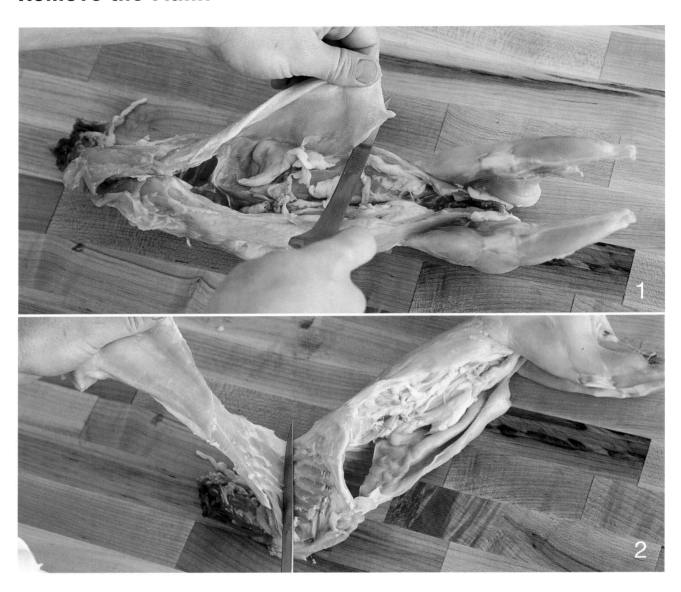

1. CUT THE FLANK AWAY FROM BODY.

Place the rabbit on its back with the hind legs splayed out. The flank is the flabby, thin muscle attached to the hind legs that continues up the loin and connects to the belly, which then overlaps some of the rib cage. Note the location of the larger hind-leg muscles that attach to the flank, as well as the location of the loin and tenderloin muscles, which run along the back and interior of the spine.

Hold the flank and cut along the hind leg, following its contour, until you are about an eighth of an inch from the loin muscle. Continue cutting along the loin muscle toward the rib cage, being careful not to nick the loin as it tapers.

2. FILLET THE FLANK OFF THE RIB CAGE.

When you reach the rib cage, continue removing the flank by filleting it off the rib cage. Avoid cutting into the loin meat running along the spine and upper rib cage. Finish by trimming the edges of inedible and unappetizing bits.

Remove the Hind Legs

Removing the hind legs is very similar to removing the legs of a chicken. Place the rabbit on its back with the posterior end facing you. The hind leg is attached to the pelvis at the ball-and-socket joint of the femur.

1. CUT THE CONNECTIONS AROUND THE BALL-AND-SOCKET JOINT.

Make a shallow cut with the tip of your knife, running along the interior edge of the leg, keeping the blade tip against the pelvis. You should hit the ball-and-socket joint; repeat shallow cuts with the blade tip until connections around the joint are severed.

2. POP THE JOINT.

Grab the leg and pop the femur out of its socket, taking care not to fracture the fragile pelvis. If disjointing does not happen with relatively little effort, cut more of the connective tissue around the joint and try again. Repeat on the opposite side.

3. REMOVE THE LEG.

Once the leg is disjointed, remove it by cutting along the flat edge of the pelvis.

Trim the Loin

At this point you have seven pieces; all that remains is trimming up the loin. With the legs removed, you can clearly see the ilium, the flat bone at the front of the pelvis. You can cut through the loin at the point where the ilium starts, but you'll lose the sirloin, which, granted, is small on a rabbit. With little effort, though, you can maximize your yield.

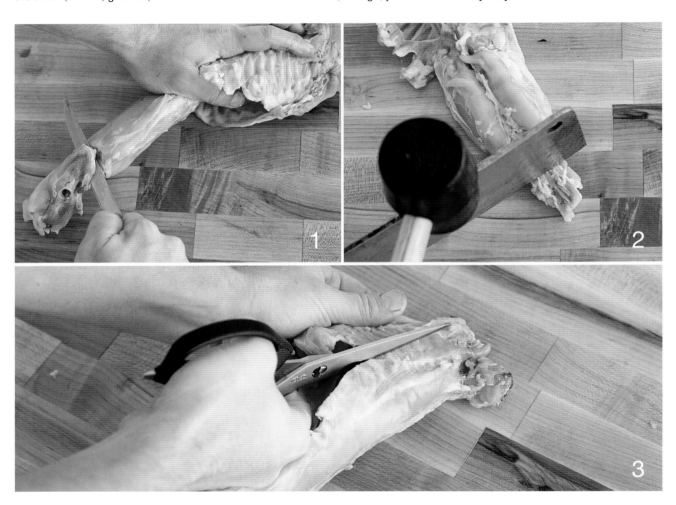

1. PRY THE ILIUM AWAY FROM THE SPINE.
Use your knife to get underneath the flat, anterior end of the ilium, staying close to the bone. Grasp the bone between your knife and your thumb and pry the bone away from the spine. Repeat on the opposite side.

2. SEPARATE THE REMAINING VERTEBRAE.
Note where the loin muscle ends and separate the remaining vertebrae using a cleaver.

3. TRIM UP THE RIB CAGE.
Cut through the ribs using shears or a cleaver, staying about a half-inch from the loin muscles. Note where the loin and shoulder muscles end, near the neck, and cleave through the spine at that point.

Split the Hind Leg

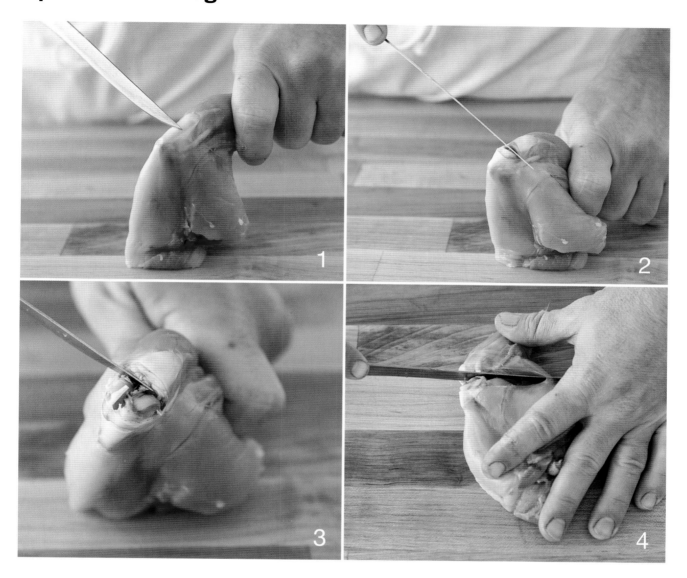

1. BEND THE STIFLE JOINT.
Hold on to the end of the hind leg and press the base against the table, bending the stifle joint.

2. SEVER THE SILVERSKIN.
Note the area of silverskin that covers the joint and cut through the middle.

3. SEVER THE JOINT CONNECTIONS.
Use the tip of your knife to sever the connections of the joint.

4. SEPARATE THE LEG.
Place the hind leg flat on the table, inside up, and finish cutting through the joint and muscle to separate the two pieces.

Bone the Forelegs

Anyone who is new to butchering should try boning the forelegs of a rabbit, because working on this micro level is a beneficial exercise in anatomy education before working on larger animals. Boning the forelegs of a rabbit is generally done if the meat is going to be ground or stewed, or if the entire rabbit is being boned.

There are two stages to boning a foreleg: scapula removal and then humerus/radius/ulna removal.

REMOVE THE SCAPULA

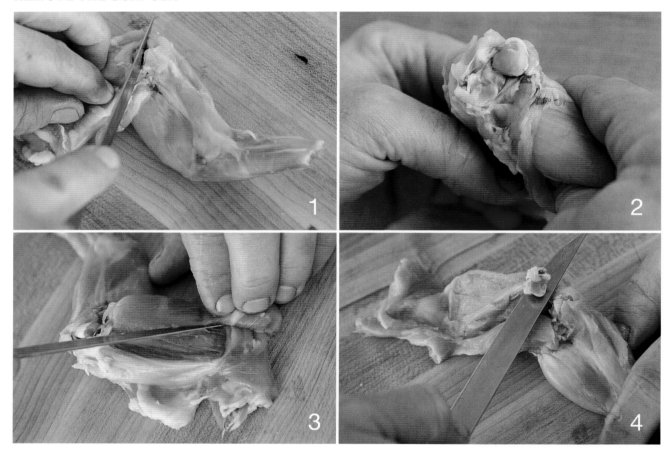

1. SEVER THE JOINT CONNECTIONS.

Place the foreleg underside-up on your work surface. Start by feeling for the joint between the scapula and humerus bones. Once you find it, make a shallow cut perpendicular to the joint.

2. POP THE JOINT.

Hyperextend the joint to separate the two bones.

3. SCRAPE THE MEAT OFF THE SCAPULA.

Once they're separated, use the edge of your knife to scrape the meat off the scapula. Start at the incision you just made, cutting down one side of the bone and scraping across.

4. PRY UP THE SCAPULA.

With the meat off one side of the scapula, you can lift and peel the bone away from the underlying muscles. Use your knife or fingers to get underneath the joint end of the scapula. Curl your finger around the joint and pull up while keeping the underlying muscle pinned to the table. Grab the bone and pull it out to finish removal. Look for remnants of the cartilaginous tip of the scapula, and remove it with a knife.

REMOVE THE HUMERUS, RADIUS, AND ULNA

The rest of the process involves tracing the bones and removing them while leaving them connected to each other. Use your hands to feel for the remaining bones – humerus, radius, ulna – and note where the edges are.

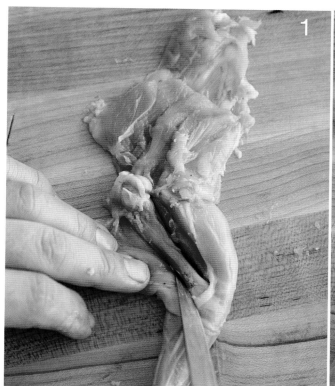

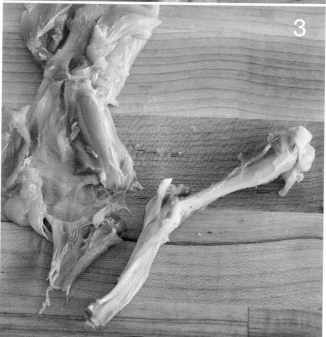

1. TRACE THE EDGES OF THE HUMERUS.

Starting at the exposed end of the humerus, use the tip of your knife to trace one edge of the bones and then repeat with the opposite edge.

2. TRACE THE RADIUS AND ULNA.

Then use your knife to get underneath the exposed end of the humerus. Follow the underside of the bones with your knife to remove them completely. The radius/ulna section will not have a substantial amount of meat, so stay close to the bones.

3. REMOVE THE BONES.

Finish by trimming away any unsavory bits.

Bone the Hind Legs

As with rabbit forelegs, hind legs are generally boned if the meat is going to be ground or stewed, or if the whole rabbit is being boned.

Boning the hind leg is all about tracing the two main leg bones, the femur and the tibia. The other leg bone — the fibula — is minor and, because it is fused to the tibia, will come out when you remove the tibia.

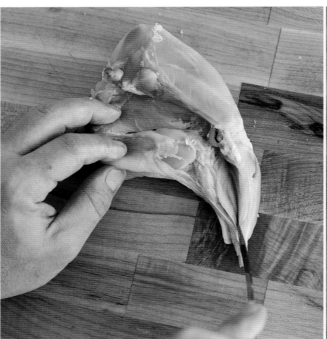 

1. TRACE THE INSIDE OF THE LEG.

Make the first cut along the inside of the bones, working from the ball of the femur all the way to the end of the shank. There are natural seams to the muscles that sit atop the femur, and it is beneficial to follow those.

2. REPEAT ON THE OUTSIDE OF THE LEG.

Follow the opposite edge of the femur, around the stifle joint, and down to the end of the shank.

3. SEPARATE THE FEMUR FROM THE MUSCLE.

Get under the femur with your fingers and then cut around the ball end of the bone.

4. SEPARATE THE TIBIA/FIBULA FROM THE MUSCLE.

Follow the underside of the bone to separate the tibia and the fibula. Trim away the patella (kneecap) that was removed, and any other inedible bits or tendons.

Bone the Saddle

A boneless loin enables you to do some truly unique preparations, especially if you keep the belly on to allow rolling and tying the whole loin. Start with the seven-piece jointed method (see page 184), and do not trim away the exterior silverskin from the loin, because this will keep the two sides of the saddle connected. The goal is to remove the tenderloin and loin muscles from both faces of the spine, keeping it all intact without piercing the thin layer of connective tissue that runs along the dorsal edge of the spine.

To do this effectively, you will need to understand the shape of the vertebrae that the muscles are attached to. Lumbar vertebrae are the only bones in the saddle. From the cross-section, you can see that there are two bony "wings," called the *transverse process*, one on either side of the column. The top muscle is the loin (*longissimus dorsi*) and the tenderloin (*psoas major*) sits on the underside. These small muscles don't leave much room for error, so keep your knife close to the bone; focus on making shallow incisions with long, smooth strokes; and remember the shape of the bone.

1. CUT ALONG THE INSIDE EDGE OF THE TENDERLOIN.

Place the loin on its backside. You can start on either side of the spine, depending on your preference. Cut along the entire inside edge of the tenderloin, keeping your knife slightly angled outward so that your incision follows the slope of the bone.

2. PUSH ASIDE THE TENDERLOIN.

Cut along the bone until just before the edge of the transverse process. Scrape the bone and push aside the tenderloin until you reach the very tip of the transverse process.

3. CUT ALONG THE UNDERSIDE OF THE BONE.

Now pivot your knife, using the outer edge of the transverse process as the stationary point, until you can begin cutting along the underside of the bone.

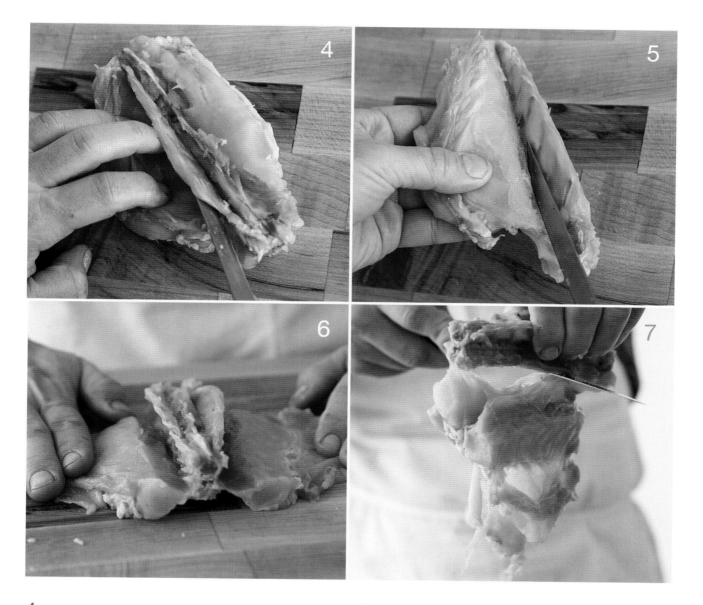

4. **SEPARATE THE MAIN LOIN MUSCLE.**
Follow the underside of the bone, separating the main loin muscle, until you reach the base.

5. **SCRAPE THE SPINE.**
Scrape along the spine to finish separating the loin muscle, stopping when you reach the connective tissue.

6. **REPEAT WITH THE OPPOSITE SIDE.**
Now you have two sides of the loin, totally off the bone and held together by a relatively thin piece of connective tissue that runs across the middle of the spine.

7. **REMOVE THE LOINS FROM THE SPINE.**
Pick up the spine and let the loins fall away. Find an area near the middle where you can insert your knife between the bone and connective tissue. Follow the spine with your knife tight against the bone, staying above the thin connective tissue to finish separation. Trim the belly edges and anything else unsavory, and what you are left with is a whole boneless rabbit saddle ready to be stuffed, tied, and cooked to your liking.

SHEEP & GOAT SLAUGHTERING

GOAT MEAT, SOMETIMES KNOWN AS chevon, is the most commonly consumed red meat in the world; it's even revered in many cultures for some supposed effects that go beyond basic nutrition. This may come as a surprise to many Americans, who may never see it as an option in supermarkets, in butcher shops, or on restaurant menus. And for the most part, goats themselves have gotten a bit of a bum rap, suffering from their reputation as animals that eat tin cans and serving as the butt end of the term *scapegoat*. Likewise, lamb is far more popular in many other countries than it is in the United States. On average, Americans consume just a single pound of lamb over the course of the year! But the demand for meat from goats and sheep is certainly changing. Large ethnic populations are creating new markets for these meats, while other consumers have begun to understand the high nutritional value and appreciate the unique taste. Furthermore, these small ruminants are perfect meat animals for small farms. For these reasons and more, meat from goats and sheep is being found more frequently at farmers' markets and other outposts of locally sourced meat.

Anatomy

THE ANATOMICAL STRUCTURE OF GOATS and sheep is so similar that it's best to cover them together while indicating the few variations along the way. It's the only case among slaughtering and butchering animals in which the likeness is so striking that once you've harvested either animal, you've effectively learned two species with one lesson. Sheep are more common than goats in America, so we will use sheep as the example. Additional details about the visceral anatomy of ruminants is found on page 201.

Domesticated sheep and goats are relatively small animals, with adults averaging just over 100 pounds. They are naturally lean, with minimal muscle development. The ratio of the skeleton and the meat to the animal's total weight is much lower than in any other common farm animal. Sheep and goats actually carry much of their weight in viscera, with a dressing percentage — assuming head-on and without hide or hanging offal — often settling around 50 percent, depending on age, breed, feed, and other inputs. (Compare this to cattle and pigs, which dress out around 60 percent and 75 percent on average, respectively.) Hence, a 100-pound sheep will dress out around 50 pounds; it might lose another 10 percent or more if it's hung for the recommended week.

SKELETON AND VISCERA OF SHEEP

THE ATLAS JOINT connects the base of the skull to the first cervical vertebra (C1). It is this connection that must be disjointed in order to remove the head. For horned animals do this early on before the skinning process reaches the skull.

THE BREAK JOINT is the lowest joint in the legs. You can remove the hooves here or at the next joint (the base of the shank). Breaks joints ossify as the animal ages, and differentiating between a break and spool formation is one of the ways to estimate the age of the animal.

THE PELVIS, OR AITCHBONE once split, is the bone you'll need to tackle when separating the bung, or rectum. Splitting the pubic symphysis, the cartilaginous midline on the bottom (ventral) edge of the pelvis, is one method for separating the bung.

THE MAIN BLOOD VESSELS, carotid arteries and jugular veins, cross as they leave the heart. This intersection occurs right around the front (anterior) edge of the sternum. Severing these vessels is the main priority when using the thoracic method of bleeding.

THE ESOPHAGUS AND TRACHEA leave the skull and run side-by-side until they enter the thoracic chamber. The blood vessels that need to be severed run above on both sides of these tubes. Severing just the blood vessels at the base of the skull is far too difficult; thus, the esophagus and trachea are severed along with the blood vessels when using the transverse method of bleeding.

THE THYMUS GLAND is largest in the "adolescent" stage of an animal's life. After that it atrophies as the animal continues to grow. In older lamb or younger hoggets the thymus gland may be large enough to save and eat. Freeze it immediately after slaughtering in order to preserve freshness.

THE NOTABLE SKELETAL difference between sheep and goats is that there are six lumbar vertebrae on a sheep compared to seven on a goat.

THE FRONT OF THE GOAT SKULL is thicker and stronger than that of sheep. Therefore, use a point on the rear top of the skull where it's thinner when stunning a goat.

BRAIN, TONGUE, HEART, STOMACH, SPLEEN, LIVER, AND KIDNEYS are the likely candidates for offal saved after slaughter. The lungs, while edible, risk contamination when the trachea and esophagus are severed during bleeding. The more reactive organs, like the brain, liver, and spleen, are best when frozen immediately after slaughter, though all offal benefits from this approach.

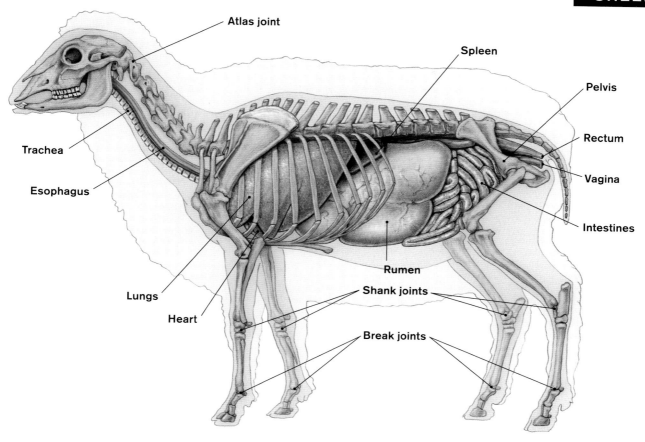

Atlas joint

Spleen

Pelvis

Rectum

Trachea

Vagina

Esophagus

Intestines

Rumen

Lungs

Shank joints

Heart

Break joints

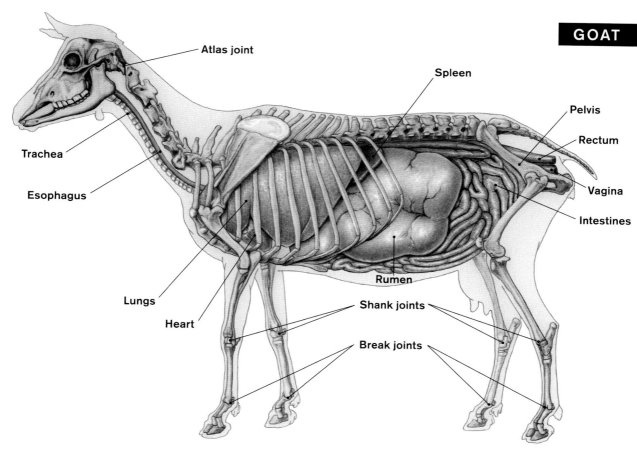

Atlas joint

Spleen

Pelvis

Rectum

Trachea

Vagina

Esophagus

Intestines

Rumen

Lungs

Shank joints

Heart

Break joints

Slaughtering

HARVESTING SHEEP AND GOATS is the best introduction to farm slaughtering, because it covers the basic steps required for any quadruped: bleed, skin, and eviscerate. Furthermore, the setup can be quite basic. There is no need for complicated hanging implements (as with cows), scalders (as with pigs), scrapers, huge gut buckets, or the like. You can do it alone; if, however, this is your first time (or close to it), having another person on hand will be helpful. Block off about 45 to 60 minutes per animal, which includes packaging offal and cleanup.

Slaughtering sheep and goats is quite easy, except for the skinning. Sheep and goats are naturally very lean animals, prone to drying out during aging and cooking. Therefore, when you are removing the hide, you must keep as much fat and fell (the thin membrane covering the carcass directly under the hide) on the animal as possible. Take your time with this part, especially as you are learning. But be forewarned: after death, the rumen of a sheep or goat will continue producing copious amounts of gas, causing it to expand and press against the other viscera. You may notice the abdomen start to enlarge if you take too long. So, while you want to take your time, the whole process of slaughtering should take no longer than an hour.

The Right Age

Truly, sheep and goats, if raised and fed right, are delicious at any age. The American diet typically exalts tenderness above all else when it comes to meat quality, but often that translates into a loss of flavor. (The older the animal, the more its muscles have worked, and the more flavor develops in the meat. For more on the science of flavor and meat, see page 9.) With sheep and goats, this often means that the spring animal — born in the spring, slaughtered six to eight months later in the fall — is the most popular. While this can make for some tender meat, the flavor will rarely compare to that of a one- or two-year-old animal, or even older. The best goat I've ever eaten was from a four-year-old animal, the age that most Americans would consider stewing, grinding, or potentially discarding the goat meat, when in fact the loins were perfectly textured — a resistant tooth without chewiness — and the flavor exquisite.

While goat goes ungraded in the United States, sheep is assigned quality grades based on age, with the three categories being (from youngest to oldest) lamb, yearling mutton, and mutton. Age is determined by the presence or absence of a spool joint on the foreleg trotter, but rough calendar estimates are under 12 months for lamb, 12 to 18 months for yearling mutton, and over 18 months for mutton. (For more on spool and break joints, see page 200.) Because of the similar anatomy and growth of sheep and goats, their muscle development rates should be considered relatively the same.

Preparation

Slaughtering sheep or goats requires some preparation that begins far before the day of the event. Read chapter 5 in its entirety, and if you're leading the slaughter, make sure not only that you know how the process proceeds step-by-step but also that you are prepared to handle the anomalies that can occur.

EQUIPMENT OVERVIEW

Goats and sheep are a good species that illustrates the potential simplicity that a slaughter can take on. Truly, all you really need is some rope and a sharp knife to get the job done well. Yet, for minimal financial investment, some equipment and tools are available that will make the process smoother. Here are some lists of options, ordered from bare-bones to ideal.

STUNNING: bullet, captive bolt gun

HANGING: rope, gambrel

BLEEDING: 5- or 6-inch knife

SKINNING: 5- or 6-inch knife, skinning knife, fist

EVISCERATION: 5- or 6-inch knife

CHILLING: ice-filled cooler, blast chiller walk-in cooler

AGING: refrigerator, blast chiller, or walk-in cooler

THE SLAUGHTER PROCESS FOR SHEEP & GOATS

Slaughtering goats and sheep is about as straightforward as you'll get for a larger livestock species. Skinning can be done with fisting or knifing, but it doesn't affect the process flow. Here is the basic rundown:

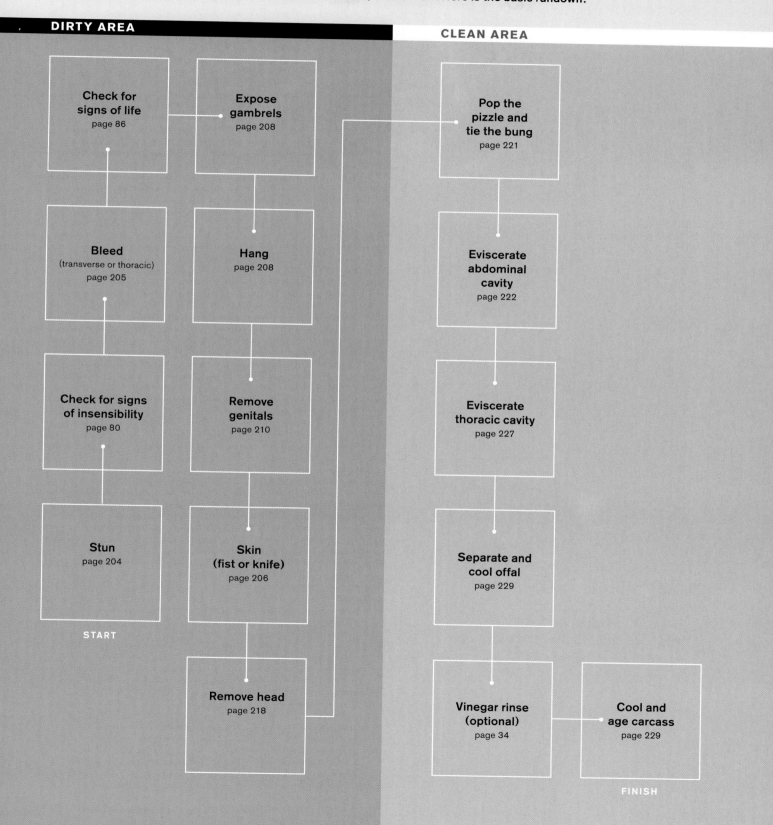

DIRTY AREA

CLEAN AREA

Check for signs of life
page 86

Expose gambrels
page 208

Pop the pizzle and tie the bung
page 221

Bleed
(transverse or thoracic)
page 205

Hang
page 208

Eviscerate abdominal cavity
page 222

Check for signs of insensibility
page 80

Remove genitals
page 210

Eviscerate thoracic cavity
page 227

Stun
page 204

Skin
(fist or knife)
page 206

Separate and cool offal
page 229

START

Remove head
page 218

Vinegar rinse
(optional)
page 34

Cool and age carcass
page 229

FINISH

Secure and Stun the Animal

The images below show the two recommended positions for stunning polled sheep. On the left, the captive bolt is positioned directly between the ears, at an angle that connects to the front of the lower jaw. There is a ridge on top of the skull that flattens between the ears; aim there. The angle is paramount: lower and you penetrate the nape of the neck; higher and you will hit the crest of the brain cavity. In either case, the stun will not be successful.

On the right I am using a point just above the eyes. Connect both eyes with a straight imaginary line. Just above that line, place the captive bolt directly in the middle and flat against the skull. The skull is thicker at this point, so ensure that you have the captive bolt against the skull so that the bolt can penetrate deep enough to produce a proper stun.

STUN POINTS FOR GOATS AND HORNED SHEEP

Sheep and goats have different cranial structures and therefore require modifications in where to aim to achieve proper insensibility; the aim will also need to be modified according to whether the sheep or goat is horned or hornless. Hornless goats have a crest or high point on their skull; find a point in the middle just behind that (toward the ears), and aim through that point toward the mouth. On horned sheep and goats, there is a ridge that runs between the horns; find the middle of that ridge and aim through that point toward the mouth.

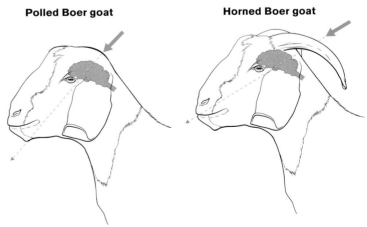

Polled Boer goat   **Horned Boer goat**

Bleed the Animal

Proceed to bleeding only after you have confirmed that the animal is insensible. (A quick test is to tap an eyeball with your finger. There should be no blinking; if the animal blinks, it is aware.)

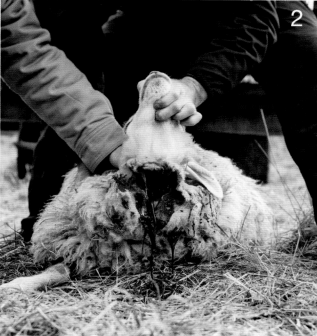

1. **CUT ACROSS THE JUGULARS.**

2. **RESTRAIN THE ANIMAL.**
After making the cut, pull back on the head to let the blood flow freely. You may either restrain the fallen animal (which I normally do) or step away; even a stunned animal will exhibit death throes. Do not be alarmed by thrashing and kicking legs, but do stay on the lookout for any signs of the animal returning to sensibility.

3. **PUMP THE FRONT LEG.**
When there are no signs of movement, take a moment to pump the front leg while placing a knee on the chest, helping expel any remaining blood from the heart and main blood vessels.

>> For those using the thoracic bleed method (see page 85), follow a confirmation of death by severing the trachea and esophagus close to the skull.

Start Skinning

Start skinning only after you have confirmed that the animal is no longer living. The majority of skinning is done while the animal is hoisted, but in order to hoist, we need to skin the hind legs to gain access to the gambrel cords.

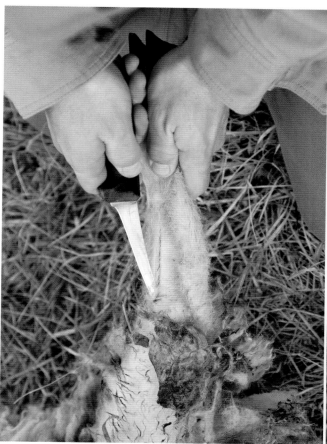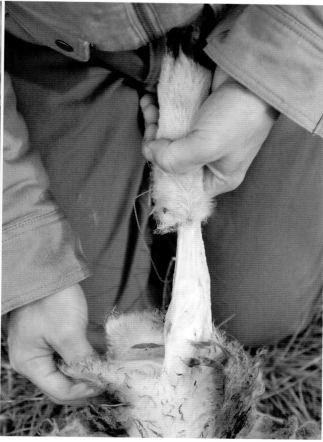

1. SKIN THE SHANK.

To expose the gambrel cords for hanging, skin around the leg, beginning at hoof-end of the leg and working toward the stifle joint. Keep the hide taut, and use just the tip of your knife to avoid cutting into the muscles. As you're peeling away the hide, stay between the hide and the ligaments that surround the cannon bone.

2. EXPOSE THE GAMBRELS.

Once the end of the leg has enough hide released to allow a firm hold, grab the end of the leg with a clean hand. With the other hand, pull the hide away at a downward angle of around 45 degrees. This will peel the hide away from the shank to expose the gambrel cord. If the hide doesn't peel, don't force it; you can tear the muscle. Instead, continue using your knife to release the hide from connective tissues and ligaments.

3. MAKE AN INCISION BETWEEN THE GAMBREL CORD AND SHANK.

With the gambrel cords accessible, it's now time to hang the animal. Make an incision, large enough for the hanging implement, between the gambrel cord and the shank muscles.

4. HOIST THE CARCASS.

A good starting height will put the rectum (also called the bung) just below your eye level.

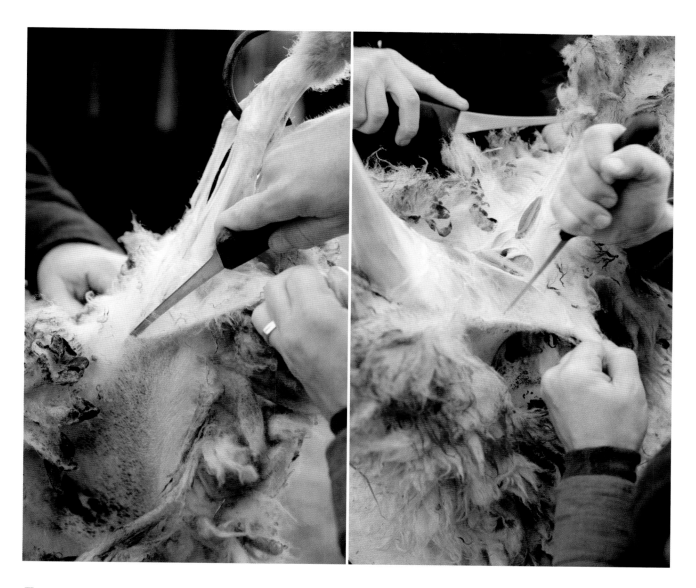

5. SKIN THE REAR LEGS.

Slide the tip of your knife beneath the hide and continue the cut you made to expose the gambrels. End at the inside point of the leg, where the leg meets the groin. Continue to skin around the leg, pulling the hide down when it's been released from the muscle. When the hide is not in your dirty hand, make sure that it drapes away from the carcass and that the contaminated exterior is not resting against muscle.

6. CONNECT INCISIONS.

Skin the opposite leg and then connect the incisions on both legs with a cut across the groin anterior to the anus. Stay shallow with your crosscut to avoid severing the genitals or puncturing the abdomen.

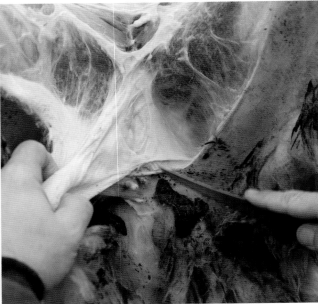

7. CUT THE CENTER SWATH.

Remove a 2- to 3-inch-wide swath of hide from the center of the abdomen, all the way from groin to neck. This provides a clean point of access to both sides of the hide, making for a less-contaminated carcass. Start at the groin (or below the mammary gland on females), pulling the hide with one hand while you cut. Avoid using a back-and-forth sawing motion; instead, slice down at an angle, drawing the knife toward you. Then push the knife across, heading at a downward angle. As you repeat this you'll find that the back-and-forth motion is similar to a rocking motion. Keep the cuts shallow and avoid puncturing the abdomen. When you reach the sternum, continue cutting through the hardened, callused fat that sits against the bone. You will take some meat off through this area. Finish by running your knife tip under the hide along the bottom of the neck, and out through the cut you made to bleed the animal.

8. SEPARATE THE GENITALS.

Before you can skin a goat or sheep, you need to separate the genitals. For males, the pizzle (penis) and testicles should be removed. Slide a knife under the pizzle at the incision you made across the groin. Cut through the pizzle, leaving the posterior portion attached to the carcass. Remove the anterior portion of the pizzle with a 2- to 3-inch swath. You can easily remove hanging testicles by cutting through the point where they join the abdomen.

Removing the mammary gland and teats from a female is a bit more difficult. Find the upper edge of the mammary gland — you can feel the softness of the gland underneath the hide, compared to the firm feeling where the hide sits adjacent to the abdominal wall. The gland is surrounded by fatty tissue and connected by a tendon (called the *symphyseal tendon*) that is less like a rope than like a sheet of connective tissue. Start removing the hide above the gland near the incision that you made between the two hind legs. When you reach the fatty tissue — it will be yellowish and distinctly different from the hide — carefully cut through it, heading at a slight downward angle. When you hit the symphyseal tendon, continue cutting but proceed slowly, ensuring that you do not cut so deep that you open the abdominal wall. Using the abdominal wall as a guide, follow the surface around both sides of the mammary gland and finish separating it, leaving a 2- to 3-inch strip at the bottom that you can use to continue cutting the midline of the hide.

Remove the Hide

Now is the time to choose whether to remove the pelt using the fisting method or the knifing method. The *fisting method* relies on the insertion of a closed hand between the pelt and the *fell*, the fascial membrane covering the carcass. Done correctly, which will inevitably take practice, fisting will provide a cleaner carcass and pelt, with less chance for damage to both muscles and hide. The downside is the physicality of the process, which some folks find to be a challenge. Furthermore, at first it may be slower than knifing, though with experience it proves to be a more efficient process.

Knifing is just what it sounds like: removing the entire hide using your knife. Long, smooth strokes are used to separate the hide. After key attachment areas are freed, you can pull on the hide to finish removing it. The downside is the high risk of carcass and hide damage, with the final carcass appearance never looking as clean as a well-done fisting job.

1. START AT THE GROIN.

Both approaches — fisting and knifing — begin the same way. The incision across the groin and the midline swath that you removed form an intersection: these corners are where you start skinning for their respective sides. Using your knife to remove the hide, work across the leg and groin and partway down the belly until you have reached the side of the carcass. Repeat this step when you need to switch sides.

2. OPEN A GAP.

Hold the flank flap of hide with your dirty hand and, near the base of the leg, use the thumb of your clean hand to start making a gap between the hide and the fell. (When fisting a hide, always use your thumb or a finger to create the beginning gap.)

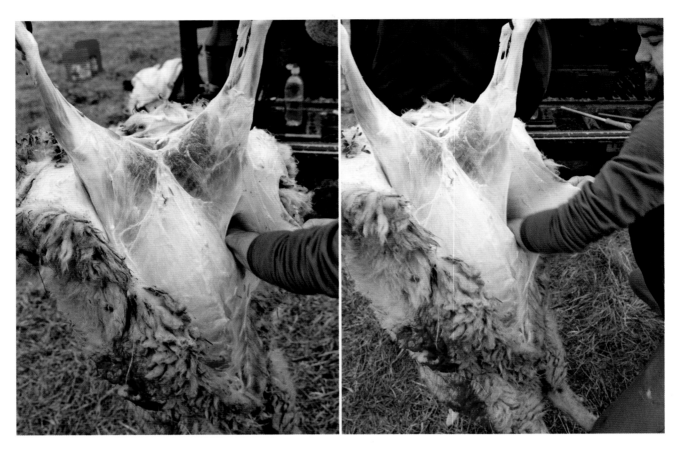

3. INSERT YOUR HAND.

Slowly open up the gap to make room for your fist and then begin tunneling your way deeper using a repetitive thrusting motion.

4. LEVERAGE DOWNWARD.

When you are elbow-deep into the hide, leverage your elbow downward to open up a larger area.

TIPS FOR EFFECTIVE FISTING

- Fisting is done in three main phases that correspond roughly to the primals: sirloin and legs first, then the loins and belly, and finally the shoulders and foreshanks. Initially knifing the flank will give you the access you need to start fisting the first stage, but take note of a couple tenets:

- Your hand should not be a clenched fist but rather a tightened shape that reduces the overall surface area so that you can insert your hand into the tight spaces between the hide and the carcass.

- You will be using your second knuckles primarily. In sliding above the fell, wet hands work better than dry hands, so keep your clean hand moistened (which should not be a problem if you have remembered to wash your hands regularly during the slaughtering process). Use your dirty hand to

 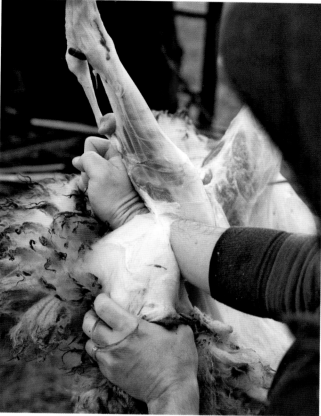

5. TUNNEL UPWARD.

Continue this process with both sides of the carcass — tunnel with your fist, leverage with your elbow — until you reach the spine, and then work your hand out the top of the hide to the near side of the tail.

6. FINISH SKINNING THE LEGS.

Either pull down on the loose upper leg hide or insert your hand below the leg and out the upper part of the back hide, near the tail; then leverage your arm upward to remove the thigh hide (pictured).

keep the hide taut, and use your clean hand to press against the hide (not against the muscle).

- The hide should feel smooth; if it starts to feel stringy or textured, then your fist has fallen below the fell. In that case, stop fisting. Use your knife to cut through the membranes attached to the hide, working your way back to the dry look of the hide. Then continue fisting.

- A mentor of mine, Eric Shelley from the State University of New York at Cobleskill, uses a metaphor to describe the process of inserting your fist against the hide and along the meat: "It's like trying to get a wet glove on." It can also be described as like trying to work your hand through the arm of a sopping-wet long-sleeved shirt.

Remove the Hide CONTINUED

7. CUT ACROSS THE RECTUM.

At this point, both legs are skinned and the hide is loosened across the back. Now pull down on the tail and cut straight across the rectum. Note the variation of ridges and white spots on the tail bones, the latter being the cartilaginous joints between the caudal vertebrae. Cut through the cartilage, wiggling your knife some to help it find its way if you are unable to easily make it through. Continue cutting down until you reach dry hide again.

8. PULL DOWN THE HIDE.

Proceed with punching down both sides of the hide, using the same tunnel-leverage approach described previously. (As you work your way down, consider raising the carcass to keep it at a convenient height.) Follow that by pulling the hide down the spine, proceeding carefully to ensure that you're not tearing meat and fat off the carcass. If so, stop pulling and use your knife to get back to dry hide.

9. SKIN THE FORELEGS.

Stop fisting when you reach the forelegs. Make a shallow incision along the inside of the leg and across the sternum. Skin the top of the front legs and sternum using your knife. Skin all the way to the hooves if you plan on removing them at the break joint.

10. FINISH SKINNING THE SHOULDERS.

Fist the rest of the shoulders and pull down on the hide to remove it from the underside of the forelegs, sternum, and neck.

Remove the Hide

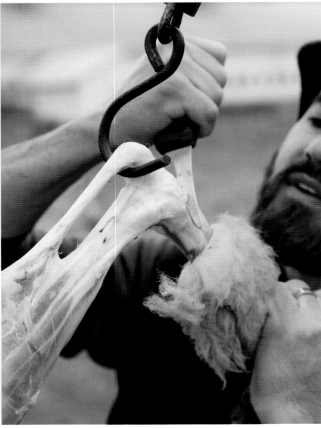

11. REMOVE THE FRONT HOOVES.

Cut at either the break joint (near the hooves) or the joint just below the shank (as shown). The break joint (with lambs, generally under 16 months) becomes a spool joint (with mutton, over 16 months) as the joint ossifies.

12. REMOVE THE REAR HOOVES.

The joint below the rear shanks is more difficult to separate than that of the foreshank. There are some strong ligaments that run across the joint and must be severed to release the two sides. Try to work your way through it with the tip of your knife. A bolt cutter or meat saw also makes quick work of lopping off hooves.

Remove the Head

At this point you have the option of removing the head, unless you plan on leaving it on the carcass to increase the hanging weight. Horns can present a problem when you are skinning the neck and shoulders, making a good case for removing the head or at least the horns, earlier in the skinning process.

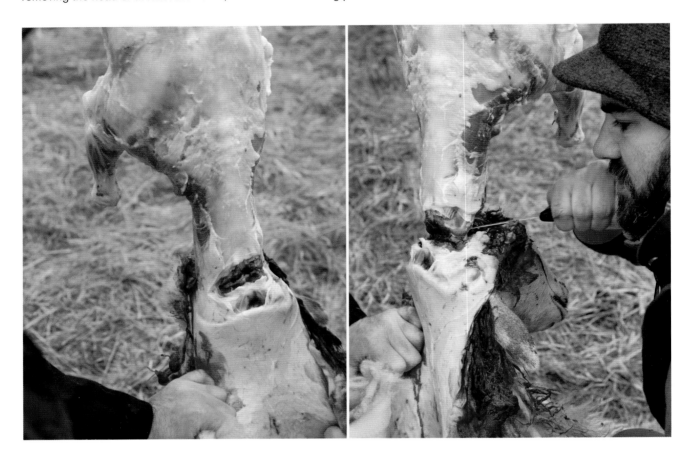

1. FIND THE ATLAS JOINT.

This is the first cervical vertebra that joins with the base of the skull.

2. DISJOINT THE VERTEBRAE.

Slice around the neck, then work from the inside out: using the tip of the knife, sever the connections in the middle of the atlas joint, wiggling it around to help find its way rather than trying to sever bone. The bones of the atlas joint are tightly joined, so make small cuts as you learn the shape and method of disjointing them.

Further Head Processing (OPTIONAL)

REMOVING THE TONGUE.

Place the head on its face, jaw up. Near the base of the skull, feel for the larynx (voice box). Insert your knife in front of (anterior to) the larynx and make an incision between the interior edges of the jaw bone, deep enough to open the sinus cavity. You should be able to insert your fingers and feel the cavity and the bony larynx. Using the interior edges of the mandible as your guide, cut all the way from your initial incision to the front, releasing the sides of the tongue. This should allow you to literally roll the tongue backward out of the bottom of the skull. Finish removing the tongue by severing the remaining connections. Wash it thoroughly, and save it.

SKINNING THE HEAD AND REMOVING THE CHEEK.

Begin with an incision along the midline of the jaw underside and then work around one side and then the other. Stay shallow around the cheeks to preserve the small nuggets of meat. Place the skinned head on its side, cheek facing up. Make an initial cut along the bottom edge of the mandible, deep enough to lift the edge of the cheek meat. Keep your blade flat and pointed slightly toward the bone. A semi-stiff blade helps here, because its slight bend allows you to apply pressure and keep the blade against the bone. Continue cutting underneath the muscle and against the mandible, using the other hand to peel away the muscle, until you reach the skull. Then cut the remaining edge, freeing the meat. Directions for cleaning the cheeks start on page 256.

EXTRACTING THE BRAIN.

Extracting brains is best done with a cleaver and after the tongue is removed. It will take some practice, but the following approach is easy to follow and will help you avoid any damage to the brain. Start by cutting down the middle of the face and through the soft cartilage at the front of the nose. Turn the skull over and cut down the middle of the palate, inside the mouth. Cleave through the front of the face along the line you cut. Then, cleave through the palate. Carefully cleave through the top of the skull and then try to pry the two sides apart. If they don't separate, cleave a bit deeper on the underside of the skull, near its base. With the skull opened, carefully pull the brain from its protective cavity and cut it from the spinal cord. Follow the instructions on page 93 for proper cleaning procedures.

Remove the Abdominal Viscera

For those who will be hanging the carcass in a well-ventilated area (where the carcass will dry out easily), take a moment to spray down the carcass now that the hide is off and before the body cavity is opened. This will help clear away any contaminants that may have settled on the carcass while the hide was being removed, reducing the amount of trimming you'll need to do later.

For males, separate the remaining portion of the pizzle by cutting on either side of it, leaving as much of the cod fat on the carcass as possible. Follow the pizzle to the opening of the pelvis where it enters the body, ventral to the rectum. Hold the pizzle straight up and taut. Keeping your knife vertical, use the tip to cut in front of (ventral to) the pizzle, staying close to the pelvic bone. Make small cuts through the connective tissue while pulling up on the pizzle until you hear it pop free.

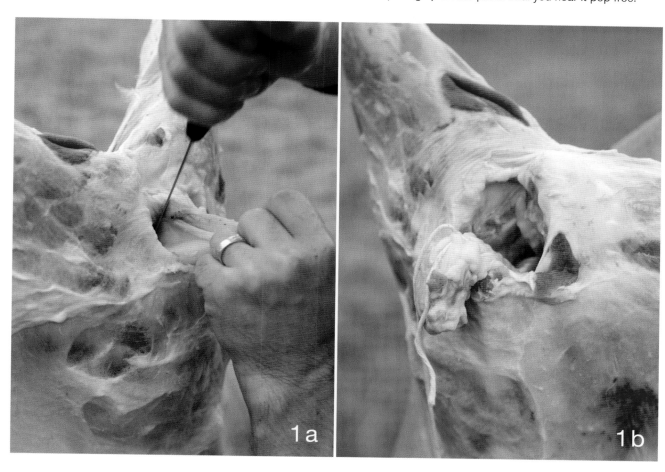

1. TIE OFF THE BUNG.

This process for separating the bung is based on using the pelvic opening as a cutting guide. (Alternately, you can separate the pelvic bones at their cartilaginous midline, known as the pubic symphysis. See Splitting the Pubic Symphysis, page 92). Follow the shape of the pelvic opening with your knife to cut around the bung, freeing it, and tie it with a length of string. Once the bung is tied, push it (and the pizzle, if the animal is a male) into the cavity for retrieval later during evisceration.

Remove the Abdominal Viscera CONTINUED

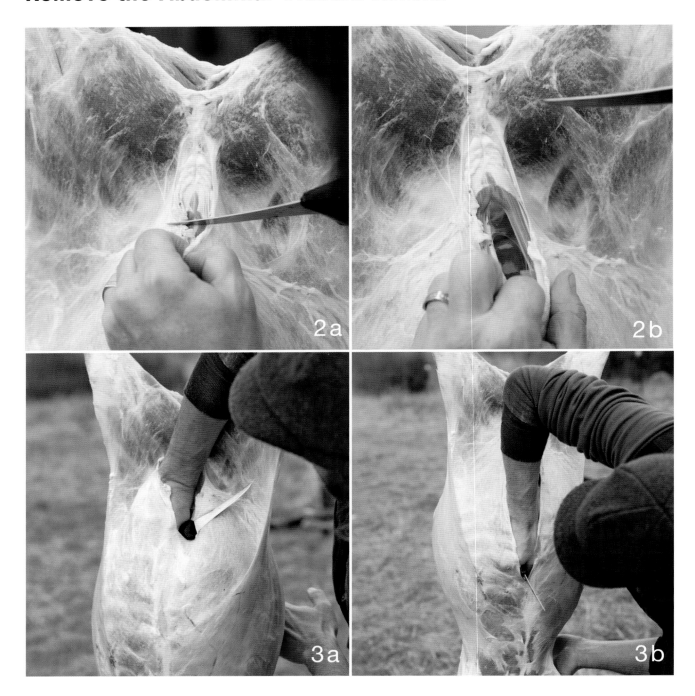

2. MAKE THE FIRST INCISION.

The initial incision into the abdomen is made by pinching the skin below the rectum, pulling it away from the bladder (which sits right behind it), and making a small crosswise cut.

3. OPEN THE ABDOMEN.

Choose one of the two methods for opening the abdomen (page 91). Here, I use the whole-handle method, inserting the knife handle into the cavity and cutting straight down until I hit the sternum, using the base of the blade to sever the abdominal wall.

Accessing the bung and genitals is simplified by splitting the pelvis at the groin along a cartilaginous line between the two sides.

 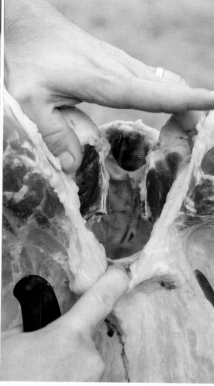 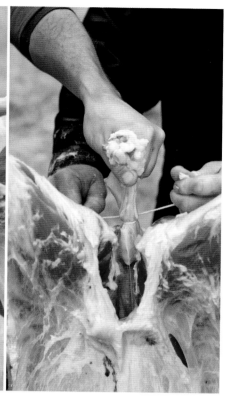

CUT BETWEEN THE LEG MUSCLES, where they join at the groin, until you hit the pelvis. Make a small incision in the pelvis by inserting the tip of your knife into the seam of cartilage.

REPEAT THIS ALONG THE SEAM until you can split the two sides, which may require a little prying with the flat of your blade. (The younger the animal, the softer the cartilage and easier it is to cut through; older animals may require some minor sawing to make your way through.)

WITH THE TWO SIDES SPLIT, the bung can be easily freed and tied. For more on splitting the pubic symphysis, see page 92.

Remove the Abdominal Viscera

4a

4b

5

6

4. REMOVE THE BLADDER, GENITALS, AND TIED RECTUM.

The guts should now be hanging out. Identify the bladder, and beware: do not tear or puncture it. Reach up into the cavity and pull down on the genitals and tied rectum. Bring them out of the cavity and sever the ureters and any other interior attachments. Notice the spleen on the left side of the stomach — the long, tongue-shaped organ with a purplish hue. Remove it, severing connections, if you plan on saving it.

5. TWIST THE PAUNCH.

Roll the stomach and entrails outward by pulling down and out. Grab the entire mass of viscera, lift up, and give it a twist. This will close the esophageal connection and prevent further stomach contents from spilling out. (Leading up to this you may see some contents spill out of the severed esophagus; this is normal.)

6. REMOVE THE CAUL FAT.

This is a lacy layer of fat covering the stomach and intestines. This can be rendered, used for culinary preparations, or draped on the carcass to stem moisture loss during aging.

7a

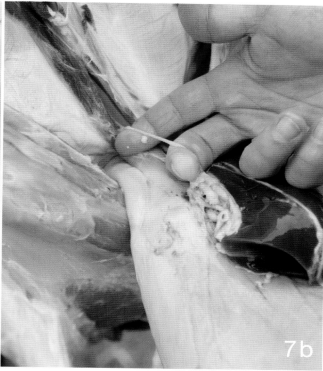

7b

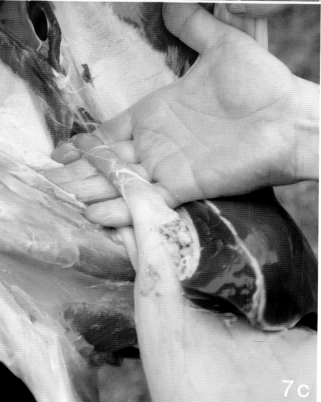

7c

7. SEPARATE THE ESOPHAGUS.

Find the esophagus, a red, fleshy tube that passes through the diaphragm and connects to the stomach at the gastro-esophageal (GE) junction. This junction is inherently weak and, if pulled on incorrectly, will break, spilling stomach contents into the carcass cavity. Use one hand to support the weight of the rumen as it hangs out of the carcass, preventing a tear of the GE junction. With your other hand, grab the esophagus. Wrap one finger around the tube and feel the thin tendon on the opposite side; it will feel like taut fishing line. Get the finger between the esophagus and the tendon, and then make space for your whole hand to grab the tube. If the process is done properly, the esophagus should feel smooth and soft.

Pulling the esophagus out may sound challenging, but it's actually quite simple. Use a motion that leverages your hand in the same way you would pull up on a fishing pole, lifting the tip and keeping force on the front to avoid tearing the GE junction. Pulling hard in this manner will safely remove the esophagus. The tube will also clear some of its contents as it passes through the restrictions of the neck and thoracic cavity, an added benefit.

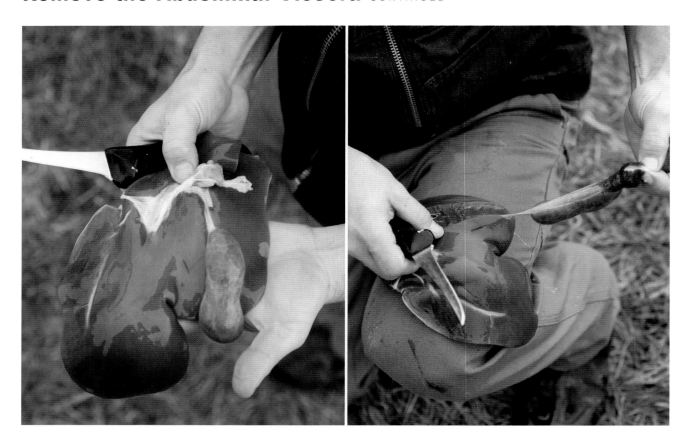

8. REMOVE THE LIVER.

Sever any remaining attachments, including those to the liver, and let the entrails drop to the ground. Take the liver and sever the blood vessels keeping it in place.

9. REMOVE AND DISCARD THE BILE SAC.

Note the bile sac (gallbladder). Pinch the duct where the sac meets the liver, and sever the connection; then carefully tear the duct off the surface of the liver, and discard it. Thoroughly rinse the liver before saving it, especially if you ended up spilling some bile while removing the gallbladder.

Remove the Thoracic Viscera

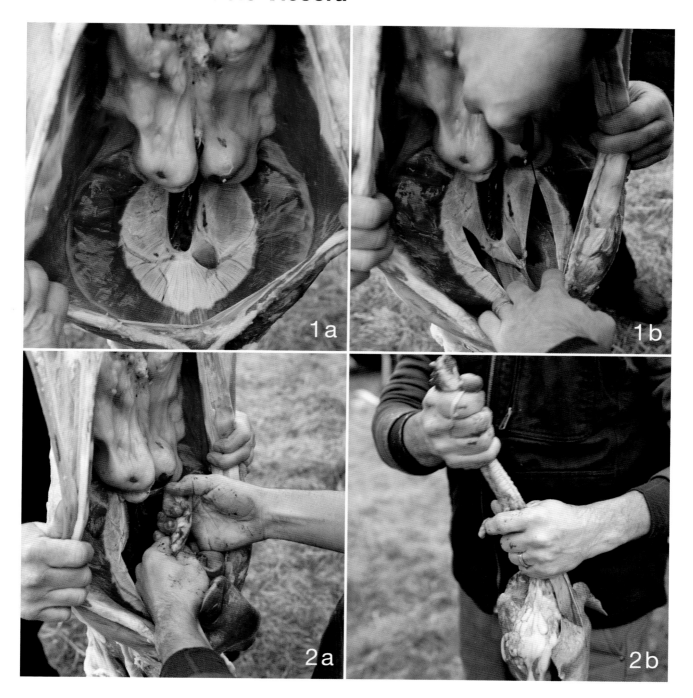

1a **1b** **2a** **2b**

1. CUT THE CONNECTIVE TISSUE.

The diaphragm has a sheet of connective tissue running along the spinal (dorsal) side. Cut along the middle of the connective tissue, keeping the muscular areas intact. Once punctured, it will retract back against the rib cage.

2. GRAB THE PLUCK.

The thoracic viscera – lungs, heart, and trachea – are also called the *pluck*. Reach into the chest cavity and, with one hand, grab the trachea. The other can grab the pluck. Pull them up through the abdominal opening. Use your knife to sever the thoracic aorta and any stubborn attachments. Trim away areas of the neck that may have been contaminated, from rumen discharge or other sources, after you free the trachea.

Clean the Heart and Kidneys

1a

1b

2

1. REMOVE AND CLEAN THE HEART.

Cut the heart loose and remove the exterior membrane (pericardial sac). Cut across the top of the heart, leaving part attached, like a hinge, and squeeze it to expel any interior blood clots. Thoroughly wash out and save the heart. (A fun trick is to blow into the trachea to inflate the lungs before discarding.)

2. REMOVE THE KIDNEYS AND "POP" THEM.

You may choose either to leave the kidneys for aging and adding to the carcass weight or to remove them for immediate freezing. In either case, score the surface membrane and squeeze them to pop.

Carcass Trimming

THE SHEEP OR GOAT CARCASS is bound to have some areas that need to be trimmed of contamination. Clean your hands and utensils and then inspect the carcass for hair, feces, digestive contents, and other grime. Trim as little meat and fat as possible while being as thorough as possible. (Inspection hooks are a great tool for removing small patches of meat.) Any part that was close to where the hide was opened (e.g., shanks, hocks, neck) is sure to need some trimming. Also inspect the interior for remnants of viscera like parts of reproductive organs, the liver, or other unsavory glands.

Finish up by spraying off the animal with a clean hose nozzle. Forelegs that still include the cannon bones can be doubled up and held in place with a loosened foreleg tendon.

Carcass Splitting, Cooling, and Aging

The sooner you start cooling the carcass, the better. Hanging the carcass whole from the gambrels to cool is ideal; alternatively, you can split the carcass if you are pressed for space. Split the carcass in half, with cuts made between the twelfth and thirteenth ribs, the same way you separate the rib and loin primals according the directions on page 233. While the hind half can be hung from the gambrels, the front half should be hung from either tendons in the foreshanks or from the rib cage.

Because sheep and goats are lean animals, excessive dehydration is a concern when aging. Wrap the legs and rumps with the gathered caul fat, and/or wrap the carcass (or carcass halves) in cheesecloth to prevent moisture loss. Hang the carcass in either a refrigerated area or, during the colder months, in a building that prevents access from hungry critters. After 30 hours, take a temperature reading of the thigh meat near the bone to ensure that the carcass is cooling at the proper rate. Following cooling should be a period of aging according to the recommendations on page 20.

Offal

Clean offal and give it a quick once-over to check for signs of infection or disease. Refer to page 93 for information on cleaning, inspecting, and preparing offal for storage. Freeze offal as soon as possible to preserve its freshness.

Cleanup

The plan for cleanup should be decided well before the slaughter. Follow the recommendations on page 96 for disposing of entrails and other inedibles, as well as for cleaning your work area and tools.

Sheep entrail identification

Large intestines

Spleen

Rumen

Small intestines

SHEEP & GOAT BUTCHERING

BUTCHERING SHEEP AND GOATS is a great way to learn the anatomy of quadrupeds, in part because the carcass size is easily manageable for a single person. The cuts of these animals often require accurate and delicate techniques, providing great practice for nimble knife work and gentle use of a saw.

The anatomical similarity of sheep and goats allows us to cover both animals in a single chapter. From this point forward, the carcass species will be referred to as lamb, the most popular incarnation of the two. Review chapter 4 before proceeding to ensure that you have a solid understanding of the foundational methods of butchering.

Butchering Setup, Equipment, and Packaging

BUTCHERING A LAMB WILL REQUIRE a good-size tabletop setup. You can comfortably break down a whole carcass on a 2-by-4-foot surface; larger is better, though a smaller surface will do if you are working with halves or quarters. The surface should be of a material that is easily cleaned and sanitized; the most common two options are wood and plastic. (For more on the benefits of using wood or plastic, see page 47.) Individual cutting boards benefit from having a nonslip shelf liner, a moist towel, or other such material underneath them to prevent movement while cutting.

Lamb can also be broken down while hanging. There are a few notable benefits to working with a carcass on the hook versus on the table. First, the hanging carcass gives the butcher 360-degree access, which allows for various angles and approaches. Second, the hanging carcass allows for pulling and other types of leverage to aid in separating muscles and bones. Finally, gravity helps make the work more efficient and less labor-intensive. All muscles are secured to the skeleton through connection points, and once these points are severed, peeling muscle from bone is easier with the help of gravity. Because lamb is relatively lightweight, a simple setup that can support a weight a few times that of the carcass weight will suffice. The working height for hanging carcasses should not be much higher than the eye level of the butcher.

REQUIRED EQUIPMENT

- Bins or lugs
- Boning knife
- Breaking knife
- Handsaw/bone saw
- Honing rod/knife steel

OPTIONAL EQUIPMENT

- Bone dust scraper
- Boning hook
- Butcher knife
- Cleaver
- Mallet (rubber or wooden)
- Paring knife

Equipment

Butchering lamb requires the same set of basic tools that butchering any other quadruped does. There are some optional tools (listed below) that are helpful, especially for some of the more tedious preparations. For example, a paring knife will help when frenching a rack. Make sure to have paper or fabric towels on hand, as well as butcher's twine. The details about particular tools and equipment, as well as recommendations on which ones to use, are found in chapter 3.

Packaging

Butchering any large carcass involves packaging a lot of frozen cuts. To get the most out of your work, make sure the packaging is effective: you want it to be skintight with as little air as possible. Under many conditions, vacuum sealing will be ideal, but for a complete list of packaging options, including recommendations and the science behind freezing, look to chapter 14.

Primals and Subprimals

WITHIN THE WORLD OF BUTCHERY all animals are broken down into primals and subprimals, distinguished often by skeletal features as well as by culinary preparations. The distinction between subprimals and primals is the guide we follow when parsing out the portions. This anatomical map may differ according to cultural traditions or local culinary habits, and regional parlance comes into play when referencing specific primals, subprimals, and cuts. Where applicable, I've provided some of the more common international names for familiar cuts. The primal definitions I've included are those used by the American Lamb Board. Scientific muscle names are provided in italics, most commonly when cuts are composed of a single muscle.

Head/Neck

The head can be removed at the atlas joint (the first cervical vertebra), like all other quadrupedal heads. From there, the neck is removed at the juncture with the beginning of the thoracic vertebrae. The cheek meat and brains can be reserved for culinary preparations.

SHEEP PRIMALS

Neck

Shoulder

Rack

Loin

Breaking down a large lamb carcass will always involve numerous decisions, because each primal can result in many different end products. It is best to have a clear idea of the resultant cuts you want before proceeding with any breakdown.

Flank

Breast and foreshank

Leg

BUTCHERING CUT POINTS

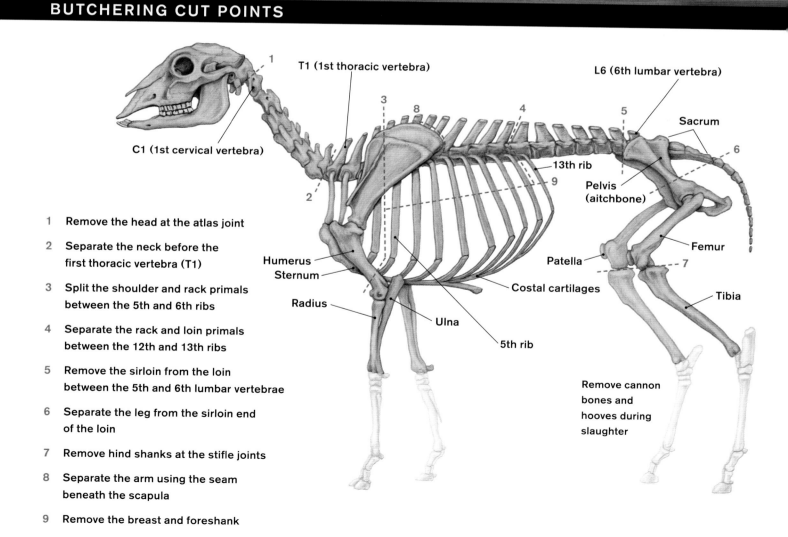

1 Remove the head at the atlas joint

2 Separate the neck before the first thoracic vertebra (T1)

3 Split the shoulder and rack primals between the 5th and 6th ribs

4 Separate the rack and loin primals between the 12th and 13th ribs

5 Remove the sirloin from the loin between the 5th and 6th lumbar vertebrae

6 Separate the leg from the sirloin end of the loin

7 Remove hind shanks at the stifle joints

8 Separate the arm using the seam beneath the scapula

9 Remove the breast and foreshank

C1 (1st cervical vertebra)

T1 (1st thoracic vertebra)

L6 (6th lumbar vertebra)

Sacrum

13th rib

Pelvis (aitchbone)

Femur

Humerus

Sternum

Radius

Ulna

5th rib

Costal cartilages

Patella

Tibia

Remove cannon bones and hooves during slaughter

ONE CARCASS, TWO APPROACHES

There are many different ways to break down a lamb carcass, and making certain cuts prevents others; you can't have both a rack of lamb and rib chops from the same side, for example. Here are two different approaches to a lamb breakdown, each utilizing a different side of the carcass. Before splitting the carcass, we'll remove the neck whole and bone it for a boneless neck roast.

	SIDE 1	SIDE 2
NECK	Remove whole before splitting; bone for boneless neck roast.	
SHOULDER	Separate shoulder and rack; bone, defat, and cut shoulder for stew meat.	Separate the arm from the shoulder ribs; bone and roll for a roast; leave shoulder attached to rack.
RACK	Remove and save fat cap; chine, trim rib bones to desired length, and leave as rustic rib roast.	Saw across all ribs (shoulder and rack), leaving flank attached to the saddle, thus allowing for Denver ribs; separate rack and shoulder; French rack and cut into two-bone chops; bone remainder of shoulder for grind.
BREAST	Bone and save for grind.	Remove from Denver ribs; bone and save.
LOIN	Bone tenderloin and loin meat, leaving flank attached; roll for boneless loin roast.	Separate flank; crosscut 4–6 loin chops.
FLANK	(Rolled with loin roast.)	Roll for cured lamb belly (pastrami or bacon).
SIRLOIN	Separate from leg; bone, tie, and crosscut for boneless sirloin steaks.	Bone, but leave attached to leg.
LEG	Seam subprimal muscle groups; treat inside round and sirloin tip as small roasts; grind bottom round and heel with saved fat from breast and rack for ground lamb.	French the shank (saving meat for grind); remove aitchbone, tie with sirloin over ball of femur for a lamb leg roast.
SHANKS	Crosscut for lamb osso buco preparation.	Bone foreshank for grind (hind shank was frenched with leg).

So, in the end this will give us almost the whole gamut: neck roast, arm roast, shoulder roll, shoulder stew meat, Denver ribs, rustic rack roast, frenched rib chops, loin roast, loin chops, cured lamb belly, sirloin steaks, small leg roasts, steamship semi-boneless leg roast, osso buco, and some ground lamb. A nice selection from just one carcass!

Shoulder

The shoulder of lamb is the second-largest primal, accounting for nearly a third of the total carcass weight. As in most quadrupeds, the forelegs and shoulder of the living lamb bear most of the burden for movement and posture. This extensive workload makes for a deeper flavor and higher collagen content than in any other primal. Although the shoulder is often characterized as the toughest primal, its higher collagen content and shorter muscle fibers lend themselves well to slow-cooking methods, especially those that use moist heat.

Lamb also has a floating shoulder, meaning that the foreleg bones have no connection to the rest of the skeleton and are held to the side of the carcass through muscle and connective tissue. Hence, this floating shoulder can be easily separated from the rib cage by cutting through the muscular attachments — no sawing or disjointing is required. The shoulder is separated from the rack between ribs 4 and 5 and is often prepared as a square-cut primal after removal of the foreshank and breast. The neck is often considered a subprimal of the shoulder, consisting of the cervical vertebrae and surrounding muscles.

COMMON SHOULDER CUTS: shoulder chops, arm chops, blade chops, round-bone chops, Saratoga chops, stew meat, shoulder roast, neck roast, neck slices.

Rack

Rack of lamb, also sometimes known as the bracelet, is a prized cut, often garnering the highest prices of any carcass cut. The rib bones are often frenched (all meat and the sheath of connective around the bone is removed) to further enhance the presentation. The main muscle of the rack is the *longissimus dorsi*, which originates at the ilium and tapers until its termination in the shoulder. The rack is separated from the shoulder between ribs 4 and 5, leaving a portion of the scapula on the rack, and then separated from the loin between ribs 12 and 13. The lean muscle, with its prominent fat coverage, lends itself to dry-heat cooking.

COMMON RACK CUTS: rack roast, frenched rack roast, rib chops, frenched chops, boneless rib roast, crown roast.

Breast/Foreshank

This primal is a combination of two subprimals: the breast and the foreshank. They are removed together by a cut that goes from the lower shoulder, near the humerus-radius joint, through to the final rib, staying relatively parallel to the spine. The flank is also sometimes included with this primal. The foreshank (made up of the radius, the ulna, and the surrounding muscles) is laden with connective tissue, since it is closest to the ground and therefore sustains the most continuous workload. The breast consists of the pectoral muscles, a portion of all 13 ribs, the sternum, and the intercostal cartilage. Both subprimals are prepared with moist-heat methods to aid in the hydrolysis of collagen into gelatin.

COMMON BREAST AND FORESHANK CUTS: boned and rolled breast, crosscut shanks, Denver ribs, riblets.

Loin

The thirteenth rib and the entire lumbar vertebrae identify the boundaries for lamb loin, a primal that includes the widest portion of the *longissimus dorsi* as well as the tenderloin. The loin, which can also be found under the moniker *saddle*, is separated from the sirloin end of the leg by a cut perpendicular to the spine at the front (anterior) edge of the pelvis. The flank is typically removed from the loin by separation from the leg and a close trim following the edge of the loin muscles.

COMMON LOIN CUTS: loin chops, English chops, saddle chops, loin roast, tenderloin.

Leg

The largest of the primals, lamb leg includes the sirloin, hind leg, and hind shank. Additional subprimals are the top round (or inside), outside round, and sirloin tip. The muscles of the leg are large and lean, offering full flavor because of their working status but minimal marbling and connective tissue because of their passive relationship to the shoulder, which leads by providing most of the locomotion and stability. Full leg preparations include bone-in, boneless, and semi-boneless, all of which include the removal of the hind shank.

COMMON LEG CUTS: sirloin steaks, sirloin roast, boneless leg roast, semi-boneless leg roast, leg steak, kebabs.

THIS SECTION will walk you through the breakdown of a lamb carcass entirely on a tabletop, resulting with a square-cut shoulder and other typical primal cuts. It assumes that you have a whole, unsplit carcass to work with, though instructions can still be followed for halves — just ignore the references to splitting or switching sides. Aside from the breast and foreshank, all primals will be removed as a pair from the adjoining sides. Any of these spinal primals — shoulder, rack, loin, and leg — can be split in half by sawing through the middle of the spine. Further processing of primals is covered on page 250.

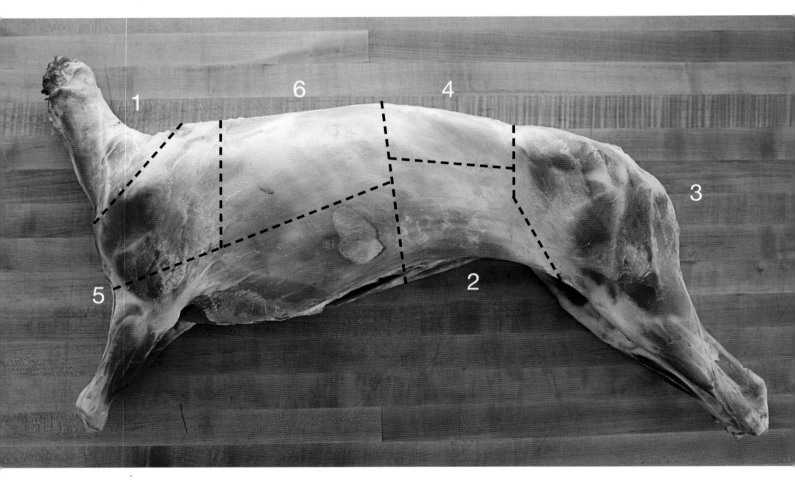

1. NECK.
Start with the neck, removing it at its base between the last cervical vertebra and first thoracic.

2. FLANK.
The boneless flank should be removed without nicking the leg muscles or cutting into the loin. Finish by sawing through the last rib.

3. LEG.
Feel for the crest of ilium, the indication of where to cut to separate the loin and leg. This point also roughly aligns with the space between the last two lumbar vertebrae.

4. LOIN.
Use the space between the last two ribs as your guide to separate the loin from the forequarter.

5. BREAST/FORESHANK.
Cut across the ribs and through the joint of the shank to separate the breast and foreshank together, with or without the attached flank.

6. RACK/SHOULDER.
Separate the rack and shoulder between the fourth and fifth ribs, leaving eight ribs on the rack.

Remove the Neck

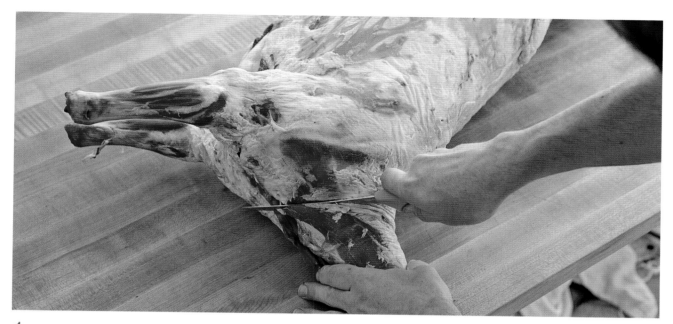

1. KNIFE AROUND THE BASE.

Cut around the base of the neck at the junction with the shoulder. Stop when you hit bone.

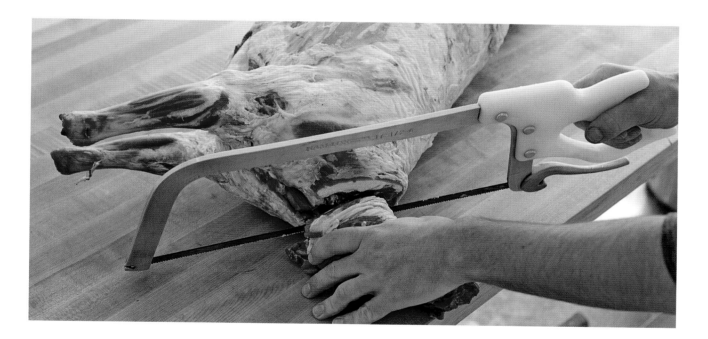

2. SAW THROUGH THE SPINE.

Using the saw, proceed through the spine until you reach meat again; then switch back to the knife and complete the cut. (Remember: saws for bone, knives for meat; cutting meat with a saw results in embedded bone shards and shredded meat surfaces.) Try to cut the neck at an angle that is roughly parallel to the anterior face, where the head was removed.

Separate the Flank

The flank consists of the abdominal walls of the animal; it is composed of muscles that are thin in profile but hefty in strength, largely endowed by copious amounts of connective tissue. The flank wraps around the upper portion of the leg, from which it can be easily separated through a natural seam. Cutting along that seam is the first step to removing the flank.

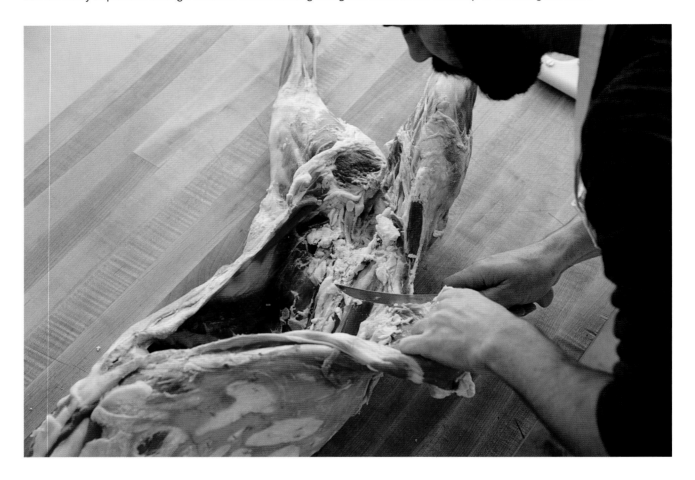

CUT ALONG THE SEAM.

To identify the seam along the leg, grab the posterior end of the flank and pull it away from the leg. You will see a separation of the thin abdominal muscles and the large sirloin and leg muscles. Follow the curvature of the leg with a boning knife, erring on the side of the flank rather than the leg. Follow this seam until you are an inch or so from the loin muscles. Turn your knife and cut parallel to the loin muscles, toward the rib cage until you hit the last rib. Saw through the last rib and leave the flank attached for removal with the breast primal. Repeat on the opposite side.

While having the breast and flank attached can be beneficial for some preparations, you can, as an alternate approach, remove the flank by itself. After sawing through the rib, follow the space between the last two ribs, cutting out and away from the spine.

Remove the Loin

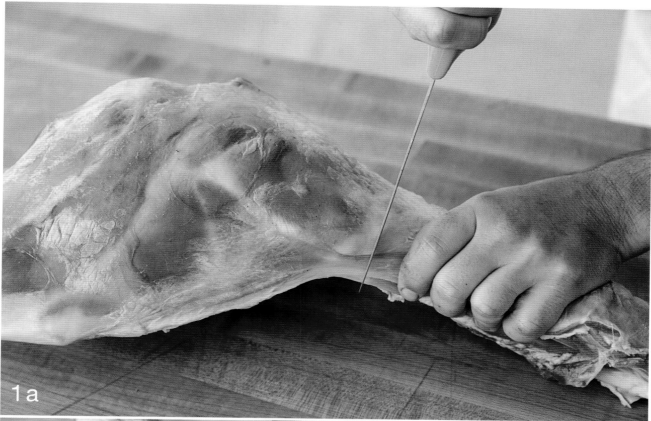

1a

1b

1. SEPARATE THE LOIN FROM THE LEG.

The loin and leg primals are split by a cut, perpendicular to the spine, at the front (anterior) edge of the pelvis. You can feel that front ridge of the pelvis, which may be visible as a slight protuberance, by squeezing along the back where the leg and loin meet. Cut adjacent to the pelvis, all the way around the loin and down to the bone. Using a saw, split the spine at your incision point to finish the separation. Use the knife to cut through any remaining muscle.

Remove the Loin CONTINUED

2

2. SEPARATE THE LOIN FROM THE FOREQUARTER.

Use the space between the last two ribs as your guide of where to split the loin and forequarter. Insert your knife between the ribs at the point where you sawed through the last rib. Cut toward the spine until you hit bone, then scrape along the bone with the tip of your knife.

Turn the carcass over and repeat on the other side. Make sure the loin muscle (*longissimus dorsi*) – the muscle that runs along the outside of the spine – is cleanly cut down to the bone on both sides. Saw through the spine where the cuts from both sides join to finish separating the loin.

Remove the Breast and Foreshank

1. SPLIT THE STERNUM.

Lay the forequarter on its spine and saw through the sternum. (Those who split the sternum during slaughter can skip this step.)

2. MEASURE THE RIB TAILS.

Cut out and along the edge of the twelfth rib until you've left around 3 inches of rib from the loin muscle. This will be the tail on your rib loin as well as the ending point for your saw line.

Remove the Breast and Foreshank CONTINUED

3. SCORE THE PATH.

Use your boning knife to trace the path that your saw will follow — a single cut through the arm and all the ribs. Start just above the upper joint of the foreshank, basically the elbow, which is the juncture between the humerus and the radius. Cut through the arm muscles to bone, and then cut across all the ribs in a straight line that connects to the end point you created in the preceding step.

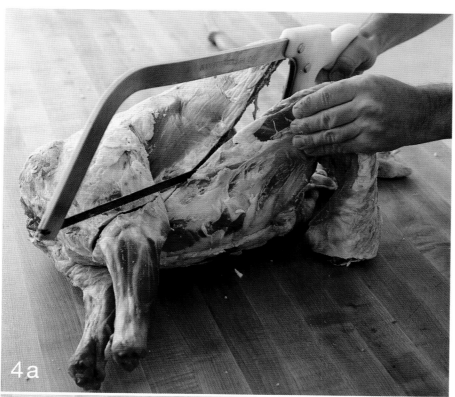

4a

4. SAW THROUGH THE RIBS.

Now follow this path with your saw, starting at the ribs. When sawing through ribs or other delicate bones, do not use force but rather allow the weight of the saw and the sharpness of the teeth to do the work. Sawing lightly will result in cleaner cuts and fewer shards. (See page 55 for more information on using bone saws with delicate bones.) Once finished, repeat with the opposite side.

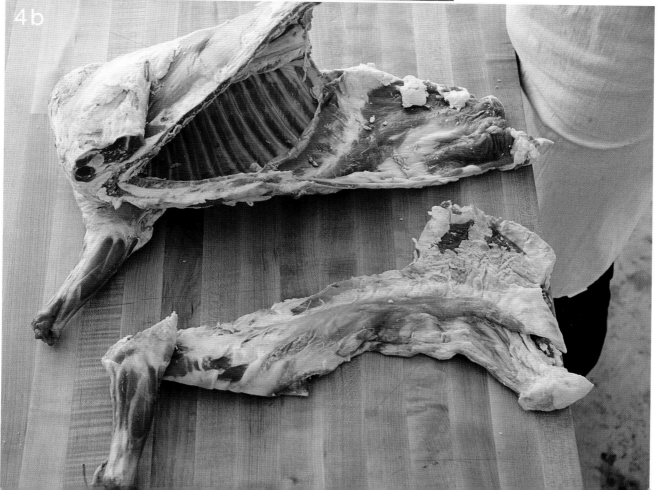

4b

Remove the Breast and Foreshank CONTINUED

5a

5. SEPARATE THE FORESHANK FROM THE BREAST.
Cut through the natural seam that is easily accessed when pulling the two apart.

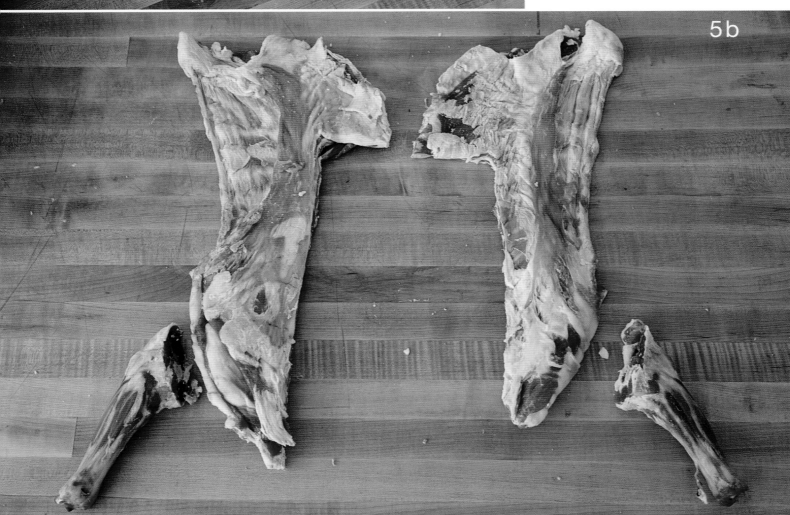

5b

Split the Shoulder from the Rack

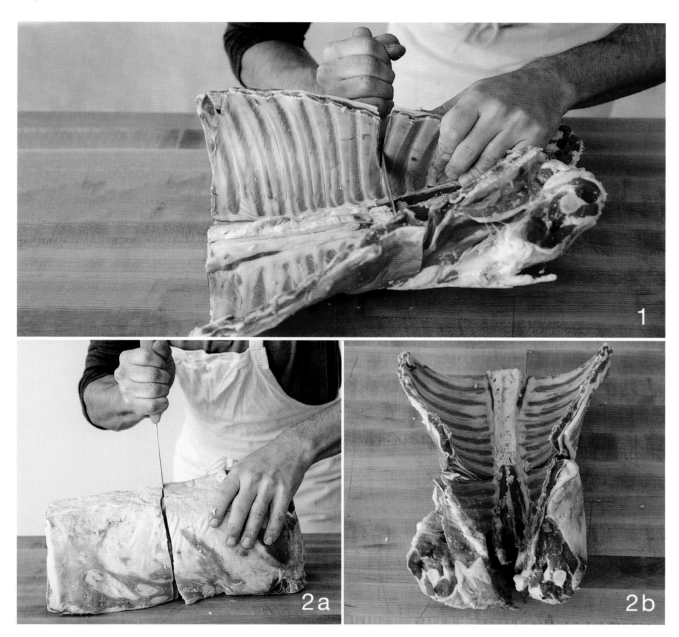

1. CUT BETWEEN THE RIBS.

Find the space between the fourth and fifth ribs; this will leave 8 ribs on your rack. Make the initial cut between the ribs, starting at the bottom (ventral) edge of the ribs and heading toward the spine. Along the way, you'll run into the scapula. On some younger animals you may be able to cut through the scapula with your knife, but on most market-ready animals the ossification is too advanced and will require a saw.

2. CUT AROUND THE BACK.

Cut over and around the scapula as you head toward the spine; then finish by cutting the loin muscle down to the bone. Repeat on the opposite side.

Split the Shoulder from the Rack CONTINUED

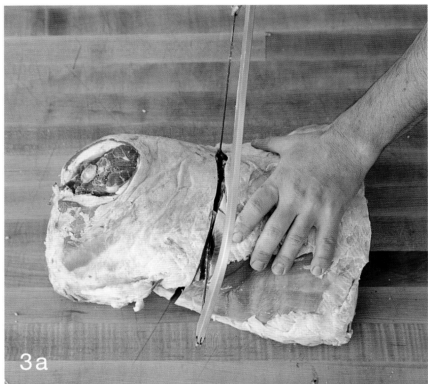

3a

3. SAW THROUGH THE SPINE AND SCAPULA.

Using your saw, connect the incisions from both sides. You now have one joined pair of square-cut shoulders and one joined pair of rack primals.

4. SPLIT THE PRIMALS.

Place the joined shoulders on its front end and saw down through the middle of the spine. Repeat with the rib primals.

3b

4a

4b

Split the Sirloin from the Leg

The sirloin, the main subprimal of the leg, can be lopped off with the bone. (Instructions for boning the sirloin while attached to the leg, creating a semi-boneless leg, are on page 302.

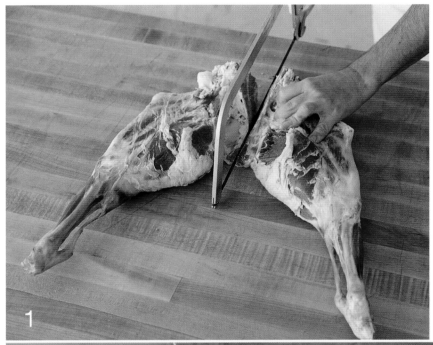

1. SPLIT THE LEGS.

First, split the two legs by sawing through the center of the spine. Notice the sacral and caudal vertebrae of the tail and where they join.

2. CUT YOUR SAW LINE.

The line you are going to saw is perpendicular to the leg bones (line it up with the visible shank bone) and sits just below the aitchbone (or the exposed edge of the pelvis, a.k.a. the pubic symphysis). Begin by cutting through the large muscle groups of the leg. Connect to the back of the spine, cutting just below the aitchbone.

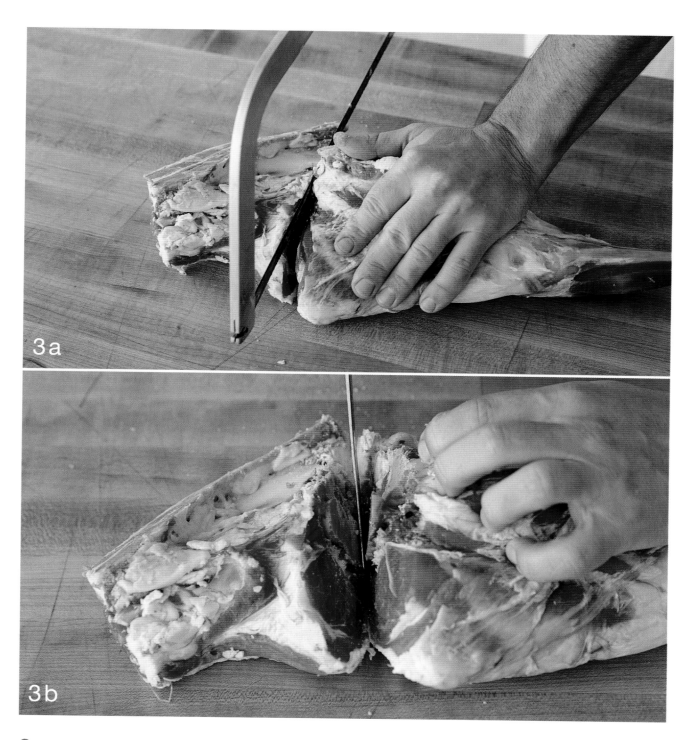

3a

3b

3. SAW, THEN CUT.

Saw through the aitchbone along the line you cut. After the bone, swap the saw for your knife and finish the separation.

THE SQUARE-CUT APPROACH is the standard for lamb breakdown, but there are some notable alternatives, mainly separating the arm from the shoulder, that may suit your culinary needs better, or just give you some options to choose from. Start with a carcass that has had the head and neck removed.

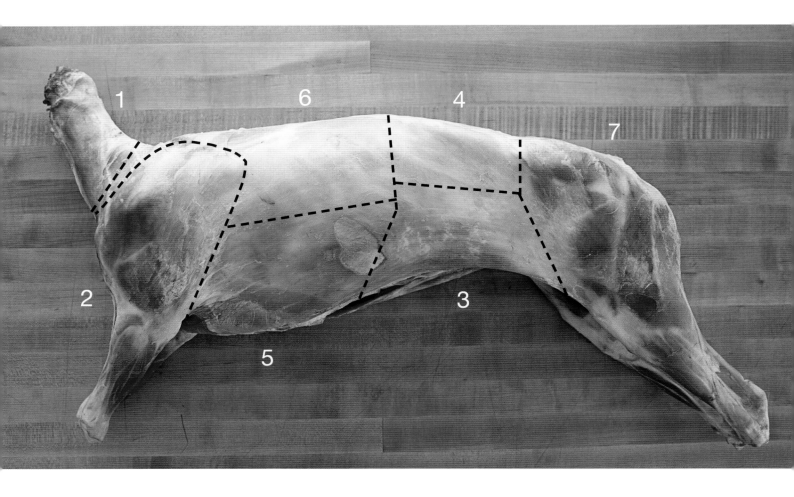

1. NECK.
Start with the neck, removing it at its base between the last cervical vertebra and first thoracic vertebra.

2. ARM.
Remove the arm using the space between the scapula and ribs, being careful not to cut into the loin muscles.

3. FLANK.
The boneless flank should be removed while avoiding nicking the leg muscles or cutting into the loin. Finish by sawing through the last rib.

4. HIND-QUARTER.
Separate the hind and forequarters, using the space between the last two ribs as your guide.

5. LEG/LOIN.
Feel for the crest of ilium, the indication of where to cut to separate the loin and leg. This point also roughly aligns with the space between the last two lumbar vertebrae.

6. BREAST AND RIBS.
Cut across the entire rib rack to remove the Denver ribs with attached breast.

7. RACK/SHOULDER.
Separate the rack and shoulder between the fourth and fifth ribs, leaving eight ribs on the rack.

Remove the Arm

Lamb has what's called a floating shoulder: its arm bones are not connected to the main body of the skeleton but rather held in place through musculature and connective tissue. This gives us an opportunity to remove the arm without needing to saw through bone. It may seem tricky at first, but it will become easier with practice.

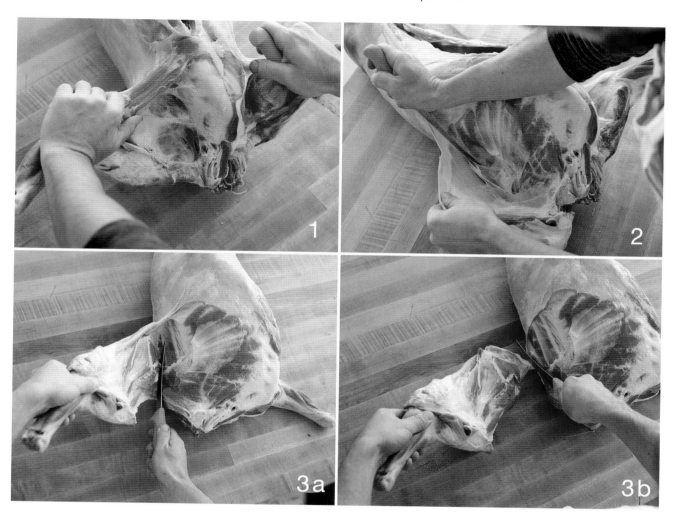

1. IDENTIFY THE SEAM.

Pull the arm away from the rib cage to expose the natural seam.

2. CUT ALONG THE SEAM.

Follow this natural seam until you reach the posterior edge of the scapula, which occurs roughly at the sixth rib. You can use your fingers to feel for the edge to help identify it: a flat cartilaginous flap that tapers as it heads toward the rear.

3. SEPARATE THE SCAPULA.

Cut between the underlying rib muscles (mainly *serratus ventralis*) and the scapula. Finish separating the arm by cutting through the thin muscle that lies atop the shoulder muscles.

Separate the Flank

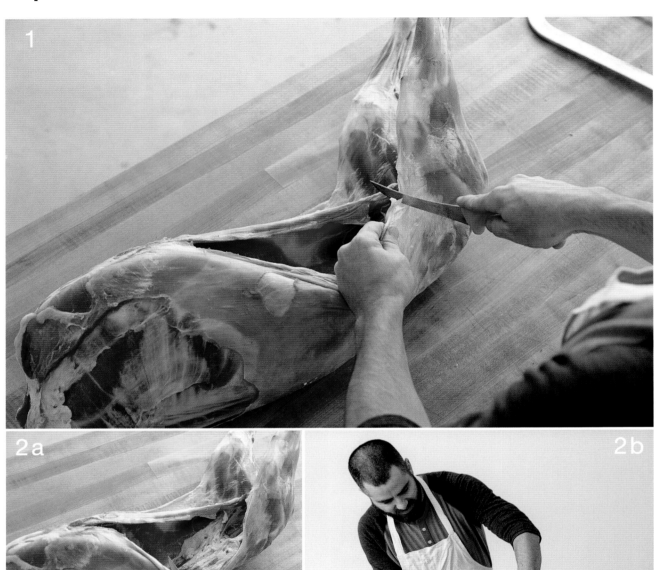

1. CUT ALONG THE SEAM.

Hold the rear edge of the flank where it joins with the leg. Make a cut at the juncture and follow the seam along the curvature of the leg muscles until you are about 2 inches from the loin muscles.

2. CUT ALONG THE RIB.

Continue cutting, staying parallel to loin muscle while maintaining the 2-inch proximity until you hit the last rib. Saw through the last rib, then repeat on the opposite side.

Split the Fore- and Hindquarters

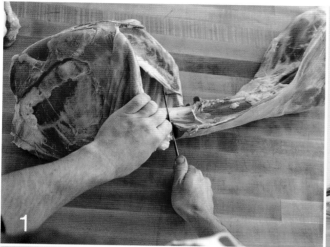

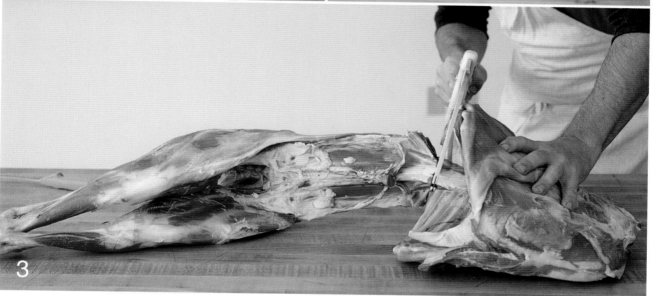

1. SEVER THE LOIN MUSCLES.

Cut between the twelfth and thirteenth (last) ribs, heading toward the spine and severing the loin muscles down to the bone. Cut across the back of the spine and connect to the opposite side between the last two ribs.

2. CONNECT BOTH SIDES.

Cut across the back of the spine and connect to the opposite side between the last two ribs.

3. SAW.

Saw through the spine following your cut line. You now have a forequarter and a hindquarter.

With the fore- and hindquarters separated, you can continue splitting the primals as instructed on pages 254 to 321. The flank and ribs can be removed together by cutting straight across the rack and shoulder primals. This version of the lamb breast includes a greater portion of the ribs than the breast from a square-cut shoulder, allowing for Denver-style ribs (see Cutting Riblets, page 262).

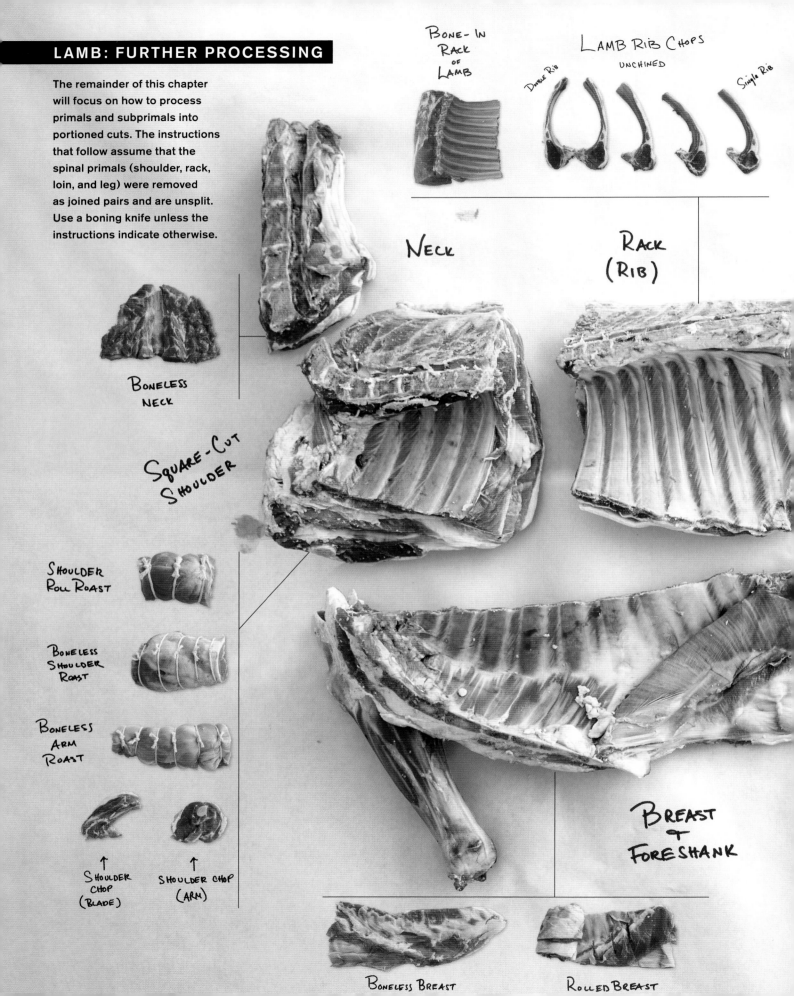

LAMB: FURTHER PROCESSING

The remainder of this chapter will focus on how to process primals and subprimals into portioned cuts. The instructions that follow assume that the spinal primals (shoulder, rack, loin, and leg) were removed as joined pairs and are unsplit. Use a boning knife unless the instructions indicate otherwise.

BONE-IN RACK OF LAMB

LAMB RIB CHOPS
UNCHINED

Double Rib

Single Rib

NECK

RACK (RIB)

BONELESS NECK

SQUARE-CUT SHOULDER

SHOULDER ROLL ROAST

BONELESS SHOULDER ROAST

BONELESS ARM ROAST

↑ SHOULDER CHOP (BLADE)

↑ SHOULDER CHOP (ARM)

BREAST & FORESHANK

BONELESS BREAST

ROLLED BREAST

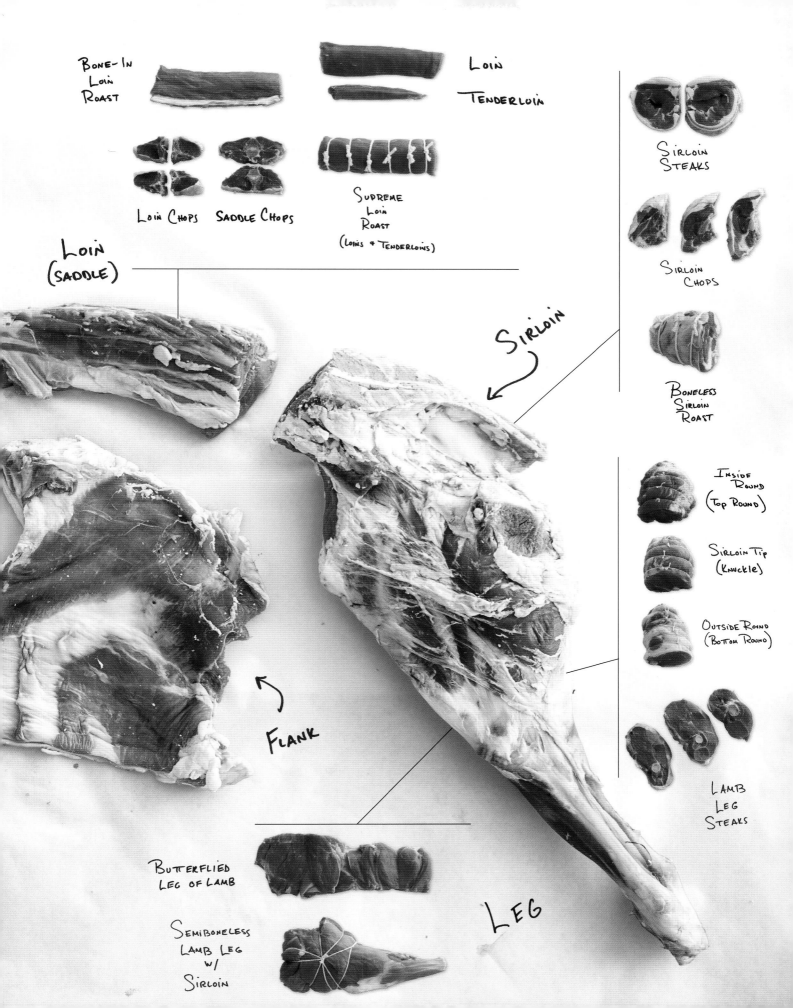

Bone-In Loin Roast

Loin

Tenderloin

Loin Chops   Saddle Chops

Supreme Loin Roast

(Loins & Tenderloins)

Loin (Saddle)

Sirloin

Sirloin Steaks

Sirloin Chops

Boneless Sirloin Roast

Inside Round (Top Round)

Sirloin Tip (Knuckle)

Outside Round (Bottom Round)

Lamb Leg Steaks

Flank

Leg

Butterflied Leg of Lamb

Semiboneless Lamb Leg w/ Sirloin

HEAD

DO NOT OVERLOOK THE HEAD as a source of delicious and nutritious cuts. While often discarded due to cultural bias, the head offers unique morsels unlike any others. I urge you to take the time to explore their value.

BRAINS

Lambs' brains, while not popular in America, are a delicious treat for anyone willing to take the time to extract them from the skull and prepare them. They provide a light, custardy texture and very creamy flavor — certainly nothing to be afraid of. See page 220 for information on how to remove the brain.

CHEEKS (MASSETER)

The cheek of any animal, but especially ruminants, is a muscle in constant use since it is responsible for the mastication of food. This means it is especially flavorful, chock-full of the connective tissue that partially enables its endurance. This makes a great case for slow-cooking methods.

Cheeks are easily removed from any skull by following the contour of the lower jaw, or mandible. Place the skull on its side, and work on the cheek that faces up. (A towel beneath the skull will help stabilize it.) Start at the bottom of the jawbone and begin cutting while applying enough pressure so that the blade stays close to the bone. Lift the muscle away from the bone as you continue to cut inward toward the top of the skull. Finish separating the cheek

when you reach the ridge of the skull just below the eye socket. Trim away any surface glands and the exterior layers of silverskin.

SUGGESTED PREPARATION: slow, moist-heat preparations like braising.

WHOLE HEAD

Lamb's head is delicious when roasted whole. Before cooking it, look over the head and trim away any glands, remnants of skin or fur, clotted blood or contaminants near the neck, or other unpalatable parts. Remove the voice box (larynx) if it wasn't already removed with the tongue. After cooking be sure to find, and eat, the muscles behind the eyes, which are constantly worked in the living animal and therefore delicious.

SUGGESTED PREPARATION: roasted whole or braised until the meat is able to be picked off the bone.

IF YOU LEFT THE HEAD on during the slaughtering process, you'll need to remove it now. Remove the head by cutting through the atlas joint, the connection point between the skull and spine. You can either saw through the spine at the base of the skull or use your knife. When using your knife, drape the animal's head over the edge of the table, which allows you to pull down on the head — a boning hook works well for that — while knifing through the joint. With your knife, cut all the way around the spine at the point where it meets the skull until you hit bone. Insert the tip of your knife into the gap of the atlas joint, and wiggle it around to sever the interior attachments. Use finesse over force, and work with the shape of the bones. Once one side is released you should be able to easily cut through any remaining attachments.

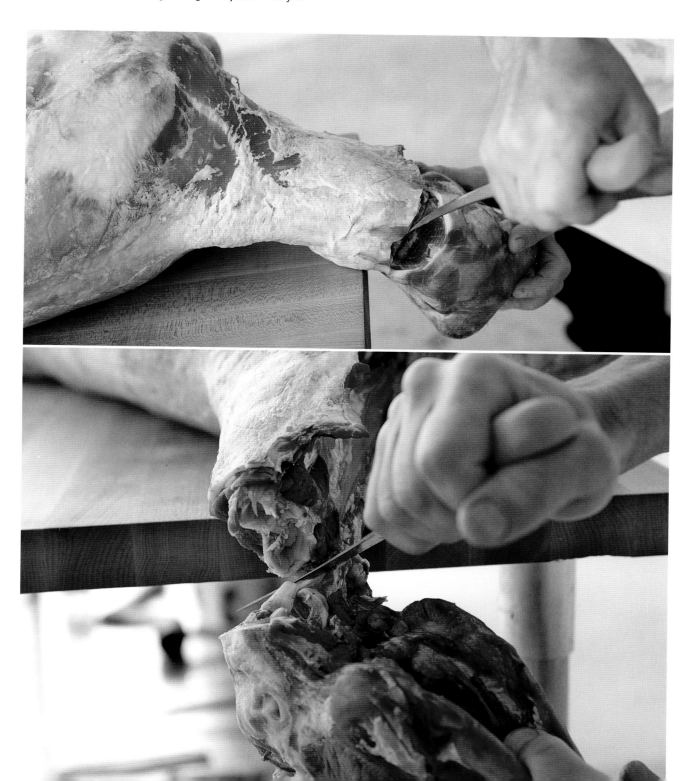

THE TOUGH MUSCLES OF THE NECK are full of flavor and copious amounts of collagen. While the recommended cooking techniques are similar (slow) for all uses of the neck, the potential preparations are vast. The neck can be left whole or further processed using the methods below.

BONELESS NECK

Lamb neck is a good primer for boning vertebrae; it's also a good way for you to become more deft with your knife-tip work. Neck bones, or cervical vertebrae, are roughly shaped like a six-pointed star, and the goal is to follow the curvature of peaks and valleys all the way around.

SUGGESTED PREPARATIONS: slow-cooking methods are required to break down the collagen; boneless neck may also be stuffed and rolled before cooking.

CROSSCUT NECK

Neck bones can be crosscut for easy portioning within stew or soups, or even added to fortify stocks. Make your initial cuts, and then follow with a saw. Clean the cuts of bone dust using a scraper or blade edge.

SUGGESTED PREPARATIONS: slow, moist-heat preparations.

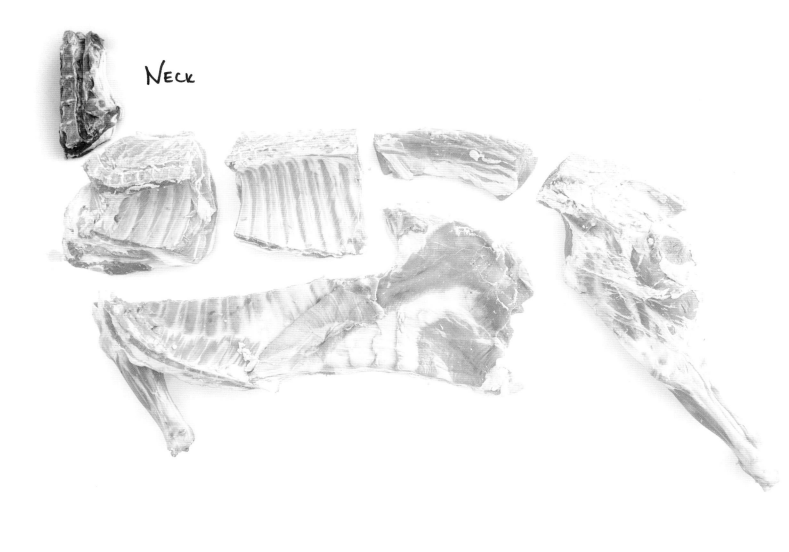

Neck

Boning the Neck

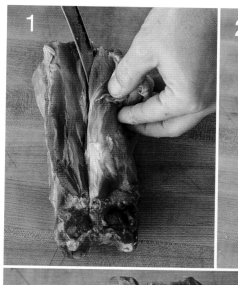

1. FOLLOW THE SHAPE OF THE BONE.

Place the neck upside (dorsal side) down with the head (anterior) away from you. Starting at the top, make a barely-off-center cut the length of the neck. Use your fingers to widen the opening, and then follow the sloping shape of the bone with the curved part of your blade, bending the blade slightly in order to stay close to the bone. (Make all your cuts long and smooth.)

2. NAVIGATE AROUND THE VERTEBRAE.

Proceed down the bone and then out as the shape changes. Slowly approach the edge of the first "point," changing direction once you pass the crest. Continue to work your way around one side and then repeat on the opposite.

3. FIND THE NECK OYSTER AND NUCHAL LIGAMENT.

Be sure to find and remove the "neck oyster," a sizable piece of solid muscle that sits next to the atlas joint at the front of the neck. Running along the top of the neck is the *nuchal ligament*, a tough band of elastin that should be removed from the boned meat prior to cooking.

THE BREAST IS A SURPRISINGLY versatile subprimal, since it includes several different muscle groups and, when prepared with an attached flank, can extend the whole length of the abdomen.

The muscles involved are mostly thin, hardworking muscles with considerable amounts of connective tissue woven throughout to help support the internal viscera. The one notable exception may be the main muscle that extends across the first several ribs, called the *serratus ventralis*. It is surprisingly tender, especially compared to the surrounding rib muscles, and is best left attached and prepared as Denver-style ribs. The foreshank, like any other shank, has the largest amount of connective tissue. (As mentioned before, areas of the carcass close to the ground will always have larger amounts of connective tissue to help provide endurance for standing and posture.) Further information about shanks and their preparations is given on page 319.

BREAST

The largest muscle group in the breast of a sheep or goat is the pectorals, which are the equivalent to beef brisket. Other muscles extend across the rib cage, including the diaphragm, the *serratus ventralis*, and the *transversus abdominis*. The bones in the breast are the sternum and the bottom (ventral) portion of the entire rib cage; the costal cartilage connects these bones. Depending on how it was removed, the flank may be attached. The most common forms of lamb breast are rolled breast (either bone-in or boned) and Denver-style ribs, which are created after the bottom portion is removed.

GENERAL CLEANING: The breast can be trimmed up before additional processing by removing the diaphragm and the thin exterior muscles (*cutaneous trunci*) that lay atop the fat. Save any clean trim for grinding. If you leave the diaphragm attached, remove the exterior membrane to make it more palatable.

FORESHANK

See the section Cutting Shanks on page 319.

Breast + Foreshank

Boning the Breast

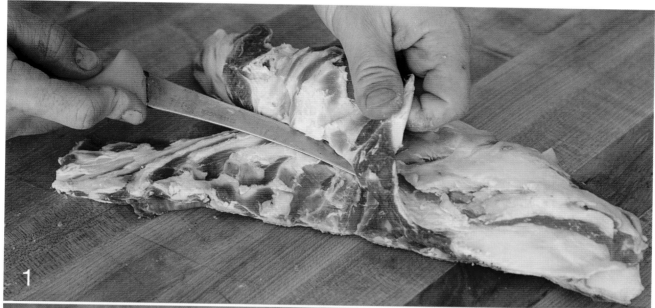

1. **CUT BETWEEN THE MEAT AND THE RIBS.**
Place the breast bone-side down. Begin removing the ribs on their exposed edge, where they were sawed through. With long, smooth motions, cut between the ribs and the meat, angling your blade edge slightly toward the bone while peeling the meat away with your opposite hand.

2. **FINISH WITH THE STERNUM.**
Work your way across the ribs toward the sternum and cartilage. After the cartilage, cut the ribs away from any surrounding muscle, and then finish with the sternum by following the curvature of the bone, leaving the calcified fat deposits with the bone.

Preparing a Bone-In Breast for Rolling

To roll a bone-in breast you will need to remove the sternum and sever the costal cartilage, allowing the ribs to lie flat. The cartilage of young animals can usually be cut with a knife, but you may need to pull out the cleaver for the more ossified cartilage in older animals.

CUTTING RIBLETS

Lamb ribs (also known as Denver-style ribs) are best when pulled from a shoulder that's had the arm removed (page 250). This way, the ribs are longer and the main muscle that makes them delicious (*serratus ventralis*) is larger. Ribs 2 to 9 provide the best meat.

Place the breast bone-side up. Mark where to remove the sternum and bottom (ventral) portion of the breast by cutting across, and between, the ribs just above (dorsal to) their connection with the costal cartilage. Follow that line with your saw to finish the separation.

Switch to your knife to finish cutting off the bottom (ventral) portion of the breast and any attached flank. Ribs can be crosscut or separated into smaller portions if desired.

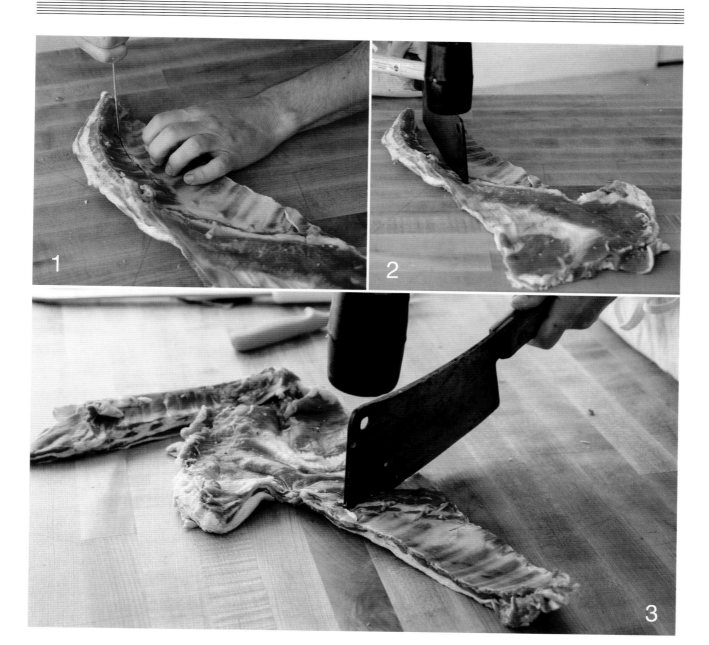

1. CUT ALONG THE STERNUM.

Place the breast bone-side down. Begin removing the sternum by making a long, shallow cut along its bottom edge, starting at the front (anterior) and working back. Use your fingers to peel the meat back, widening the cut, and continue following the shape of the sternum with your blade until you reach the ribs.

2. REMOVE THE STERNUM.

Finish removing the sternum by cutting through the cartilage that connects the ribs to the sternum, using your knife or a cleaver if necessary.

3. SPLIT THE RIBS.

Separate the remaining ribs (usually ribs 7 to 13) by cutting between the bottom of each rib, through the costal cartilage that connects them, using a knife or cleaver. Cutting through the underlying meat as well is perfectly acceptable. Once all the ribs are separated, the breast can be used for rolled preparations.

SQUARE-CUT SHOULDER

THE SHOULDERS CONTAIN the main muscles responsible for locomotion; they are working muscles, with intense flavor and generous deposits of connective tissue. Despite this, many of the most tender muscles in the body are located within the shoulder. In larger animals, like beef, these muscles are often separated and treated as individual cuts. In sheep and goats (as well as pigs) these muscles are often too small for such extraction, so they stay nestled within. No matter: we still benefit from them in other ways. The shoulder contains the last length of the loin muscles, which can be prepared as a shoulder roll. The deep flavor of the shoulder makes it the best candidate for any stew meat, and when prepared whole, bone-in or boneless, the shoulder is an ideal option as a slow-roast dish for a family.

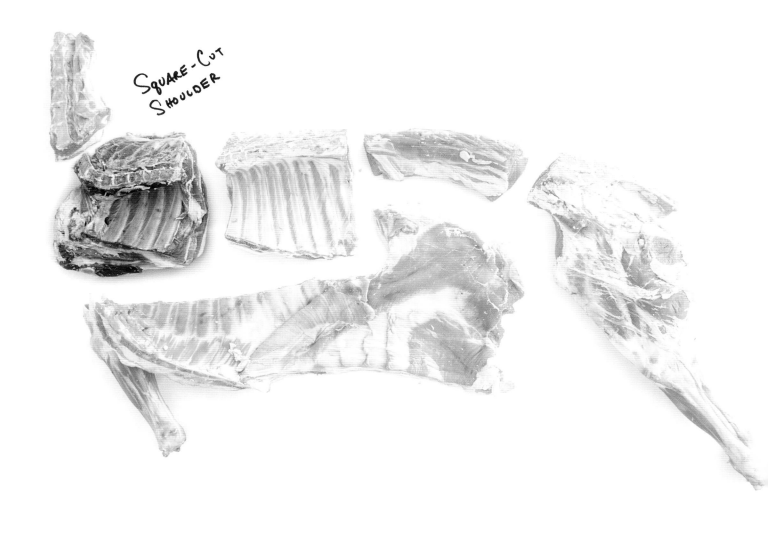

Square-Cut Shoulder

Square-Cut Option 1: Chops (Arm, Blade)

The square-cut shoulder lends itself to easy processing into chops from both sides, the arm and the blade. One reason why the breast and foreshank are separated above the humerus-radius joint is so that the resultant face includes a nice cross-section of the round arm bone.

1. SPLIT THE SHOULDERS.
Before processing chops, split the pair of square-cut shoulders by sawing down the center of the spine, starting inside the cavity.

ARM CHOPS

The muscles in arm chops are moderately tender and benefit from a sear-to-oven method of cooking (searing on the stove top and transferring, in the pan, to a 300°F oven to finish cooking in an even, dry-heat environment). Thus, cut them at least 1 inch thick.

1. MAKE THE INITIAL CUT.
Place the square-cut shoulder rib-side down, and start at the bottom (ventral edge) of the arm. Measure out the thickness of the chop and use your knife to make the initial cut down to the arm bone, staying parallel to the bottom face of the arm.

Square-Cut Option 1: Chops (Arm, Blade) CONTINUED

ARM CHOPS CONTINUED

2. SAW THROUGH THE BONE.

Pull out your saw and make your way through the arm bone, switching back to the knife until you reach the ribs, and then back to saw to complete the cut. Repeat up the arm until you reach the scapular joint. Rib bones are, more often than not, removed from each chop.

BLADE CHOPS

The remaining portion of the shoulder can be cut into blade chops, starting at the rack (posterior) end. The best blade chops are the first three, while the rest, being closer to the neck, have more densely packed collagen and tougher muscle groups. You may consider cutting three or four chops and then boning the remaining portion of the shoulder for grinding or stew meat. With home fabrication it's easiest to cut chops between the ribs.

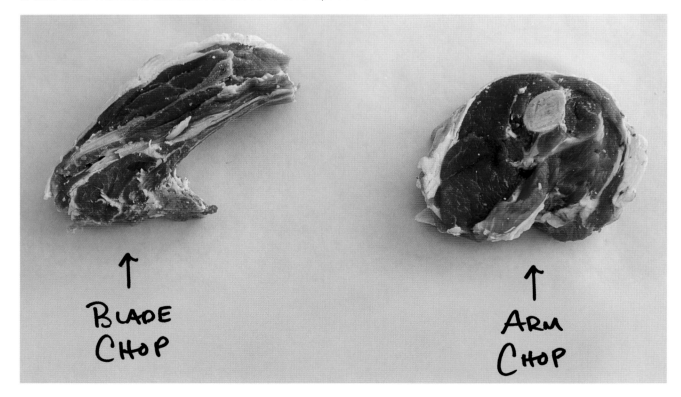

BLADE CHOP

ARM CHOP

1. MAKE THE INITIAL CUT.

Place the square-cut shoulder bone-side up. Starting at the rack (posterior) end, cut between the first two ribs; the tip of your knife should scrape against the scapula.

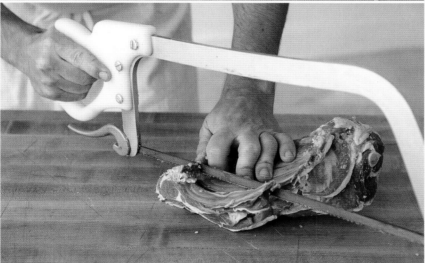

2. SAW THROUGH THE SPINE.

Separate the spine and finger bones between the ribs, stopping before you hit the loin muscle.

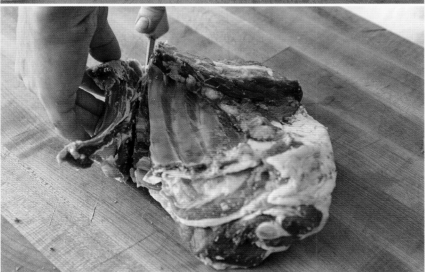

3. CUT THROUGH MUSCLE, SAW THROUGH BONE.

Pull out your knife and cut through the loin muscle, saw through the scapula, then finish the cut with your knife.

Square-Cut Option 2: Boneless

Boneless lamb shoulder either can make for one large roast or can be split along a natural seam to provide two roasts, the arm and shoulder roll. For two roasts, split the shoulder along the natural seam, starting between the base of the arm and ribs, and then follow the previous directions for processing the arm and shoulder roll separately. The shoulder also happens to provide the best stew meat in the entire carcass, another worthy reason to bone it out.

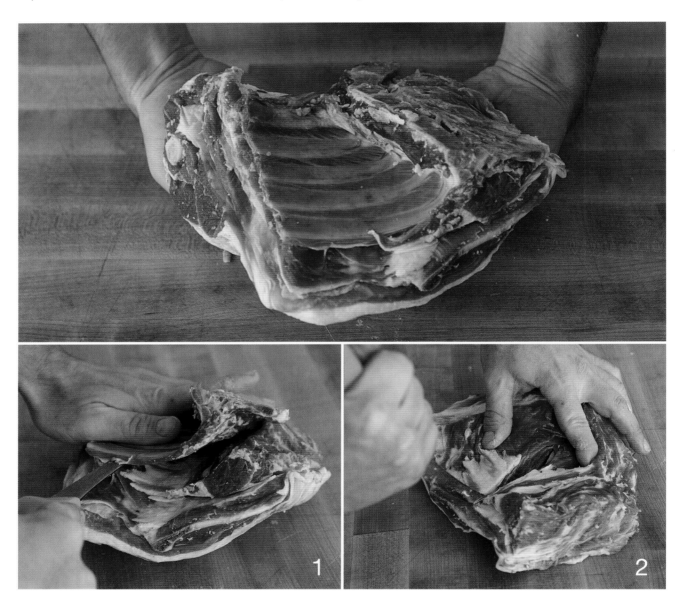

1. REMOVE THE SPINE AND RIBS.

(Follow the directions on page 275 for boning a lamb shoulder roll, starting at the feather bones and working back along the ribs.)

2. CUT DOWN THE CENTER OF THE SCAPULA.

Place the shoulder fat-side down, and use your hands to feel for the boundaries of the scapula, wiggling the arm back and forth to help identify the scapular joint. Make a cut down the center of the scapula using enough force to scrape along the bone, starting near the joint and drawing the knife toward you.

3a

3b

4

3. PEEL THE MEAT AND EXPOSE THE SCAPULA.

Work from this center point to peel the meat away from both sides of the bone, keeping the flexible blade close to the bone.

4. REMOVE THE SCAPULA AND HUMERUS.

Once the flat of the bone is exposed, remove the scapula and humerus following the instructions for a boneless arm on page 272.

Square-Cut Option 2: Boneless CONTINUED

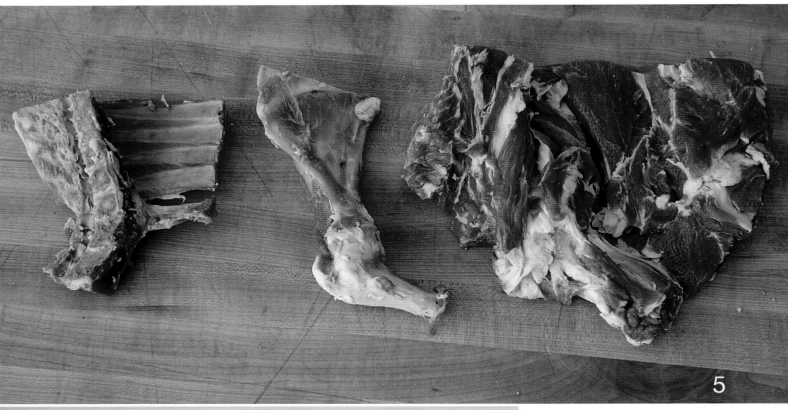

5

BONELESS
SHOULDER
ROAST

6

5. TRIM THE MEAT.

After boning the shoulder remove the thick band of yellow elastin from the top of the shoulder. Then make a pass through the interior and remove pockets of fat, connective tissue, and any blood clots or glands. Stew meat is best when it's mostly lean, so trim away any excessive fat while cubing the meat.

6. ROLL AND TIE THE MEAT.

When tying the whole shoulder into a roast, start at the loin end and roll the meat toward the shank.

SHOULDER ROLL AND ARM

AFTER THE ARM IS REMOVED from the shoulder, what is left is the shoulder roll — the lamb equivalent to the beef chuck roll. There are two main muscle groups: the shoulder eye muscles (a continuation of the loin muscles) and the rib muscles. The two groups can be boned together and rolled into a roast or separated and cooked according to their individual virtues. You can bone the shoulder roll from an attached pair of shoulder primals, as shown here, or just the same from a single shoulder.

The main muscle of the shoulder eye is the *longissimus dorsi*. The shoulder section of this muscle group is more varied in texture because of the smaller size of the *longissimus dorsi* and the more prominent representation of other muscles — such as the *spinalis dorsi* and *complexus*. This makes for a wonderfully diverse eating experience.

The main muscle that spans the shoulder ribs is the *serratus ventralis*. In beef, this muscle is the main feature of all cuts of short ribs. While smaller on lamb, the muscle offers the same benefits: deep flavor and a toothsome texture. It can be left on the bone or removed and prepared as boneless stew meat.

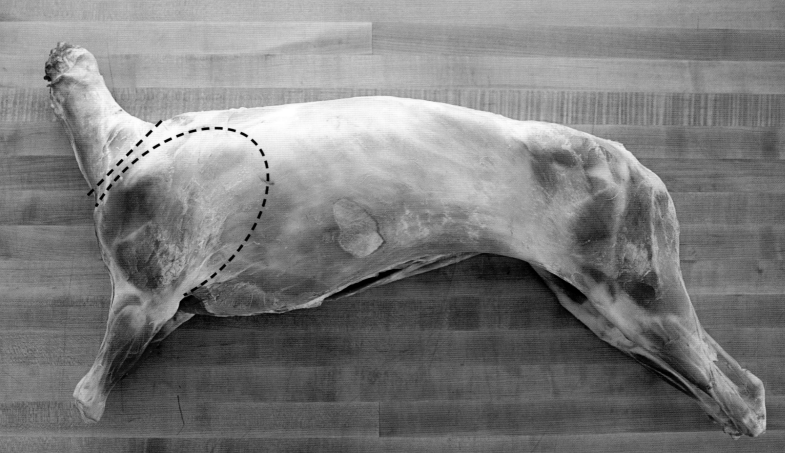

Boning the Arm

The lamb arm is removed, as illustrated on page 250, by taking advantage of the floating-shoulder condition that is common among quadrupeds. The arm can be cooked whole but is more commonly prepared by separating the foreshank and boning the remainder. The foreshank can be removed using either a saw or a knife. I like to use the knife because it forces me to better understand the joint structure.

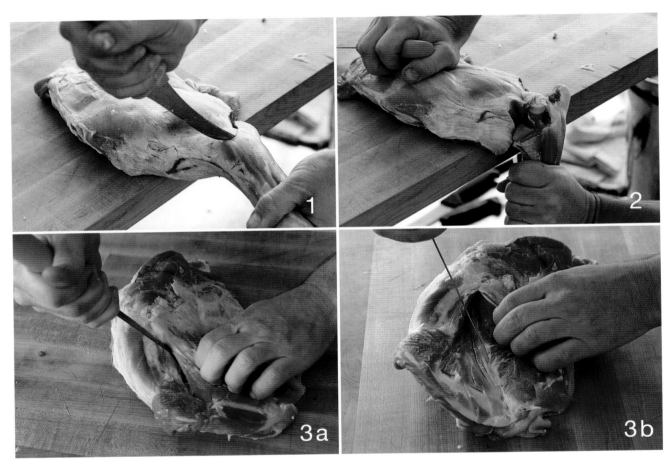

1. SEVER THE FORESHANK.

Place the arm fat-side up, with the rear (posterior) edge facing you. Hang the foreshank over the edge of the table, and bend it back and forth to get a feeling for where the joint is. The ulna is fused to the radius but has a protrusion on the back of the elbow. (Note the skeletal diagram on page 201.) Cut above the protrusion and then down toward the joint and across it, cutting down to the bone.

2. SEPARATE THE FORESHANK FROM THE ARM.

Apply downward force on the foreshank while using the tip of your knife to sever the strong ligaments that lie atop the joint and hold it together. Once severed, the joint should open up, allowing you to finish the separation with easy cutting. The foreshank can be processed further with the instructions on page 319.

3. PEEL BACK THE MUSCLE.

The remaining arm has two bones, the scapula and humerus. Lay it fat-side down, with the shank (ventral) end facing away. The muscle lying atop the scapula needs to be peeled away while being kept attached to the other muscles. Make a cut along the exposed edge of the scapula, and then cut underneath the muscle to the other side, keeping the blade close to the bone.

Boneless lamb arm can be rolled into a roast, with an optional stuffing, and slow-cooked with either moist- or dry-heat methods. It also makes excellent stew meat when cut into bite-size cubes.

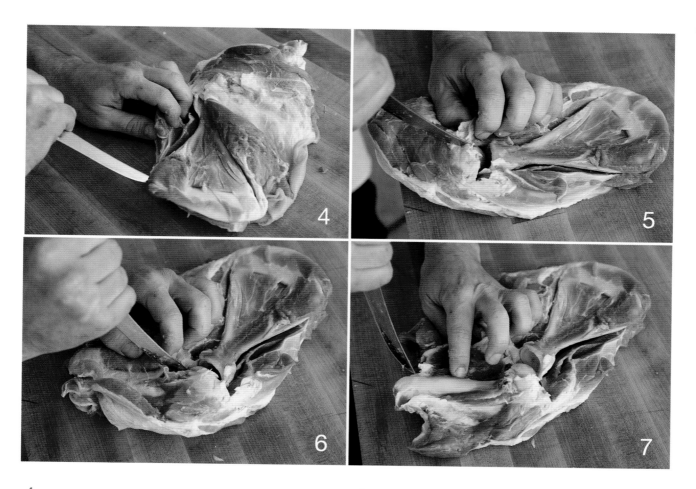

4. CUT ALONG THE SCAPULA.

Slide the tip of your knife under the edge of the scapula and scrape it along the bone, starting near the joint and pulling toward you. Repeat on the opposite edge.

5. CUT THROUGH THE JOINT.

Find the joint between the scapula and the humerus, and cut through it with your knife.

6. CUT ALONG THE SEAM.

Find a seam between the arm muscles that surround the humerus. Using just the tip of your knife, follow that seam down to the bone.

7. UNCOVER THE HUMERUS.

Continue to peel the muscles away until you can see the whole bone, using long strokes, starting at the top of the bone and drawing the knife toward you.

Boning the Arm CONTINUED

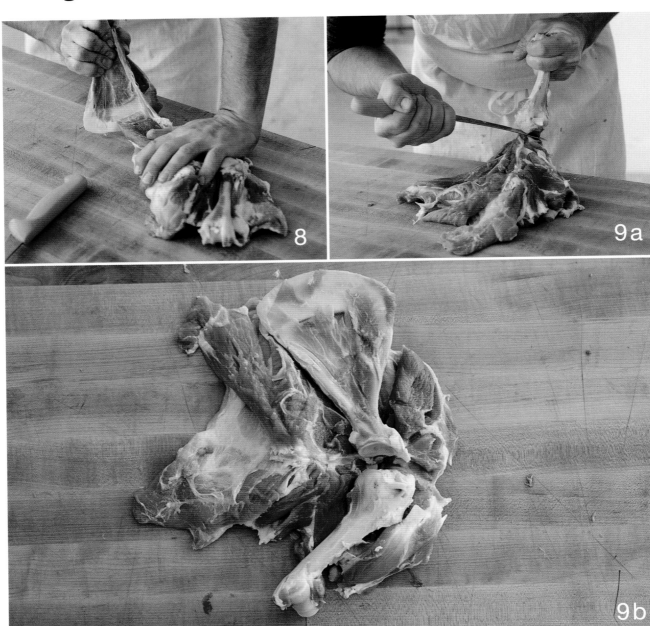

8. PULL UP THE SCAPULA.

Cut around the joint end of the scapula enough to get a boning hook or your fingers underneath it. With one hand, hold the meat against the table while pulling up on the bone with the other, peeling it away from the underlying muscles. Use your knife to help sever any stubborn attachments.

9. REMOVE THE HUMERUS.

Cut around the joint end that was attached to the shank. Hold the joint end of the humerus, and pull up on the bone while using your knife tip to cut away the underside muscles, following the curvature of the round bone shaft. Finish removal by cutting around the bottom joint end that used to be attached to the scapula. Trim away any excessive connective tissue, thick tendon remnants, or other unpalatable bits like exterior fell or bone shards.

Boning the Shoulder Roll

1. CUT ALONG THE FEATHER BONES.

Place the shoulder roll bone-side down. Find the feather bones, the middle wing of the thoracic vertebrae that sticks straight up. Make an initial, shallow cut on either side of the feather bones. Trace over the initial incisions with a deep cut down to where the vertebrae join with the ribs, all the while keeping your blade edge against the feather bones.

2. NAVIGATE OVER THE VERTEBRAE.

There is an outcropping at the base of the vertebrae (*transverse process*). Cut down to it and then carefully cut over and around it, staying close to the bone.

3. PEEL THE RIBS.

After navigating the transverse process, begin peeling the meat away from the ribs. Hold the meat with one hand while closely following the flat of the ribs with your knife. Continue to peel back the muscles with one hand while cutting with the other, keeping the blade edge slightly angled toward the rib bones until you reach the vertebrae and join with the initial cuts you made, thus releasing the shoulder roll muscles.

4. TRIM THE CONNECTIVE TISSUE.

Trim away the thick band of yellow elastin at the top of the shoulder and any other heavy deposits of connective tissue.

5. SEPARATE OR ROLL AND TIE.

The boneless shoulder roll can be rolled and tied as is, or the shoulder eye and rib muscle groups can be separated through natural seams. Find the seam near where the ribs and vertebrae were connected, and separate the two muscle groups, staying parallel with the shoulder eye muscles.

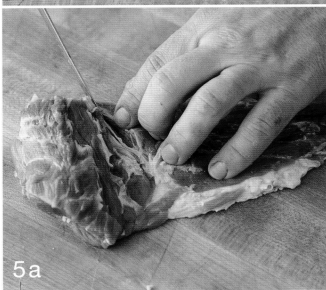

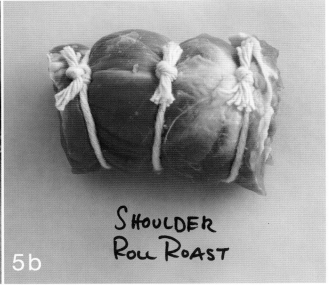

SHOULDER ROLL ROAST

CUTTING RIBS FROM A SHOULDER ROLL

Rather than bone the shoulder roll, an alternate preparation is to cut ribs. Place the shoulder bone-side up on the table and saw through the ribs just below (ventral to) where they begin curving to connect with the vertebrae. Once you're through the bone, use your knife to finish cutting. Ribs can be left as a group or cooked individually.

RACK

THE RACK IS THE SO-CALLED money meat of the animal, representing the most popular high-priced cuts. The traditional rack of lamb is by far the most popular form, though you can also cut individual chops or bone the rack for a boneless rib roast. The main loin muscle (*longissimus dorsi*) is present in this primal, as are other notable muscles like the *spinalis dorsi*.

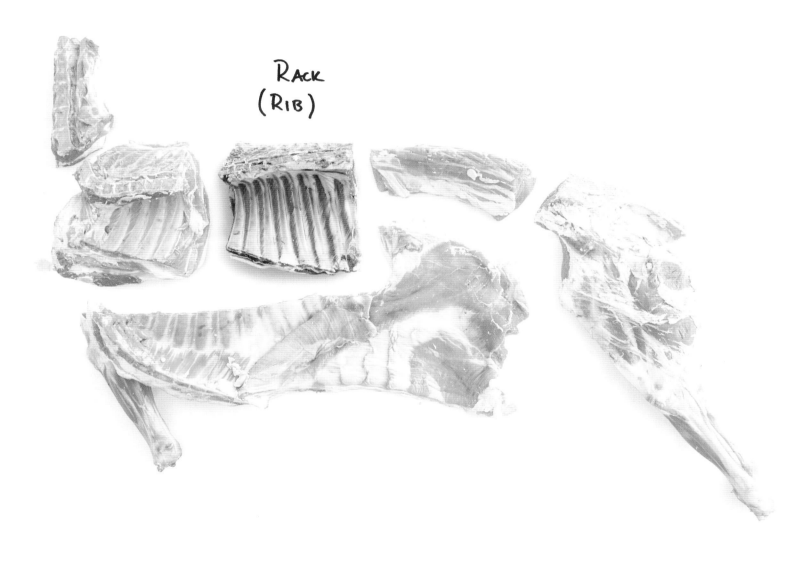

Rack
(Rib)

Rack Option 1: Boneless Rib Roast

Working with loin muscles requires a delicate approach with both knife and saw because those muscles are much leaner and finer grained, and therefore easily damaged, than those found in the shoulder or leg. I find it helpful to have the teeth of my saw blade facing toward me when working on lamb, but especially with the rack. (Refer to page 55 for more on saw techniques.)

1. SPLIT THE RACK.

Split the pair of rack primals if you plan on boning only one side; otherwise feel free to leave them connected. To split, place the pair on either end and saw through the center, being careful not to let the saw wander. (Scoring a line down the center of the rack can help keep you inline.)

2. PEEL THE FAT CAP.

Enveloping the loin muscles is a cap of fat and muscle that can be peeled away by hand. Start on the shoulder end and press your thumb into the natural seam between the fat cap and the loin muscle. Use one hand to keep the loin muscle secured, to avoid any tearing, while peeling away the fat cap, using your knife to sever any stubborn connections.

3. REMOVE THE SCAPULA.

There is a small muscle (*rhomboideus*) that sits atop the main loin muscle at the shoulder end of the ribs. Peel that away from the loin, starting at the feather bones, and continue with that seam to remove the scapula and attached muscles.

4. CUT ALONG THE FEATHER BONES.

Make the first cut along the flat of the feather bones, cutting while keeping the blade flat against and slightly angled toward the bone. Stop when you reach where the ribs and vertebrae join.

Rack Option 1: Boneless Rib Roast CONTINUED

5. **PEEL THE MEAT.**
Turn the rack around and make your second cut along the ribs, 2 or 3 inches above the loin muscles. Using long strokes, peel the meat off the ribs by cutting along them toward the spine, pulling the meat back with the opposite hand while keeping the blade flat against and slightly angled toward the bones.

6. **CONNECT THE CUTS.**
There is a small ridge of bone (called the transverse process) where the ribs and vertebrae connect; carefully navigate the blade over the ridge and back toward the bone, connecting the two cuts.

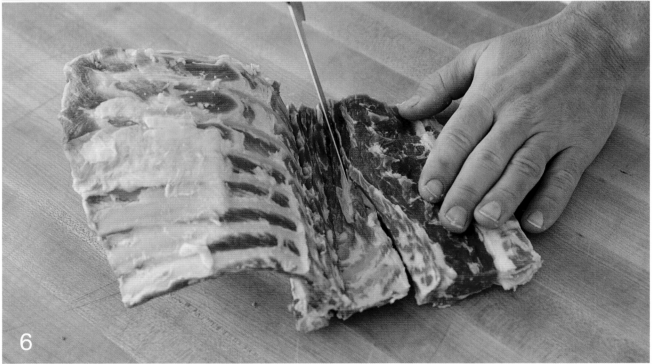

NOTE: This boneless rack can be tied as-is, with the rib fat wrapping around the underside of the loin muscle. It can also be further trimmed of all exterior connective tissue. To achieve the cleanest appearance, trim with the grain direction, which on the main loin muscle (*longissimus dorsi*) goes from the shoulder end toward the loin. (It is the opposite direction on the underside of the muscle that was previously attached to bone.)

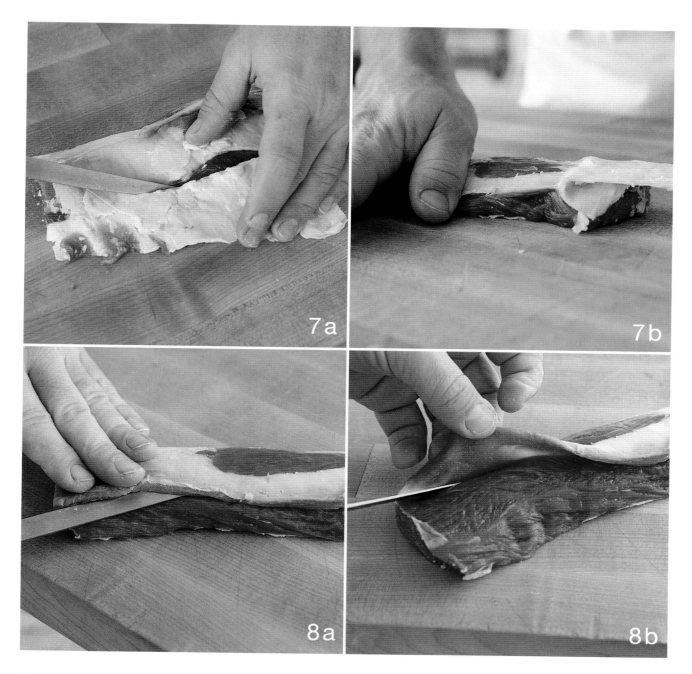

7a

7b

8a

8b

7. REMOVE THE FAT.

Trim away the fat that was formerly attached to the ribs at the seam above the loin muscles. Then peel away the thin layer of fat that sits atop the loin, securing the loin muscle to the table with your opposite hand.

8. TRIM THE SILVERSKIN.

The silverskin sits on top of the main loin muscle, but runs beneath the rib cap (*spinalis dorsi*); you'll want to remove the visible portion while leaving the rib cap on. Starting at the shoulder end, cut beneath the outer edge of the silverskin. Continue cutting underneath the silverskin, across the loin, avoiding the rib cap.

Rack Option 1: Boneless Rib Roast CONTINUED

9. CUT ALONG THE RIB CAP. The silverskin should only be attached
at the edge of the rib cap. Pull the silverskin up and cut along that edge, following the shape of the rib cap muscle, to finish silverskin removal.

NOTE: The loin muscle can also be barded (wrapped in fat). Trim the previously removed fat cap to a thin layer and wrap around the loin muscle, securing with twine at 1-inch increments.

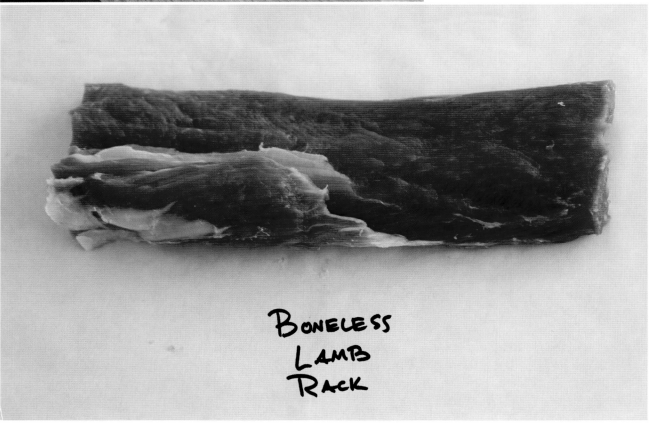

BONELESS
LAMB
RACK

Rack Option 2: Roast and Chops

The rack of lamb may very well be the most esteemed cut from the carcass. When presentation is a priority, it's often frenched, meaning that the rib bones are stripped clean. (I prefer eating that meat off the bone.) This portion-ready rack is trimmed to the loin muscles and the chine bone is removed, allowing you to cut bone-in portions before or after roasting it whole. Having already split the joined rib primals, this example works with only one side; you can do this same process with them joined if you wish.

PORTION-READY LAMB RACK

1. CUT ALONG THE FEATHER BONES.

Place the rack bone-side down. Cut along the feather bones, keeping your knife flat against the bone, until you reach the junction with the ribs.

2. NAVIGATE THE RIDGE.

Use your knife to navigate up and over the ridge, being careful not to cut into the loin muscle. Stop just after reaching the ribs.

3. HOLD THE MEAT AWAY FROM THE SAW.

Set the rack on end (either the loin end or the shoulder end). Use one hand to pull the loin muscle away from the bone to avoid damaging it while sawing.

Rack Option 2: Roast and Chops CONTINUED

4a

4b

4. NOTE WHERE THE RIBS AND VERTEBRAE JOIN.

Saw through the ribs, one by one, just below (ventral to) that juncture, until you are through all eight ribs. Saw slowly to avoid splintering the ends of the ribs. The result is called a chined lamb rack.

5. **PEEL AWAY THE COVER.**

Remove the layer of fat and thin muscles from the ribs, including the scapula and the small attached muscles that continue on top of the loin muscles. The entire grouping can be peeled away with minor knife work. Follow that by removing the yellow band of elastin that sits at the base of the loin muscles.

6. **TRIM THE CONNECTIVE TISSUE.**

There's a thin layer of connective tissue that runs from the ribs across the loin. It *can* be left on, but the whole thing is more delicious with it removed. Carefully get underneath this translucent layer with your knife, blade facing away from the ribs. Cut out, releasing enough to get a good grip, then follow the far edge of the tissue along the loin. Peel away and finish removing along the edge of the ribs.

Rack Option 2: Roast and Chops CONTINUED

VARIATION: FRENCHING THE PORTION-READY LAMB RACK

1. SCORE THE TOP SIDE.
Decide on the length of your rib tails (the average is around 3 inches from the loin eye) and trim the bones if necessary. Peel away the fat cap (page 279) and place bone-side down. Estimate around 1 to 2 inches from the loin eye and make a corresponding mark on the edges of the ribs at both ends of the rack. Connect the two marks with a straight cut, down to the bone, and when passing the gaps between the bones insert the tip of your knife through to the other side.

2. SCORE THE UNDERSIDE.
Flip the rack over and notice the hashed line from the previous cuts. Connect the hashes, making sure to score the bones along the way.

3. REMOVE THE INTERCOSTAL MEAT.

The meat between the ribs is known as intercostal meat. Remove the meat between each rib by cutting down one side, across your score line, and back up. Really cut against the bones as you go down and up.

4. SCRAPE AWAY THE MEMBRANE.

Scrape the underside membrane, starting at the score line and working your way toward the cut end of the rib. Make sure to scrape along the sides as you peel the membrane as a whole.

Rack Option 2: Roast and Chops CONTINUED

VARIATION: FRENCHING THE PORTION-READY LAMB RACK

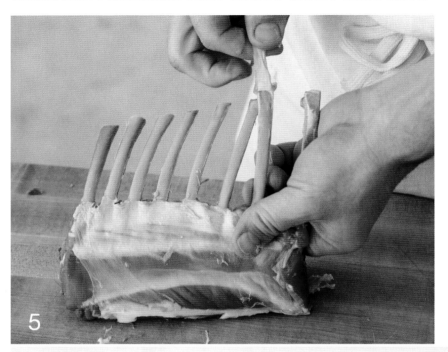

5. **PEEL THE MEMBRANE.**
Flip the rack over and scrape the membrane of the first rib up from the score mark. Once the membrane, front and back, is loosened you can peel it off the rib. It helps to use a towel, which grips the membrane as you pull.

6. **REMOVE THE CHINE AND TRIM.**
Finish processing your frenched rack by following the preceding instructions for a portion-ready roast (page 283).

FRENCHED
RACK OF
LAMB

Rib Chops

Chops are the portioned cut from a rack of lamb. As you cut them, you will notice a distinct difference in the look of chops from the loin end and shoulder end of the rack; the loin end is almost a solid single muscle, while the shoulder-end chops show varied muscle groups with more intramuscular fat deposits. This is due to the tapering of that main loin muscle, *longissimus dorsi*, as it goes from the pelvis (its widest end) to the shoulder (its thinnest end). If you haven't already shortened the ribs to your target tail length — a typical measurement is around 3 inches — do so as the first step of processing. For frenched chops, follow the preceding directions for frenching (see page 286) prior to cutting.

Chined rib chops

BONE-IN RIB CHOPS (CHINED)

With the chine bone removed, as with the portion-ready roast, chops are simple to cut. Decide how many ribs per chop (typically one or two). Place the rack meat-side down, and use a breaking or butcher knife to cut cleanly between the ribs.

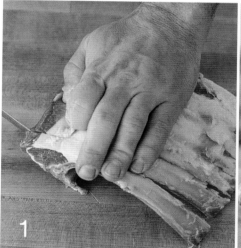
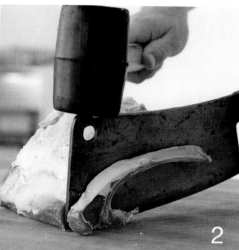

BONE-IN RIB CHOPS (CHINE-IN).

The presentation of chined chops is wonderful, but I actually prefer a more rustic chop that leaves the chine bone in place and has more rib length and fat coverage. Cutting these is easy, too: there's just one extra step. (Plus, it's a good excuse to use a cleaver.)

1. CUT THE LOIN MUSCLE.

Decide if your chops will have one or two ribs. Cut between the ribs, through the loin muscles, and down to the chine and feather bones.

2. CLEAVE THE CHINE.

Place your cleaver on the chine bone between the cut ribs. Hit the spine of the cleaver with a mallet to separate wthe chop. Repeat with the remaining ribs, starting at step 1.

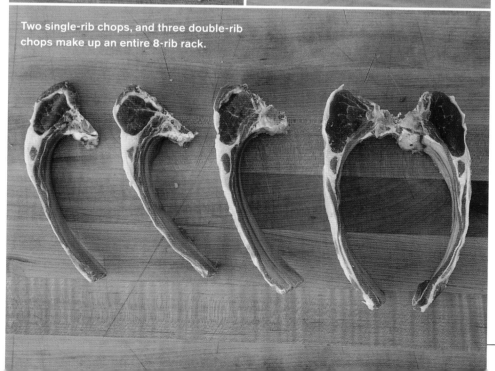
Two single-rib chops, and three double-rib chops make up an entire 8-rib rack.

THE LOIN, ALSO SOMETIMES referred to as the saddle, is the other middle-meat primal, after the rack. It lacks the presentation of the rack, and its bones are a bit more challenging for a diner to navigate, so it hasn't received the same notoriety. Nonetheless, the loin offers its own benefits — the widest section of the main loin muscle (*longissimus dorsi*), which originates at the pelvis and tapers as it heads toward the shoulder. It also contains the tenderloin (*psoas major* and *psoas minor*), which sits on the inside of the spine.

Loin

Option 1: Boneless, Including Tenderloin

Boning the loin primal will leave you with two separate muscle groups: the main loin group (*longissimus dorsi* and *multifidus dorsi*) and the tenderloin (*psoas major* and *psoas minor*). Start with a whole (unsplit) loin.

1. REMOVE THE LAST RIB.

Cut along the inside of the last rib down to the spine. Pry it away using your hand.

Option 1: Boneless, Including Tenderloin CONTINUED

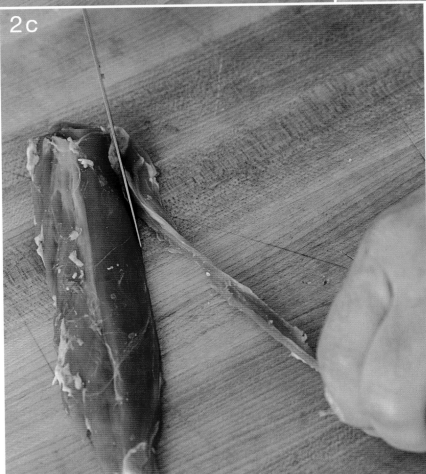

2. SEPARATE THE TENDERLOIN.

Separate the tenderloin (the smaller of the two loin muscles groups) by first cutting along the flat of the finger bones. Then, navigate over the bottom edge of the vertebrae. The tenderloin can be further trimmed of the small side muscles and silverskin.

3a

3. RELEASE THE MUSCLE FROM THE FINGER BONES.

Flip the primal around and start separating the main loin muscles on the opposite side of the feather bones. Pull the loin muscle away from the bones as you cut along the flat face, keeping your knife close to the bone. Once you've hit the base of the finger bones, navigate over the ridge of vertebrae. Trim any excess flank and exterior fat until only the loin eye is left.

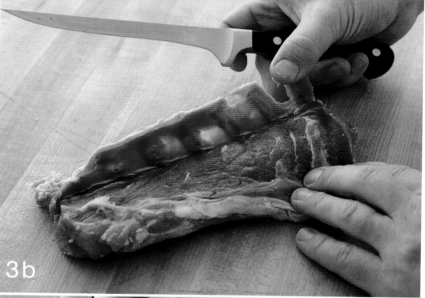

3b

3c

Option 1: Boneless, Including Tenderloin CONTINUED

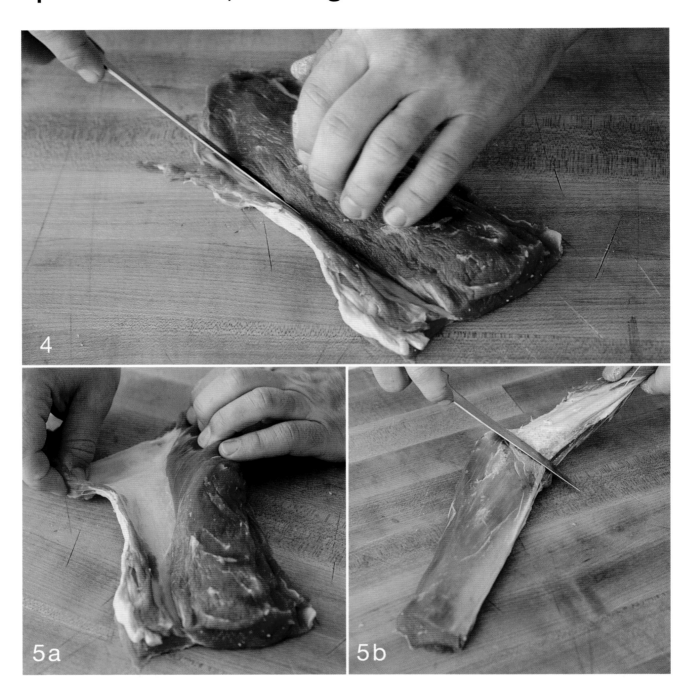

4 | 5a | 5b

4. TRIM THE LOIN EYE.

Trim the underside of the loin eye, working with the muscle grain (start at the larger end). Cut along both sides of the muscle, and release the connective tissue without cutting all the way through it; you want to keep it as one continuous sheet for easier removal.

5. PEEL AWAY THE FELL.

With both sides released, the fell (the first layer of connective tissue that sat just beneath the hide) should peel away from the muscle. Begin to pull the muscle away, then turn it over and secure the muscle to the table while peeling the fell. Use your knife to cut any stubborn connections.

6a

6. FILET THE SILVERSKIN.

Place the loin on its silverskin. Trim away any edges of the silverskin that would prevent your knife from lying flat against the table. Filet with the muscle grain: start at the thinner (anterior) end. To get your knife parallel with the table and silverskin, come in just above it and quickly curve your knife until it's lying flat. Then, with the muscle pinned flat with one hand, cut along the silverskin, keeping your knife flat. Avoid angling your knife toward the silverskin.

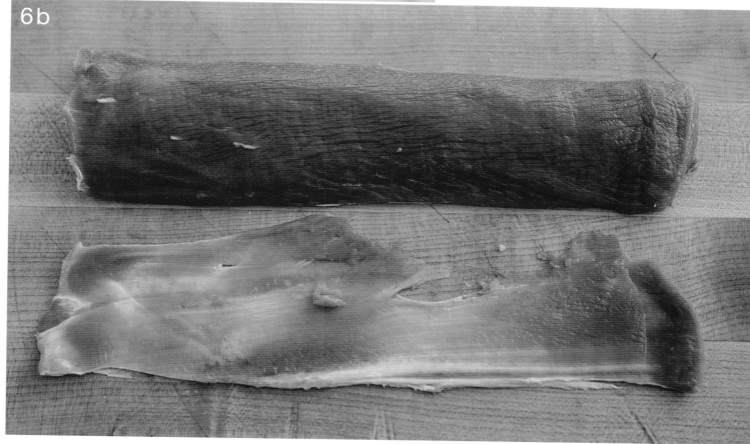

6b

BONELESS LAMB TENDERLOINS don't provide much meat on their own; better to combine them with the loin eye for an aptly named supreme loin roast. Lay one loin eye on the table. Follow by laying the two tenderloins, side by side, on top of it, then sandwich them together with the remaining loin eye. Tie the bunch together and roast to your liking.

As with a boneless rack roast, the loin and tenderloin can be barded and tied for roasting using a thin layer of fat pulled from the flank or elsewhere on the carcass. They can also be crosscut into boneless chops or medallions.

Loin

Tenderloin

Boneless Loin Roast

IN MOST CASES, processing loin primals into chops or bone-in roasts will require you to split the joined primals. If you plan on boning both sides, then don't bother: just bone one side, then the other.

SPLIT THE SPINE. Secure the joined loins on the table, top down. Saw through the middle of the vertebrae, starting at the rear end.

FINISH WITH A KNIFE. Once you're through the bone, stop sawing. Pull out your boning knife and finish the split by cutting through the connective tissue.

Option 2: Roast and Chops

Preparing a loin roast can involve almost no work: you can cook it with great results as is, flank, fat, and fell (exterior membrane) on. But if you want a little more refined preparation, then taking a few moments to trim it up will help.

TRIMMING A BONE-IN LOIN ROAST

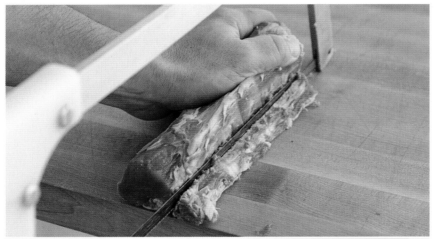

1. REMOVE THE SPINOUS PROCESS.
Start with a split loin primal. Remove the tenderloin, based on the preceding instructions. Then, sit the loin on its bone and saw off the bony outcropping (*spinous process*) that's left after the tenderloin is removed.

2. TRIM AWAY THE FELL.
Trim any remaining flank up to the loin eye. Peel away the fell by pulling up on one end while cutting its connection at the base of the loin eye. Cut away any other stubborn connections.

3. REMOVE THE SILVERSKIN.
The remaining silverskin is thickest at the base of the muscle where it connects to the vertebrae. Use the tip of your knife to sever that connection along the entire length of the loin, being careful not to cut too deep and into the muscle. Remove the silverskin by carefully cutting under it along the curvature of the muscle.

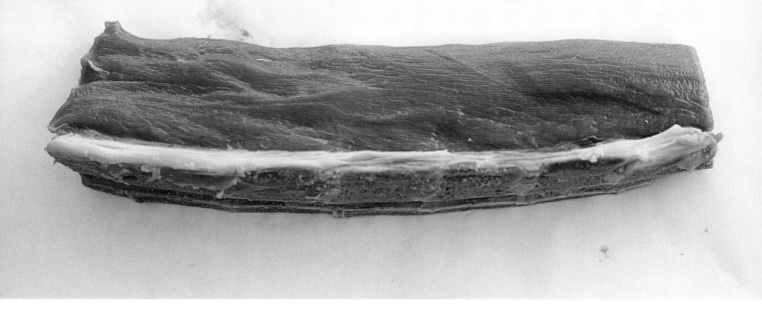

Bone-In Loin Roast

LOIN CHOPS

Lamb loin chops are analogous to beef porterhouse steaks: they include a portion of the loin eye and the tenderloin. When fabricating without a bandsaw it's best to portion them based on the gaps between the finger bones, which are fragile and difficult to cleanly saw through by hand.

1. MARK YOUR CHOPS.

Trim the flank, fat, and connective tissue based on how rustic you want the chops. (Here, we cut them how I like them: rustic and mostly untrimmed.) On the tenderloin side, feel for the tips of the finger bones. Cut between each of the finger bones, all the way through both sides of muscle and down to the bone.

2. CLEAVE (OR SAW).

One by one, place your cleaver between the finger bones and hit the cleaver's spine with a mallet to separate the chop. Alternatively, you could use your handsaw to finish the separation.

ONE FUN VARIATION of the loin chop is to give it wings: cut double-sided chops, known as saddle chops, from joined loin primals. This follows the same premise as single-sided chops: cut between the finger bones, making sure to sever the loin muscles all the way down to the bone. Cleaving through the whole vertebrae can be messy, so it's best to finish the separation using a handsaw.

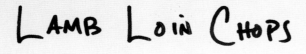

LAMB LOIN CHOPS   SADDLE CHOPS

LEG AND SIRLOIN

A WHOLE LAMB LEG PROVIDES us with a number of preparations, especially when considering the attached sirloin. It can be roasted whole, bone-in or boneless; muscle groups can be seamed for smaller individual roasts; and crosscut leg steaks can provide a great grilling option. The sirloin is my personal favorite for ground lamb that is destined for burgers, sometimes adding additional fat from the flank.

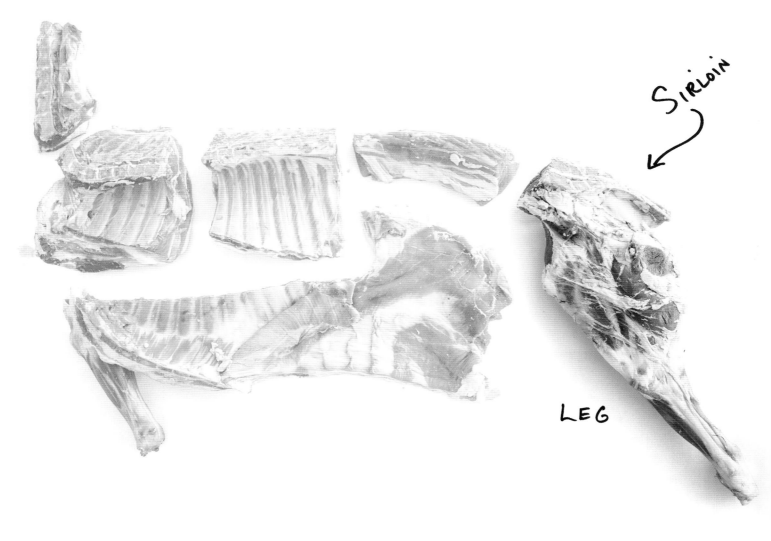

Sirloin

LEG

Sirloin Option 1: Boneless

The sirloin strikes a great balance between the rich flavor of working muscles and the tenderness of sedentary fibers. It can provide a small boneless roast or boneless steaks. Separate the sirloin from the leg following the directions on page 248. The portion of the aitch bone (the pelvis) in the sirloin is small but varies greatly in shape, especially near the loin end, where it slopes dramatically. Thus, use just the tip of your knife and stay close to the bone.

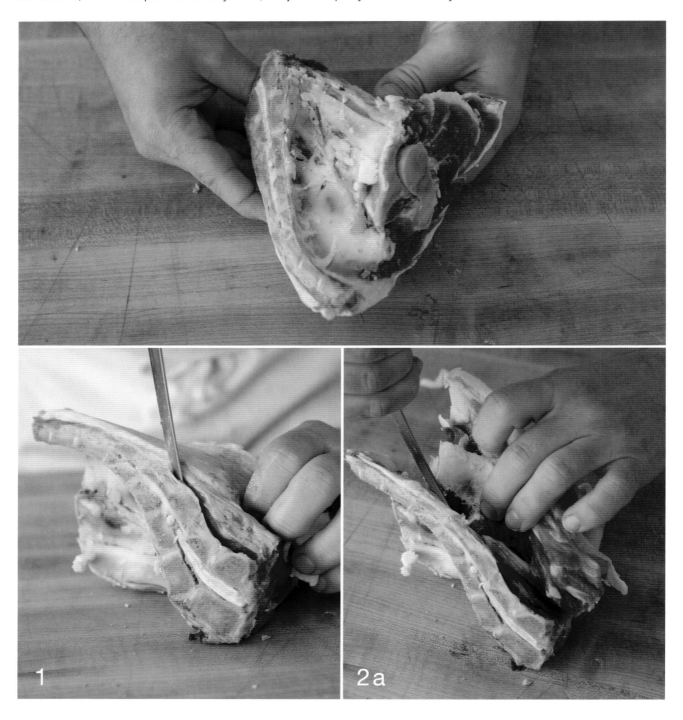

1

2a

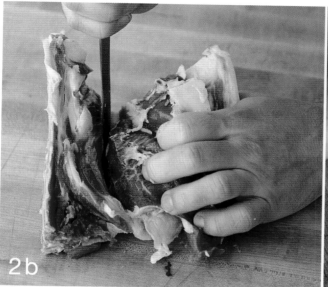

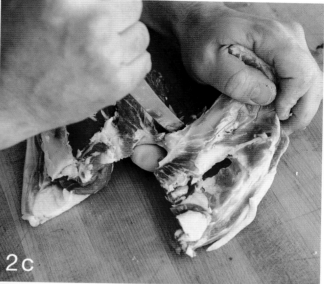

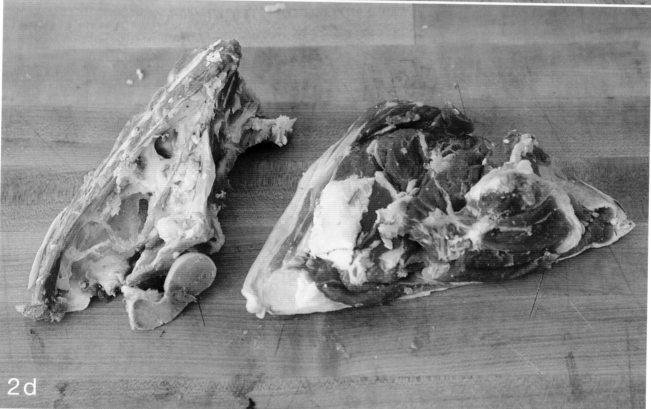

1. SEPARATE AT THE SPINE.

Sever thick connective tissue that runs along the vertebrae and connects at the base of the muscles. A small portion of the loin eye sits protected under a lip of bone; navigate over it for now. You can remove it later for grind, as it doesn't connect well to the other sirloin muscles.

2. FOLLOW THE CONCAVE SLOPE OF BONE.

Pull the sirloin away while severing its connections to the flat sloping face of the pelvis as it heads down and away toward the joint. To finish the separation, cut around the joint and the remnants of the femur's ball-head.

Sirloin Option 1: Boneless CONTINUED

3a

BONELESS
SIRLOIN
ROAST

3. TRIM AND TIE.

Trim away any remaining connective
tissue, and tie in 1-inch increments
for a roast. Or tie in 2-inch increments
and cut between the twine for boneless sirloin steaks.

3b

Sirloin Option 2: Chops

Sirloin chops are fabricated from a bone-in sirloin subprimal. Remove the sirloin from the leg following the instructions on page 248.

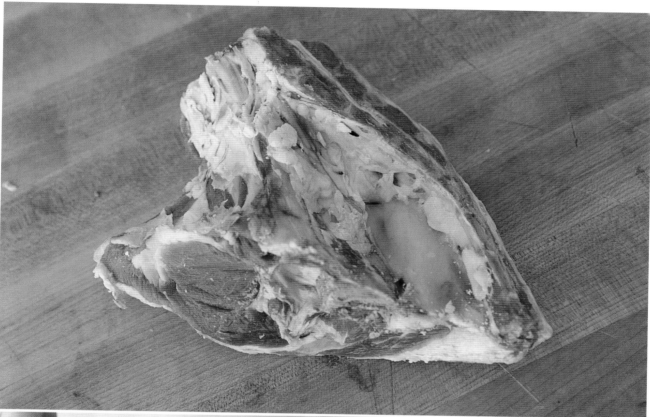

1. KNIFE THROUGH THE MUSCLE.

Decide on your chop thickness (I usually cut them 1–1.5 inches thick). Starting at the leg end of the sirloin, use a breaking or butcher knife to mark the first chop, cutting down and around the bone.

Sirloin Option 2: Chops CONTINUED

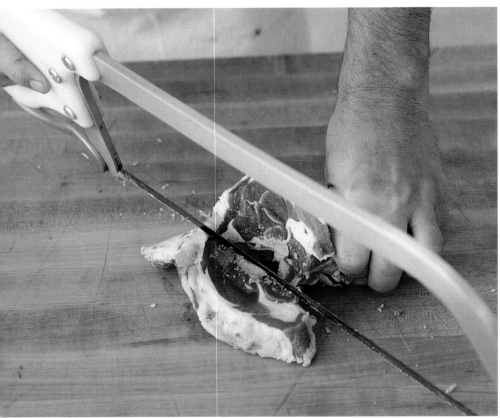

2. **SAW THROUGH THE BONE.**
Switch to a saw until you are through the bone, and switch back to a knife to finish the cut. Trim away any remnants of the vertebrae on the chop.

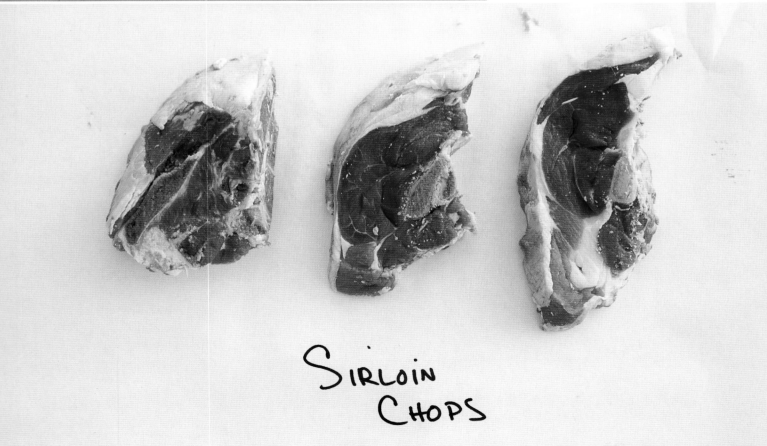

SIRLOIN CHOPS

Hind Leg Option 1: Boneless

A boneless leg of lamb provides a wide variety of preparations, including subprimal muscle groups to a bone-in/bone-out presentation. This can be done with or without the sirloin removed. Start with a single, bone-in leg.

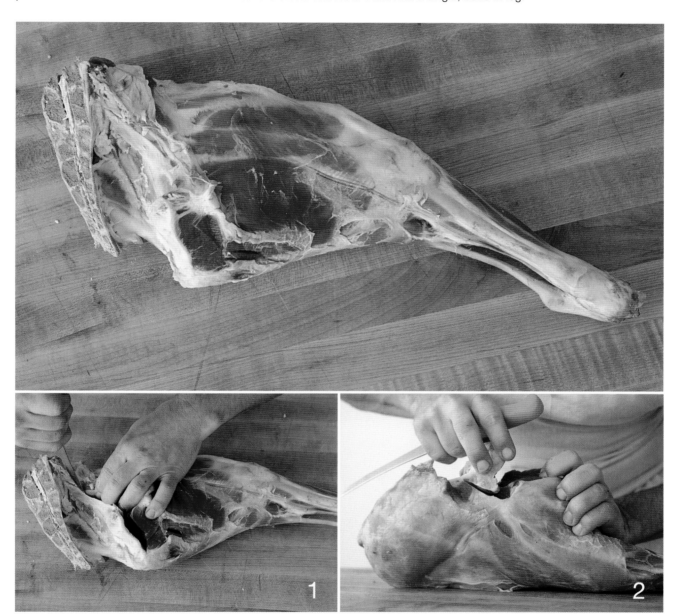

1. CUT ALONG THE AITCHBONE.

The first bone to remove is the pelvis, or aitchbone. It is attached to the femur through a ball-and-socket joint. Place the leg on the table with the inside facing up. Cut along the upper edge of the aitchbone, using your opposite hand to pull the bone away from the meat.

2. OUTLINE THE NOTCH.

The tail end of the pelvis has a bony protuberance that needs to be traced in order to remove the aitchbone. Feel for it with your hand, then make a shallow cut around it.

Hind Leg Option 1: Boneless CONTINUED

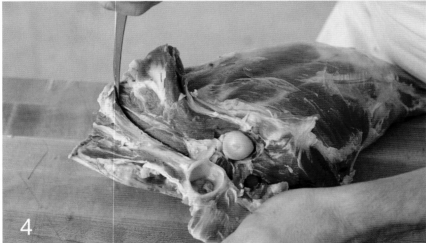

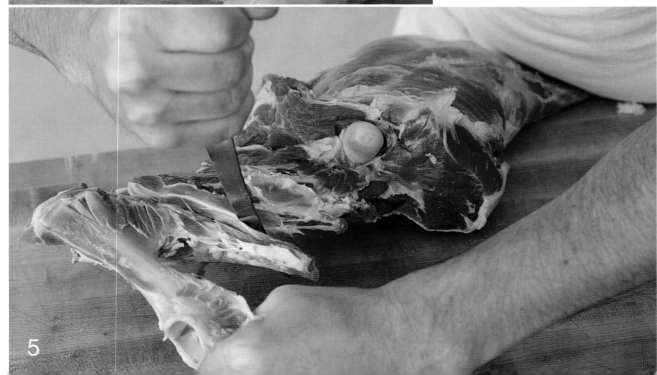

3. CUT THROUGH THE JOINT.

Find the ball-and-socket joint, and cut through the surrounding connective tissue and the ligament that runs inside the joint. Once the tissue and ligament are severed, you should feel the aitchbone release, providing you better access to the underside.

4. TRACE THE AITCHBONE.

The underside of the aitchbone is concave, sloping down and away. Pull the bone away from the meat as you follow the contour with the tip of your knife.

5. REMOVE THE AITCHBONE.

Trim any meat that was left on the bone, and save for grinding. Remove the sacral and caudal vertebrae with the aitchbone.

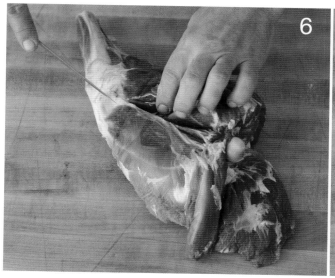

6. CUT ALONG THE SEAM.

Find the seam between the top round and sirloin tip and use it to access the femur. (The photo on page 313 will help identify the seam location.).

7. TRACE THE FEMUR.

Then trace the shape of the femur across where the sirloin tip attaches, using long strokes with the tip of the blade.

SEMI-BONELESS LEG

A lamb leg without the aitchbone is considered a semi-boneless leg, with or without a boneless sirloin. You can remove the boneless sirloin by cutting along the same line as indicated on page 302 and then process it according to the preceding directions.

SEMIBONELESS
LAMB LEG
w/
SIRLOIN

Hind Leg Option 1: Boneless CONTINUED

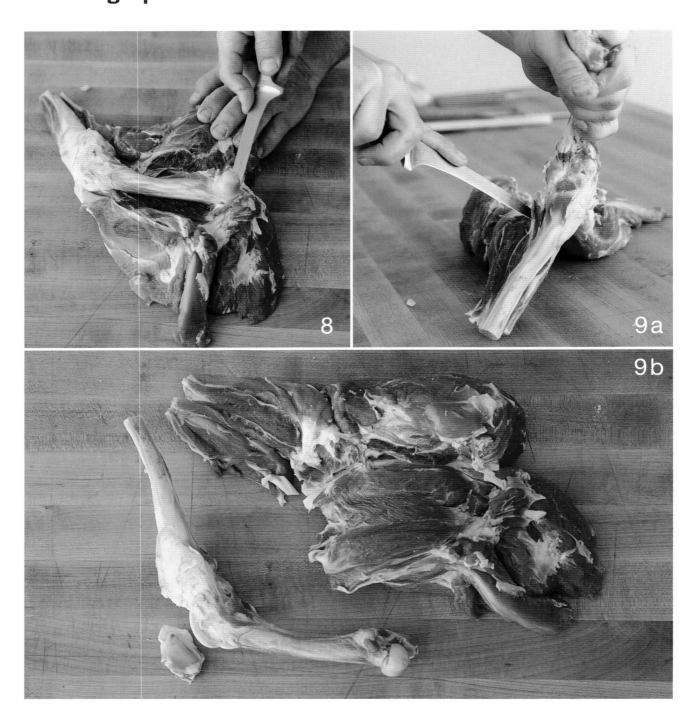

8

9a

9b

8. CUT ALONG THE UNDERSIDE OF THE FEMUR.

Cut around the ball-and-socket end of the femur, and then trace the underside of the bone.

9. SEPARATE THE SHANK.

Cut around the stifle joint, leaving the patella attached to the muscle. Continue down both edges of the shank muscles where the tibia is exposed. Remove the tibia and femur together, then the patella.

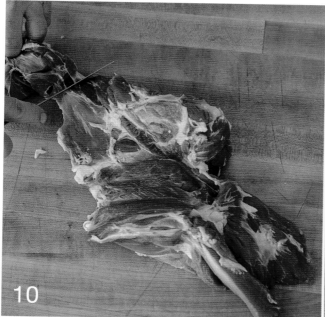

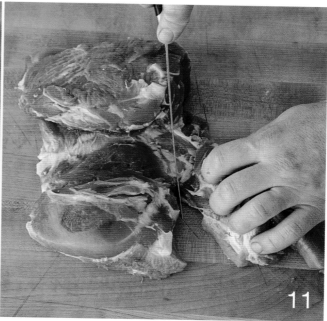

10. SEAM THE SHANK AND HEEL.

Find the seam between the heel and the outside round and use that to remove the heel and shank muscles. Trim any interior pockets of fat from the leg.

11. SEPARATE THE SIRLOIN.

Find the seam at the base of the sirloin tip and follow it to remove the boneless sirloin. Make a couple of cuts across the sirloin tip and top round subprimals, in the same direction that the femur ran, to allow the leg to flatten out. You may also cover the leg with plastic wrap and pound it to help even the muscle thicknesses.

NOTE: Butterflying is purely optional; you can tie and roast the boneless leg as it is now. But butterflying helps to even out the rolling of the roast, especially if you are stuffing it. (It's also a great way to grill a whole leg.)

BONE-IN/BONE-OUT LEG

Another option for the hind leg is to fabricate a bone-in/bone-out leg. This preparation has the flavor of cooking a bone-in roast without the hassle of carving it off the bone. To create a bone-in/bone-out leg, start with a boneless leg, but save the femur. Don't butterfly the leg: the bone fits back in place much more easily when the muscles are kept whole. Take advantage of the situation and stuff the inside of the leg with some combination of herbs and seasonings. Then, place the bone back inside the leg, approximately where it was removed, and tie the leg to secure the bone inside. After cooking, untie the leg, remove the bone, carve, and serve.

Hind Leg Option 1: Boneless CONTINUED

BONELESS
LEG OF LAMB

12. BUTTERFLY IT.

Make a couple of cuts across the sirloin tip and top round subprimals, in the same direction that the femur ran, to allow the leg to flatten out. You may also cover the leg with plastic wrap and pound it to help even the muscle thicknesses.

Hind Leg Option 2: Subprimal Roasts

A lamb leg has the same muscle structure as any other quadrupedal leg and can therefore be seamed into similar subprimal groupings. In this case they are the top round, outside round, and sirloin tip. Start with a semi-boneless leg (see page 309).

Use the image below to identify the top round. One edge of the top round sits next to the sirloin tip; the other edge sits adjacent to the outside round; the top is a thin tapered area and lies across a portion of the heel. These are the three adjoining muscle groups from which you will be separating the top round. Learning this process will involve exploration; the heel is the least valuable of the three adjacent muscle groups, so do not fret if you slice into it while finding your seam.

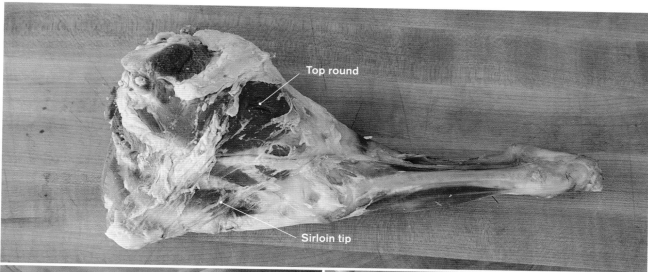

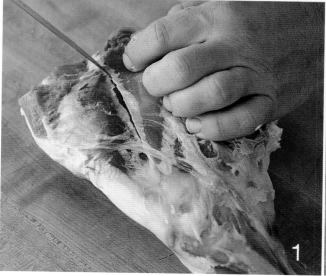

1. SEPARATE THE TOP ROUND.
Start by separating the edge next to the sirloin tip. Use small exploratory cuts at first, keeping between muscle groups while pulling the top round away from the sirloin tip.

2. PEEL AWAY FROM THE HEEL.
When that edge is largely released, follow the contour of the top round toward the heel and continue to follow the seams.

Hind Leg Option 2: Subprimal Roasts CONTINUED

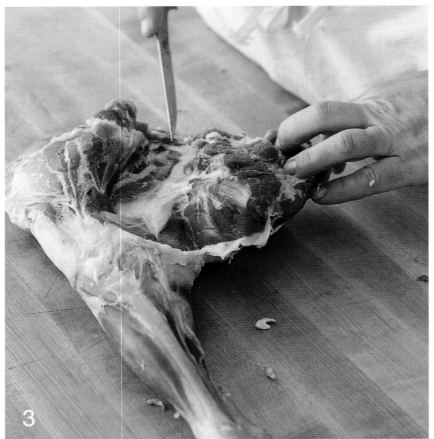

3. SEVER INTERIOR CONNECTIONS.

After you've passed the juncture with the heel, peel the top round outward by severing connections near the interior of the leg where you will find a grouping of connecting tissue and blood vessels.

4. FOLLOW THE FLAT.

Finish separating the top round by following the flat edge of the outside round. Reposition the leg as you wish to give better access to the outside round.

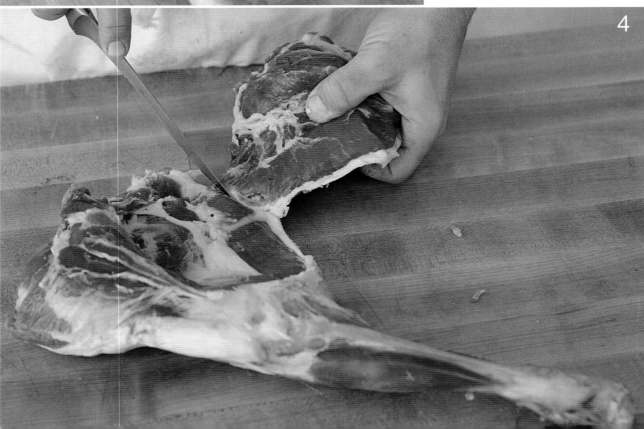

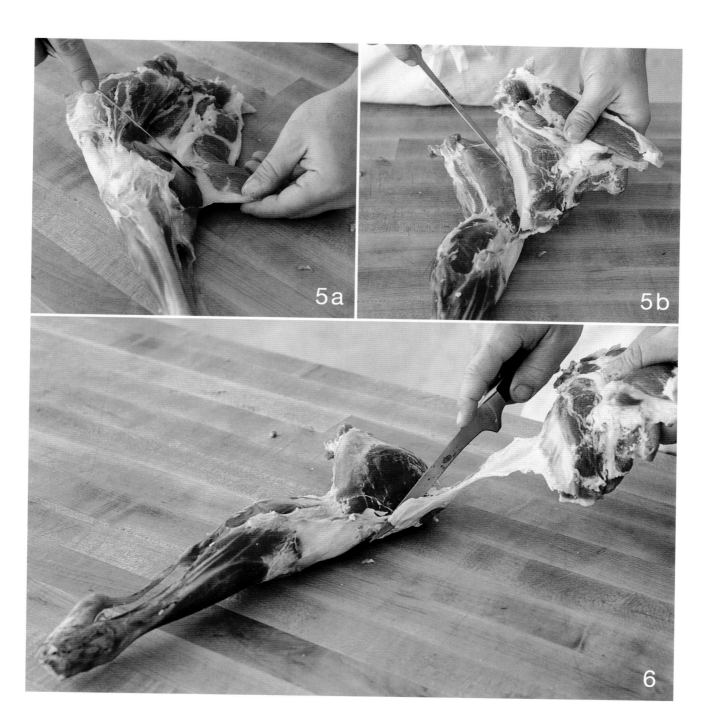

5a

5b

6

5. PEEL THE OUTSIDE ROUND AWAY FROM THE HEEL.

The top edge of the outside round, like that of the top round, is thin and lies atop the heel. Release the top edge and follow the contour of the heel, after which you will encounter a thick layer of silverskin. Sever it, and continue following the seam along the edge of the sirloin tip and femur.

6. PEEL THE SIRLOIN TIP.

When you remove the bottom round from the sirloin tip, take with it some of the connective tissue that wraps across the sirloin tip.

Hind Leg Option 2: Subprimal Roasts CONTINUED

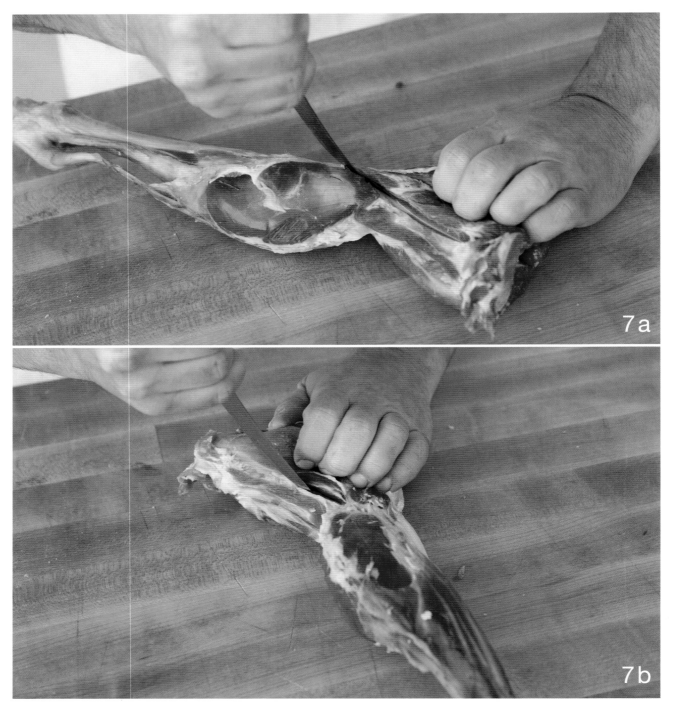

7a

7b

7. SCORE BOTH SIDES OF THE TIP.

The sirloin tip is now the remaining muscle group attached to the femur. You can see how it tapers as it literally wraps around the bone. Cut at the edge of where the muscle connects to the bone, pressing hard enough to score the bone itself. Repeat on the opposite side.

8. CUT BENEATH THE PATELLA.

Connect the two score lines with a deep cut right above the patella (knee cap). Like the others, make sure it's deep enough to score the bone.

9. PEEL AND REMOVE.

Hold the bone against the table while you pull the released end of the sirloin tip toward you. If all three sides were adequately scored it should peel away from the bone, as shown. Finish removal by cutting the connections around the ball-end of the femur.

Hind Leg Option 2: Subprimal Roasts CONTINUED

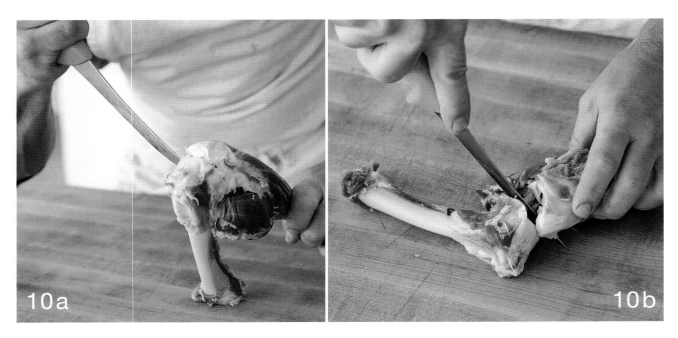

10a 10b

10. POP THE STIFLE JOINT.

Pull down on the hind shank, as shown, and cut across the stifle joint. You should feel it release as you sever the exterior connections. Use the tip of your knife to sever the interior central ligament, cut through the joint, and remove the heel and hind shank as one.

Lamb leg subprimals can all be treated as small boneless roasts, tied to maintain shape or cut into kebab meat. Boneless steaks can also be cut from either the sirloin tip or the top round, preferably after the exterior connective tissue is trimmed.

INSIDE ROUND (TOP ROUND) SIRLOIN TIP (KNUCKLE) OUTSIDE ROUND (BOTTOM ROUND)

LAMB LEG ROASTS

Sawing through individual bones or separated shanks is easier when the saw is braced against a wooden table, as shown. Let the saw teeth do the work as you move the bone back and forth on top of the saw blade.

A RULE THAT holds true in all situations is that the parts of a carcass closest to the ground will have the most pervasive deposits of connective tissue. These parts — the foreshank and and hind shank — are therefore suitable for moist-heat methods of cooking or boning and grinding. When cooking shanks whole, I prefer to split or crack them in the middle to allow the marrow to seep out during cook-

ing. To do so, cut through the meat in the middle of the shank until you reach bone. Use your saw the split the bone. You can either finish the separation with your knife or leave the two sides of the shank connected by the remaining muscle.

SHANKS can also be easily boned for grinding. Place the shank with the larger end facing you and start boning

at the trotter (distal) end. Cut along the exposed part of the bone, using the tip of your knife and drawing it toward you, keeping it against the side of the bone. Work your way around the shank bone, peeling the meat back with one hand while tracing the shape of the bone with the tip of your knife, in long strokes. Remove any ligaments or tendons, and then cut the shank into pieces for grinding.

Hind Leg Option 3: Steaks

Lamb leg steaks are a cross-section of many muscle groups, offering a diverse array of textures. They are best cut thick, at least to 1.25 inches, to allow for searing and a slow-cooked finish. They may also benefit from a quick tenderization with a Jaccard. Start with a semi-boneless leg that has had the sirloin removed.

1. POP THE STIFLE JOINT.

Place the leg cut-face down on the table. Hold the hind shank and apply some downward pressure while cutting through the middle of the stifle joint. When you hit the right spot, you should feel the joint release dramatically.

3. MARK YOUR STEAK.

Decide on a thickness and, using a breaking knife, cut a line across the leg parallel to the exposed ball-joint face. Cut above the bone first, then over it and behind it.

2. REMOVE THE SHANK.

With the stifle joint popped, cut around the side and back of the joint. Then navigate your way through the center of the joint. A thick ligament runs through the center of the joint; when you can get to it, sever it with the tip of your knife and the joint will open up easily, allowing you to finish the separation.

A good leg steak includes a nice, round cross-section of the femur. Therefore, we need to lop off this ball end first. The surrounding meat can be boned and used for grind. The process is exactly the same as cutting a steak. Hence, your first steak should be as thin as needed in order to just remove that ugly ball end of the bone; from there, commence with cutting more presentation-worthy steaks.

4a

4b

4. SAW AND FINISH.

Make your way through the femur using a handsaw. Once through, use the breaking knife to finish cutting the steak.

5. REPEAT.

Work your way down the femur with this cut-and-saw method. Based on the recommended thickness, you can get 3 to 4 good leg steaks. Bone the rest and use for grind; save the joint ends of the bone for stock.

PIG
SLAUGHTERING

PIGS WERE FIRST DOMESTICATED because they are omnivores. They are recyclers of sorts, allowing a family to supplement the animals' outdoor diet with kitchen scraps and other sources of nutrition otherwise inedible to humans. Alive, they are no-waste consumers; slaughtered, they are no-waste carcasses. Pigs are the true nose-to-tail animal: you can use every part of the carcass, along with most of the viscera. For centuries, people worldwide have realized this, turning every part of the animal into charcuterie and other preserved forms of flesh.

Today the commercial pig is raised without thought toward nose-to-tail use, but at home we can look to cultural traditions to show us how to maximize the return from the carcasses we raise. Pigs are the perfect species for learning whole-animal utilization: how to render the fat, fry the skin, smoke the hocks, braise the head, fry the brains, stuff the intestines, cure the loin, and so forth. Take the time to properly slaughter your animal, ensuring that the skin is clean, the head cleaned, and that every step is performed with an expectation that all parts of this animal will be consumed by you and those around you.

Anatomy

PIGS ARE ONE OF THE FEW ANIMALS whose head and skin are left on after slaughtering. That fact, combined with the fact that a pig has a monogastric digestive system (i.e., a stomach with only one compartment), gives pork a high dressing percentage of around 75 percent on average. The details of the pig's digestive system and other visceral anatomy are found on page 90.

Setting Up for Slaughter

CONSIDERABLE PREPARATION is needed for pig slaughter and should start far ahead of the actual event day, especially if you're planning on scalding and scraping the carcass (which is recommended). The general considerations to be taken into account are covered in chapter 5, with additional requirements and information provided here.

An average market-ready pig is the largest animal covered in this book, and the carcass must be moved often during the slaughter process. The process can be physically draining, so gather some help before you attempt to slaughter a pig; it is not a task to be done alone. (This is one of many reasons why it was traditionally an event that brought neighbors together.)

Stunning and Sticking

Although a projectile can kill the animal, a captive bolt pistol is the preferred method of stunning any animal that you can get close to, enabling a more complete exsanguination. Pig boards — the large plastic handheld barriers that are used for corralling pigs — are essential in separating and containing the animal for stunning, so you should have at least two on hand. Refer to page 80 for detailed instructions on methods of stunning and pages 86 and 332 on the chest stick method of bleeding a pig.

EQUIPMENT OVERVIEW

Pigs can be slaughtered, scalding included, with minimal, even rustic, equipment. These lists encompass the main options, ordered from barebones to ideal.

STUNNING: bullet, captive bolt gun

HANGING: paracord or gambrel

BLEEDING: 6–7-inch sticking knife

SKINNING: 5–6-inch knife, skinning knife

EVISCERATION: 5–6-inch knife

CHILLING AND AGING: modified chest freezer, chest refrigerator, blast chiller, or walk-in cooler

SCALDING: towels, 50-gallon drum, old cast-iron or metal tub, temperature-controlled vat

SCRAPING: dull knife, hog scrapers, bell scrapers

If you choose to scrape with a knife (not recommended) or your scrapers don't have a hook on one end, you'll need a boning hook to remove the toenails. Also, make sure to have a propane torch on hand for final singeing.

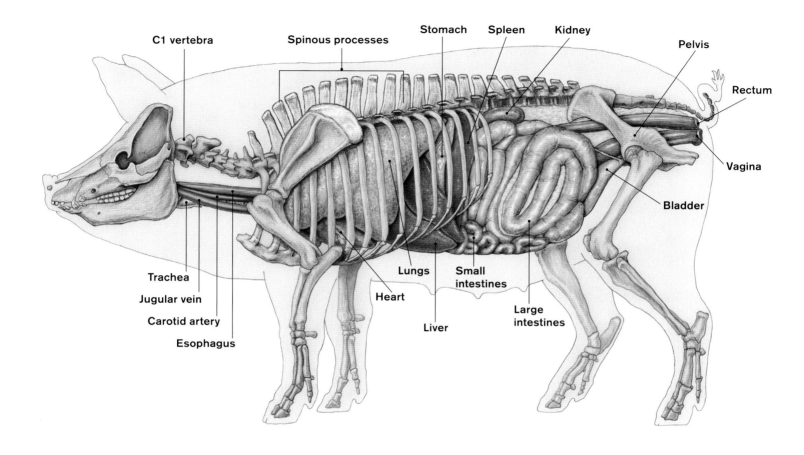

C1 vertebra · Spinous processes · Stomach · Spleen · Kidney · Pelvis · Rectum · Vagina · Bladder · Large intestines · Liver · Heart · Lungs · Small intestines · Esophagus · Carotid artery · Jugular vein · Trachea

THE SKELETON OF A PIG is quite small in relation to the weight it needs to carry. This makes for a low skeleton weight, increasing the percentage of muscle (meat) you get from a carcass. Combine that with the high dressing percentage, and you realize that pigs return some of the largest amounts of meat from any species of livestock.

THE HEAD OF A PIG has many culinary purposes and is therefore often left on after slaughter. Should you choose to remove it, leave the jowl on the carcass and remove the head at the atlas joint, which connects the skull to the first cervical vertebra (C1).

THE THIN, BONY SPINOUS processes on the thoracic vertebrae make even splitting difficult in the shoulder. This is where it may be best to pull out a cleaver, following the edge of the bones, until you reach the skull, where you can continue sawing.

THE PELVIS (or aitchbone once split), is the bone you'll need to tackle when separating the bung, or rectum. Splitting the pubic symphysis – the cartilaginous midline on the bottom (ventral) edge of the pelvis – is one method for isolating and tying off the bung.

THE MAIN BLOOD VESSELS, carotid arteries and jugular veins, cross as they leave the heart. This intersection occurs right around the front (anterior) edge of the sternum. Severing these vessels is the main priority when sticking a pig for bleeding.

BRAIN, TONGUE, HEART, STOMACH, SPLEEN, LIVER, AND KIDNEYS are the likely candidates for offal saved after slaughter. Intestines are the common casing for sausage, though home cleaning of them is an arduous process; because of that, they aren't often saved. The more reactive organs, like the brain, liver, and spleen, are best when frozen immediately after slaughter, though all offal benefits from freezing.

Make sure your scalding setup is able to reach and maintain an ideal temperature.

Scalding Setup

Removing pig hair and scurf can be tedious, but it is fully worth the effort. Pigskin not only tastes great but also protects the carcass and the valuable, hefty layer of subcutaneous fat during transport, aging, butchering, and even in some cases cooking. Skinning, while it may seem more convenient, has some considerable disadvantages. Along with a loss of fat from the actual skinning and the inevitable trimming, the fat of a skinned pig will also oxidize and develop rancid flavors more quickly. This is because the fat is exposed to air and other elements immediately rather than sitting protected underneath a layer of delicious skin.

The main requirement for scalding is a vat that can maintain a temperature of 145°F to 175°F and is large enough to submerge either the entire pig or at least half of it, while allowing enough space for you to rotate the carcass, ensuring that all areas are exposed to the scalding water. The exposure to hot water loosens the grip of hair follicles by breaking down (denaturing) proteins that hold them in place. If the water is too hot, you risk burning the skin and "setting" the hairs, making scraping arduous; if it's too cold, the hairs won't loosen enough to get them off without shaving or torching. The right temperature is often a source of debate, though the average falls around 155°F. In essence, the right temperature is the one that does the job, and that will change depending on conditions and breeds. Aim for a starting temperature above the goal temperature because the moment the carcass is in the water, the temperature will drop considerably. But, be wary of going too high: you can always increase the water temp and scald again, but if you "set" the hairs there's no going back. If the weather is frigid, take that into account as well when figuring out how effective your efforts at maintaining a temperature will be. Also, keep in mind that the pig will be occupying a considerable amount of space in your vat. Don't fill it too much; allow room for the pig to avoid spilling water over the edges and onto your heat source below.

Using a vat that enables the pig to be laid on its side is often better. Along with some other advantages explained on page 335, having half of the pig submerged allows you to work on removing hair from the other half while the follicles on the submerged half loosen. You can then rotate the pig and repeat. One common item that can serve as such a vat is an old bathtub; another is a 55-gallon drum cut in half. Large drums can be used intact, though lifting the carcass in and out can prove to be an incredible effort. (Some folks choose to partially bury whole barrels at a 45-degree angle to help with this.) Commercial items for maintaining accurate water temperatures include electronic coils and thermostats. If you need to scald only a few pigs, any of these coils and thermostats work. But if you're frequently scalding multiple pigs, it's helpful to have a unit that can bring the water back up to temperature quickly after each carcass.

You can heat the water with fire, propane, or electricity, the first two being the most frequently used in home processing. Heating is often done directly underneath the dipping vat. You can use propane with either burners or handheld torches placed amid wood to provide a continuous flame for the fire. The water temperature drops quickly after the carcass is introduced, so having extra hot water available to maintain an ideal temperature is helpful. You can do this by boiling water in a turkey fryer

SCALDING, SIMPLIFIED

While I highly recommend using a vat for scalding — submerging the carcass provides a better scald, easier hair removal, and will enable more complete usage of the carcass — the setup can often feel like too much effort, especially the first time. And for others, procuring the necessary equipment might just be unfeasible. Fortunately, there is a secondary method for scalding which requires minimal setup.

You will need to have access to ample amounts of hot water at a temperature slightly above the target temperature, as cooling will immediately occur when you start pouring the water. Ideally this is one or more pots that are capable of quickly reheating water during the scalding process.

Place the carcass on a stable working surface. A sturdy table is ideal, as it allows you to work at a comfortable height, but a pallet on the ground or a similar, relatively clean surface will work. Place the carcass on the work surface and cover one half of the upward side with towels. Pour the scalding water over the towels, making sure they are in close contact with the skin of the carcass. Allow it to sit for a few minutes, pouring more water if necessary. After, pull on some of the hair. If it comes out easily, peel back a section of the towels and begin scraping. When the hair and scurf become difficult to remove, replace the towels and reheat the area.

This works because the towels allow the hot water to stay in contact with the skin, making it more efficient and effective than just pouring water over the carcass. Repeat this process across the entire side of the carcass. The head and trotters will be the most difficult areas. If possible, submerge them in buckets of scalding water. Otherwise, do your best using the towels. Finish this process with a propane torch, singeing stubborn hairs and helping to clean the more pigheaded areas.

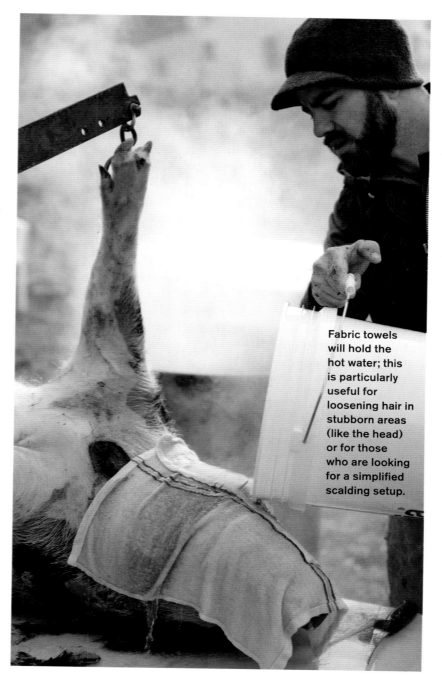

Fabric towels will hold the hot water; this is particularly useful for loosening hair in stubborn areas (like the head) or for those who are looking for a simplified scalding setup.

Make sure to prepare your scalding tub in an area that allows you to move the pig easily.

or any extra-large pot. Also, you will need a large bucket to gather water from the scald tub to keep the carcass wet and warm while you are scraping it.

You will need a way to lift the carcass in and out of the water. A small group of strong people can do it by hand if you are using a tub. If you are having a group work by hand, use towels wrapped around the scalding-hot trotters to help the lifters grip the wet skin. A tractor or another mobile method of hoisting can be helpful when moving larger boars and sows or when working with an upright drum.

While in the water, the carcass needs to be kept moving to help disperse the heat and prevent any over-scalding or light cooking of the skin and muscles. This can be done by hand, using the pig's legs as leverage, but an easier method is to use two chains or ropes placed underneath the carcass. This allows two people to agitate the carcass and then easily rotate it in the water.

Rosin is frequently used during the scalding process to aid in hair removal. The rosin causes the hair to clump and adds texture to the hairs, making removal of large swaths easier because it sticks to the scrapers. Commercially sold mixes are available through butchering resources, although recipes for concocting your own mixes are available online.

There is a concern about scalding multiple pigs in the same tub, a practice that can propagate and transfer contaminants and microorganisms. One method of counteracting this is to add a couple bottles of hydrogen peroxide into the scald tank before you begin scalding, which will help prevent transmission of microorganisms from one carcass to the next. Moreover, singeing hair after scalding will help eliminate surface pathogens.

Hair Removal

A properly scalded pig should be rather easy to scrape clean of hair and scurf (the porcine equivalent of dandruff); the hard part is getting the exposure time and temperature of the scald right, without which the scraping becomes an immense effort. A bell or hog scraper is the main tool for clearing the carcass of hair, scurf, and dirt. Hooks are needed to remove toenails, and some bell scrapers are sold with hooks at the end. You may also use boning hooks for toenail removal. Have some knives on hand that can be used to shave the headstrong hairs, especially those around the head, ears, and trotters. A propane torch will be useful for singeing any remaining hairs and helping release areas of scurf that were not properly scalded. Finally, coarse-haired brushes will help clear the last bits of scurf and grime from the carcass, so have a few on hand.

You will need a surface near the scalding tank to transfer the carcass for scraping. The closer the surface, the shorter the distance you will have to carry a sopping-wet, steaming-hot, cumbersome carcass of deadweight. Tables work best, and one with the right height allows for you to work easily while standing. But the table needs to support the weight of the pig, and then some, so make sure it's strong enough before you find out otherwise. A pallet or clean tarp is another option, though you'll be working on your knees at ground level.

Slaughtering

PIG SLAUGHTERING IS the most involved process covered in this book. Make this a group effort and be sure to take your time; the pig isn't going anywhere. Pigs aren't ruminants, so dawdling after a pig's death does not mean a rumen full of gasses and an ever-increasing visceral size. Take your time with the stick, which is the most difficult method of killing any animal. And spend ample time on scalding and scraping, because clean skin will benefit carcass condition and culinary preparations.

The Right Age

Pigs are slaughtered and consumed at every stage of life, from on-sow sucklers to aging sows who fail to produce effective litters. That said, the typical market-ready pig is usually a barrow or gilt weighing between 240 and 260 pounds. A carcass of this size from the common industrialized pig will provide cuts similar in size to what you see on supermarket shelves. On a farm, things tend to be much different: you will determine what the final weight of the pig should be and the right time to slaughter it according to factors such as its genetics and age, as well as the cost of its feed and the season.

After scalding, the carcass should be hung by an appropriately sized gambrel at a comfortable height.

Preparation

Preparing the right setup for pig slaughter is an undertaking of its own. Make sure you've read chapter 5 in its entirety before attempting a slaughter. You should complete additional research far ahead of time, and the lead person should have a firm, step-by-step understanding of the process and be prepared to deal with gaffes as they happen.

THE SLAUGHTER PROCESS FOR PIGS

Pig slaughters do require a couple up-front decisions, primarily whether you are going to scald or skin the carcass. I recommend scalding, for a variety of reasons (page 326), but for some the setup required may be too much work. The remainder of the process is quite straightforward:

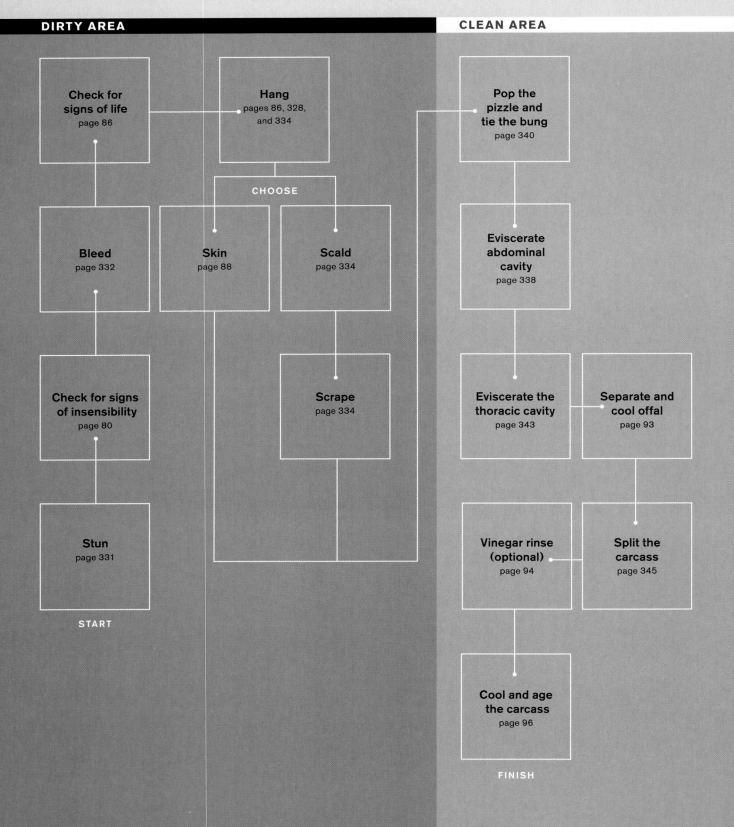

DIRTY AREA

CLEAN AREA

Check for signs of life
page 86

Hang
pages 86, 328, and 334

Pop the pizzle and tie the bung
page 340

CHOOSE

Bleed
page 332

Skin
page 88

Scald
page 334

Eviscerate abdominal cavity
page 338

Check for signs of insensibility
page 80

Scrape
page 334

Eviscerate the thoracic cavity
page 343

Separate and cool offal
page 93

Stun
page 331

Vinegar rinse (optional)
page 94

Split the carcass
page 345

START

Cool and age the carcass
page 96

FINISH

Secure and Stun the Animal

Stunning should be done only when the animal is under minimal duress, relatively speaking. Work to contain the animal with methods that prevent excessive startling, agitation, or aggression. Begin stunning when these conditions are adequately met, and then proceed with bleeding only after you have ensured insensibility.

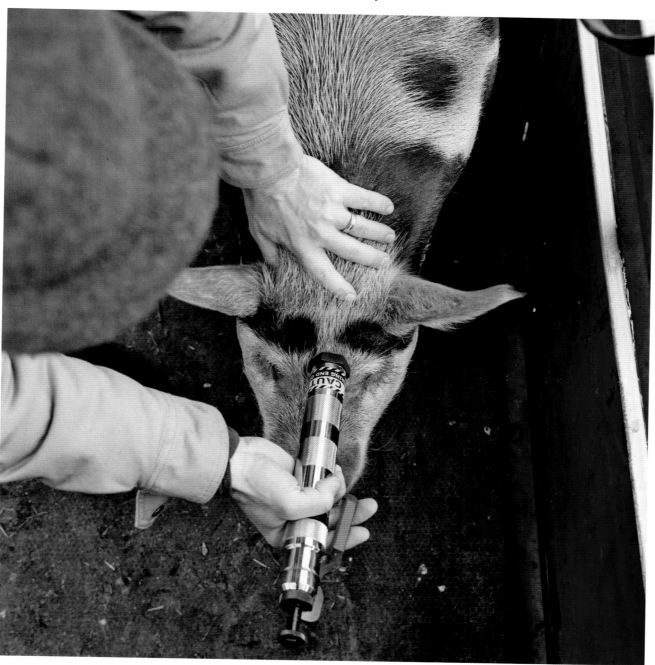

The correct area to aim for when stunning pigs, whether you are using a projectile or captive bolt, is in the middle of an X made between the ears and eyes, connecting each ear with the opposite eye. When using a captive bolt, ensure that it is positioned flat against the skull prior to firing.

Bleed the Animal

Proceed to sticking only after you have confirmed that the animal is insensible. Sticking a pig is no easy task. There are many methods to doing it, but the intention should be to make as small an incision as possible while allowing for the blood to exit after severing blood vessels. Be wary of an off-center stick that will cause damage to the shoulder meat and cause excessive clotting within the neck and shoulder.

1. FIND THE STICKING SPOT.

Roll the animal on its back. Feel for the depression just above its sternum and position your knife there, blade facing forward.

2. STICK THE PIG.

Push the knife straight down in the center of the depression until you hit the spine. If needed, push the handle toward the tail to increase the interior incision. (Blood should immediately begin exiting through the incision; if not, stick again.)

3. GATHER THE BLOOD.

If you're planning on using the blood, have a bowl ready and in place when the sticking happens. As the blood collects in the bowl, have someone continually mix it as it cools down, so that it doesn't coagulate.

4. DISCHARGE THE LAST OF THE BLOOD.

The animal will continue to kick as it loses blood; this is normal, though be aware of any indications that sensibility is returning. (Signs include vocalizations or arching of the back.) When the animal has stopped moving, help discharge the last of the blood by pumping the front leg while placing pressure on the chest with your knee.

Scald and Scrape the Carcass

The first step of hair removal is submerging the carcass in the scald tub. Drag the pig near the scalding tank. Put some hot water in a bucket and splash it on the hind leg, looking for movement and signs of life. If there are no signs of movement, splash some water in the ear, again checking for motion. Proceed only after the animal is unresponsive; otherwise repeat in a few minutes. If you are using rosin, pour hot water over the animal to moisten the hairs. Liberally dust the carcass with rosin, rubbing it into the hair with your hands; then roll the carcass over and repeat. (Follow the manufacturer's directions for rosin application.)

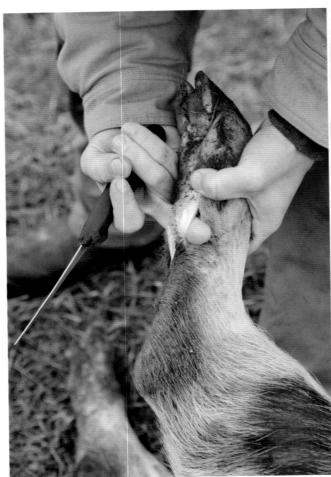

1. EXPOSE THE GAMBRELS.

Release the tendons in all four trotters by making deep cuts on either side of the hind dewclaws and using your fingers to make a gap behind the tendon. (In this case, we hung the pig from the hind legs first, then switched to scald the whole carcass.)

2. HOIST THE CARCASS.

Insert the gambrel and hoist the carcass above the scalding vat.

ALTERNATE METHOD: CHAINS AND A LOW SCALDING VAT

The ideal vessel for scalding is an old tub. Being lower to the ground means that you can skip the need for hoisting and dipping; being able to roll it over allows you to work on one side while the other is scalding, and vice versa. Having chains beneath the carcass makes rotating easier and also helps scrub hair off the submerged side. Place the chains (or ropes) across the scalding vat; they should be positioned between the fore and hind legs. Carefully place the pig in the vat and roll it to one side, submerging it as much as possible. Use the chains to agitate the pig and the water, preventing overscalding while also helping to loosen hairs on the underside. After 60 to 90 seconds, use the chains to roll the carcass over. Grab a few clumps of hair and test to see if they release easily. If not, let the now-submerged side go for a bit longer; then rotate and repeat. Once you can easily remove a handful of hair with little to no resistance, start scraping while the opposite side heats up. Repeat this pattern — scrape the top while the bottom heats up — until you have removed a large portion of the hair.

3. DUNK THE CARCASS INTO THE SCALDING TANK.

Lower the carcass into the tank as far as it will go.

Scald and Scrape the Carcass CONTINUED

4a

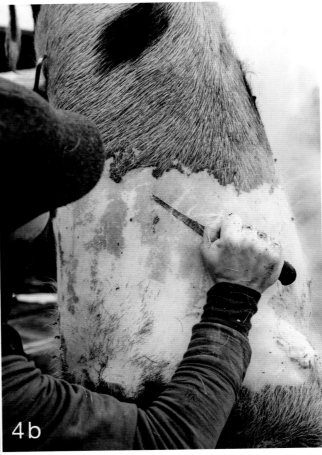

4b

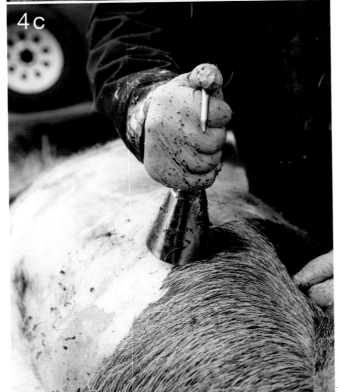

4c

4. SCRAPE THE CARCASS.

When the submerged portion of the carcass has been properly scalded, move it to the scraping table and begin scraping immediately. (You can even begin scraping while the animal is suspended, to be sure the carcass has been properly scalded.) While the carcass is on the table, periodically pour hot water from the scalding tub over the carcass to keep it hot and moist. Bell scrapers are easy to use: pull them across the skin at a roughly 45-degree angle while applying enough pressure to remove the hair and scurf. The action is similar to shaving. The most resilient areas will be the face, ears, and trotters. Give them extra attention while the animal is steaming. When bell scrapers fail to work, move to a sharp knife and start shaving off the remaining headstrong hairs. Scrape until removing the hair becomes difficult, then scald again or switch the gambrel to the forelegs and scald the opposite end.

 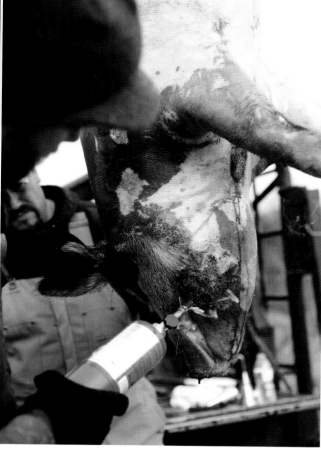

5. REMOVE THE TOENAILS.

Have someone remove the toenails while the legs are hot, using the hook end of a scraper or a boning hook. Insert the hook into the top of the toenail. Once hooked, pull hard to pop the nail off. Repeat with all the feet and then clean away the hair and scurf from between the toes.

6. SINGE THE REMAINING HAIRS.

Use a torch to singe remaining hairs and any broad areas that you could not scrape. Hair singes easily, so you shouldn't need to keep the torch focused on any single area for too long. Keep the torch moving to prevent overheating the skin, which can cause bubbling and light cooking of the underlying muscle. Follow singeing with knife shaving to remove hair remnants. Run your hand along the skin and feel for areas where the coarse hairs are still present. Finish by rinsing the carcass thoroughly with hot water and scrubbing with coarse brushes. Areas that prove impossible to get completely cleaned can be trimmed off after the carcass is cooled.

Eviscerate the Carcass

Pigs have copious amounts of fat; that is the biggest difference between eviscerating a pig and eviscerating any other animal. The fat can make for some confusion: the genitals, bung, and internal viscera in a pig are not always obvious, because they are surrounded by huge deposits of adipose tissue. Proceed through it carefully, and avoid any cavalier cutting.

1. **SEVER THE SKIN.**
Walk to the front side and make an initial shallow cut from the rectum to the snout, just deep enough to sever the skin.

2. **CUT AROUND THE RECTUM.**
Stand behind the carcass and begin by cutting around the rectum, just deep enough to sever the skin and begin loosening its connections.

3. CUT ALONG THE PIZZLE.

When working with a male pig, make a shallow cut to one side of the pizzle. Follow that with another shallow cut on the opposite side of the pizzle, connecting it to the first cut just below the urogenital opening. Carefully cut along one side of the pizzle until you have identified the shaft that runs from the urogenital opening to the rectum.

4. SEPARATE THE PIZZLE.

Continue to cut and separate the entire length on both sides, without severing the pizzle, and then toss it between the legs so that it hangs over the back. Be wary of leaking urine, an indication that the interior bladder is extremely full.

Eviscerate the Carcass CONTINUED

5. SPLIT THE ABDOMEN.

A few inches below the pelvis, make an incision about 6 inches long into the abdominal cavity. You cannot pinch the skin of pigs to help prevent accidental punctures of the bladder or other viscera; therefore, make safe, shallow cuts until you have broken through. Increase the initial incision using just the tip of your knife. Split the abdomen all the way to the sternum using either method described on page 91. If you are splitting the pubic symphysis, do so now, starting with cutting between the hams down to the pelvis.

6. TIE THE BUNG.

Pull up on the pizzle and finish separating it with the bung. Tie off the genitals and bung with string.

7. REMOVE THE BUNG AND GENITALS.

Pull the tied bung and genitals down and begin severing ureters and other connections, being careful not to puncture the bladder, until you can let them hang outside the cavity.

8. REMOVE THE VISCERA.

Reach in and pull up and out on the large mass of viscera, which includes the stomach, intestines, spleen, and liver.

Eviscerate the Carcass CONTINUED

9a

9b

9. RETRIEVE THE CAUL FAT.

Let the viscera partially hang out of the cavity and take a moment to remove, and save, the caul fat from around the stomach and intestines.

The caul fat, which surrounds the viscera, should be saved for future use; it can be rendered or used for other culinary preparations.

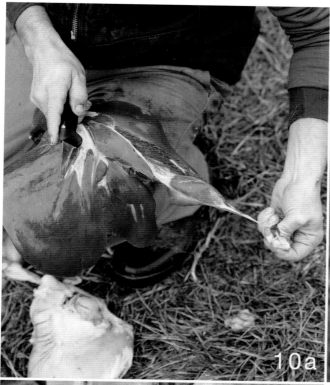

10. REMOVE THE LIVER AND SPLEEN.

Remove the bile sac from the liver (photo 10a). Consider removing and saving the spleen (photo 10b), the long, purplish organ on the left side of the stomach. (I removed it with the caul fat in the photo on page 342.) Hold the viscera mass with your non-knife hand, and begin to sever any remaining attachments, including the connection with the esophagus called the gastro-esophageal (GE) junction, while pulling the mass out of the cavity and depositing it into a waste container.

11. SPLIT THE DIAPHRAGM AND SEPARATE THE STERNUM.

Cut through the white connective tissue, following its curvature from one side to another. The muscle will retract once severed, exposing the thoracic cavity.

Separate the sternum by running your knife slightly to either side of the very middle, moving through the cartilage of the ribs. If there is too much resistance, as with older animals, use a meat saw to split the sternum down the middle.

Eviscerate the Carcass CONTINUED

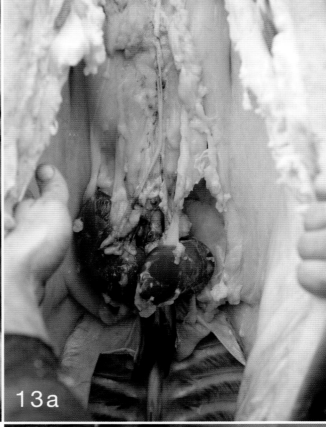

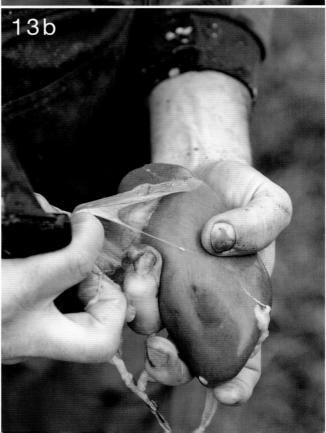

12. REMOVE THE PLUCK.

Grab the pluck (the heart and the lungs) with one hand, and pull down and out through the bottom of the sternum, using your knife to cut blood vessels and other attachments. Remove the heart from the group; cut open the top, remove the clots, rinse, and save. Reach back inside and, with your knife, follow the remaining portion of the trachea and esophagus to the tongue and remove all three together. Inspect the interior for visceral remnants, glands, or other items that require removal.

13. REMOVE THE KIDNEYS.

You may choose either to leave the kidneys for aging and carcass weight or to remove them for immediate cleaning and freezing. Pop the kidneys by scoring the surface membrane and squeezing them.

Split and Cool the Carcass

Rinse the carcass thoroughly with cold water. Begin splitting by standing in front of the carcass and using a meat saw to separate the two sides, beginning with the pelvis. Use long strokes on the saw (rather than short strokes), which will help prevent the saw blade from wandering to either side. I often switch from a saw to a cleaver when I reach the ribs and feather bones of the shoulder. The thin feather bones are hard to split cleanly, and cleaving often produces a cleaner split. There is no need to smash your way through the spine with a cleaver; rather, let the weight of the cleaver do most of the work. A heavy cleaver will easily split the spine and the face of its blade will follow the side of the feather bones, leaving them on one side of the carcass and cleanly separating the two shoulders. Saw through the skull until you reach the teeth, then use a cleaver to finish the separation. Immediately follow splitting with cooling according to the instructions on page 96.

Thoroughly rinse all offal and check for signs of infection. Process the offal according to the directions on page 93. Freezing offal soon after slaughter will ensure freshness.

PORK BUTCHERING

FABRICATION OF A PIG is truly an exercise in zero-waste processing. The popular term *nose-to-tail* was derived from the fact that literally everything from the pig — including the viscera — can be used with delicious results, a claim rarely applied to any other animal. While you may not have taken the time to clean the intestines, the paunch, and other edible offal that would otherwise be destined as packaging for sausage or other concoctions, these parts are traditionally processed throughout the world for our culinary benefit. Certainly the easiest reminder that this was a sentient, sauntering animal prior to arriving on our butcher block is that the skin, head, feet, and tail are all left on. Humility and respect are at the foundation of a zero-waste approach to slaughtering and fabrication with any animal that has given its life for our sustenance, and pork is the easiest animal to achieve this with.

Butchering Setup, Equipment, and Packaging

FOR PIG BUTCHERING, YOU'LL NEED a sturdy tabletop about 6 feet long. (Shorter tables can work, though you may end up doing a bit of rearranging to the suggested order of butchering and carcass breakdown, thus shortening up the carcass to fit in the space.) The surface should be made from a material that is easily sanitized and kind to blade edges, like wood or food-grade plastic (see page 47). Because pork fat is largely unsaturated and soft, it easily greases up surfaces, so have a scraper on hand for frequent clearing of buildup. Pork also tends to exude more moisture during processing than other meats do, so a good number of towels will help manage dampness.

Equipment

The basic tools of butchery apply to pork, though there are some optional tools that will help along the way. All the tools mentioned below are described in further detail in chapter 3.

Packaging

Any whole-animal fabrication will result in a lot of frozen or otherwise preserved cuts. Pork fat is mostly unsaturated and thus becomes rancid more quickly than the fats found in red meat, which are saturated (page 10). Therefore, it is imperative that you take the right measures to ensure that the meat maintains freshness. The two main priorities in freezer packaging are making the package skintight and removing as much air as possible. For most applications, vacuum sealing will often provide the best solution, though it is sometimes cost-prohibitive. For an explanation of the science behind freezing and a complete list of packaging options, see chapter 14.

Primals and Subprimals

AS WITH OTHER ANIMALS, the pig carcass is first broken down into primals and subprimals, the boundaries of which are defined by anatomical features. The instructions in this book for pork primals are adapted from the guidelines of the National Pork Board. This map of the carcass is accurate for most regions on the United States but will differ dramatically for other parts of the world. Fortunately, the scientific names of muscles are universal, and those are provided in italics where applicable.

Shoulder

The shoulder muscles in pigs, and most other quadrupeds, are strong working muscles responsible for directing the movement of the animal. Thus, the muscles require a serious resilience and generous amount of energy, provided by copious amounts of connective tissue and healthy fat deposits, respectively. The pervasive collagen and fat make pork shoulder ideal for slow-cooking methods, and therefore a popular cut in the world of barbecue. It's also the most popular primal for use with sausage.

The pork shoulder primal is typically subdivided into two subprimals: the Boston butt and the picnic. The foreshank (or hock) and front trotter are separated from the traditional picnic subprimal. Where to divide the shoulder from the loin is a subject of debate, with points ranging from the second rib all the way to the sixth; I like to make the split between the fourth and fifth ribs, making for nice shoulder chops or a lengthy shoulder roll (the pork equivalent of a chuck eye roll).

COMMON SHOULDER CUTS: shoulder roll, shoulder chop, blade steak, country-style ribs, pork brisket, bone-in butt, picnic cushion, smoked hocks, trotters.

REQUIRED EQUIPMENT	OPTIONAL EQUIPMENT
▪ Bins or lugs	▪ Bone dust scraper
▪ Boning knife	▪ Boning hook
▪ Breaking knife	▪ Butcher knife
▪ Handsaw/bone saw	▪ Cleaver
▪ Honing rod/knife steel	▪ Mallet (rubber or wooden)
	▪ Paring knife
	▪ Rib puller
	▪ Twine

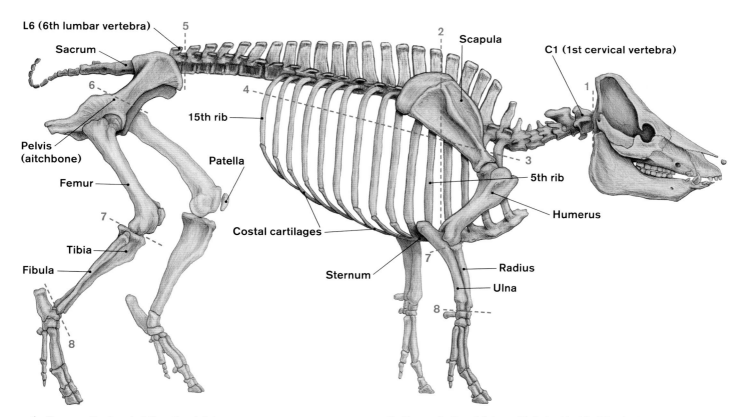

Loin

Boston butt

Head

Leg

Belly

Picnic

Breaking down a large pig carcass will involve many decisions, since each primal can result in several different end products. It's best to have a clear idea of the resultant cuts you want before proceeding with any breakdown.

BUTCHERING CUT POINTS

L6 (6th lumbar vertebra)

Sacrum

Pelvis (aitchbone)

Femur

Tibia

Fibula

Patella

Costal cartilages

Sternum

Scapula

C1 (1st cervical vertebra)

15th rib

5th rib

Humerus

Radius

Ulna

1. Remove the head at the atlas joint

2. Split the shoulder and loin primals between the 4th and 5th ribs

3. Separate the butt and picnic subprimals by cutting across the base of the ribs

4. Cut across the ribs to separate the belly primal from the loin

5. Separate the sirloin and loin just behind the last cervical vertebra

6. Separate the leg primal from the sirloin end of the loin

7. Separate the rear and fore hocks at the stifle and arm joints, respectively

8. Remove trotters at the base of the leg and arm bones

Loin

The loin on a pig extends all the way from the shoulder to the leg. On most other animals, the rib and saddle (or loin) primals are separated — for example, the rib and short loin on beef, the rack and loin on lamb — whereas the pork loin includes both of these, along with the sirloin and tenderloin subprimals. The loin is a lean primal composed primarily of the muscle *longissimus dorsi*, known as the loin eye. The most common loin cut is the bone-in center-cut pork chop, which is one of the more valuable cuts on the carcass. The loin also includes the back fat, a thick layer of adipose tissue that is often removed for curing, rendering, or sausage making.

COMMON LOIN CUTS: center-cut pork chops, loin chops, crown roast, boneless loin roast, back bacon, back ribs, riblets, button ribs, tenderloin.

Side (Belly)

A side of pork contains both the belly and the spareribs, and makes up about 23 percent of the headless carcass weight. The side is removed from the loin by a parallel cut made a few inches from the loin eye. It is characteristically fatty, with large swaths of adipose tissue among streaks of abdominal muscle. This lends itself to curing, which is the most common use for belly, as bacon or pancetta. Spareribs are exclusively barbecued, sometimes whole or after further processing into St. Louis–style or Kansas City–style ribs. Spareribs can also be removed as the entire rib cage with an alternate approach to splitting, covered on page 362.

COMMON SIDE CUTS: spareribs, St. Louis-style ribs, Kansas City ribs, bacon, pancetta.

Leg (Ham)

The hind leg of a pig is the largest primal. It is commonly referred to as fresh ham, largely because of its most common domestic usage as cured ham. It contains a number of the same hind-leg subprimals found in other animals: leg tip (or knuckle), inside round, and outside round. The muscles of the leg are large and lean, providing structure and movement to the animal but only secondary to the shoulder, resulting in less connective tissue. The hind shank (or ham hock) and trotter are removed from the leg and often cured, smoked, or boned for grinding.

COMMON LEG CUTS: smoked ham, baked ham, cured ham, ham steaks, fresh ham, smoked ham hocks.

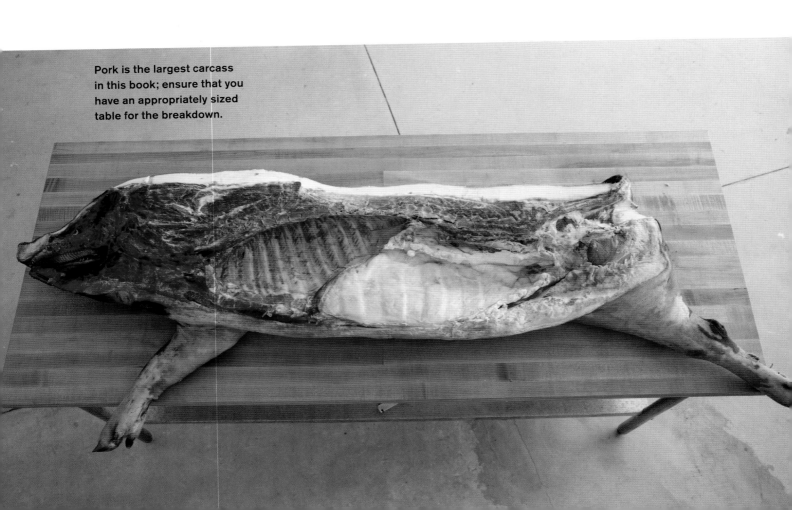

Pork is the largest carcass in this book; ensure that you have an appropriately sized table for the breakdown.

ONE CARCASS, TWO APPROACHES

There are myriad methods to breaking down a carcass of pork, but you can't have it all every time; removing the tenderloin prevents the cutting of pork porterhouses, for example. Fortunately, there are two halves to every carcass, which affords you the opportunity to break down each side differently, thus maximizing the diversity of cuts. A breakdown like this covers all your bases, accommodates most of the common cooking preparations and portioned cuts that come from pork, and does it all with a single carcass. Here's an example of how you can do just that:

	SIDE 1	SIDE 2
HEAD	Remove jowl and cheek separately; leave remaining meat on the bone for headcheese.	Remove jowl and cheeks separately; bone face for face bacon.
TENDERLOIN	Remove full tenderloin.	Leave tenderloin attached.
BELLY	Separate belly and picnic together from loin and butt; remove extra-long spareribs; split belly from picnic; save boneless belly.	Split belly and loin from shoulder primals; separate belly from loin; leave spareribs on belly for bone-in belly preparations; separate sirloin end of belly for boneless preparations.
BOSTON BUTT	Separate from loin between fifth and sixth ribs, giving you a long coppa (shoulder roll); seam-out coppa; bone the rest for sausage.	Separate butt from loin between second and third ribs, leaving shoulder chops on loin; save butt for pulled pork.
PICNIC	Separate hock and trotter together; bone remaining picnic for sausage.	Separate hock and trotter separately; leave picnic bone-in, skin-on for cracklin' shoulder roast.
LOIN	Separate from the sirloin; bone entire loin, creating baby back ribs, riblets, and a boneless loin roast that can be split into smaller sections.	Separate from the sirloin, leaving tenderloin attached to both sides; run chops, anterior to posterior, creating shoulder, rib, and porterhouse chops.
SIRLOIN	Separate from ham; bone and tie for sirloin roast.	Separate from ham; crosscut for thick sirloin steaks.
HAM	Remove hock and trotter separately; skin ham, then seam out smaller roasts.	Remove hock and trotter together; skin ham; crosscut a few ham steaks from the sirloin end; reserve distal end for a bone-in ham roast.

Remove the Leaf Lard

Leaf lard is the internal cavity fat found along the underside of the abdomen and around the kidneys. It makes the cleanest lard and, when not removed during slaughter, needs to be removed on the table. It's easy: get a good hold of the front (anterior) portion of the fat, and then quickly pull up and back, peeling it away from the flank. Use your knife only if parts are stubborn, but cut carefully because the tenderloin sits below a portion of the fat.

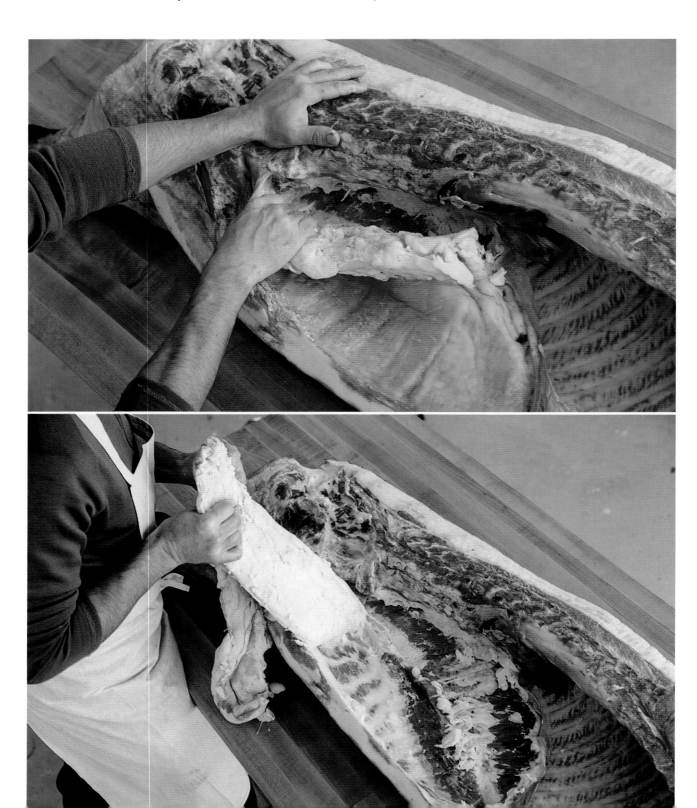

Remove the Head

The head, while not a primal, is completely edible and contains a cut prized for its curing ability: the jowl. It's a fatty cut that, especially on traditional pig breeds, is perfect for curing. The head and jowl are removed by severing the connection between the atlas joint (the neck vertebrae) and the skull. The area of the skull where the atlas joint occurs is quite flat and can be scraped clean, keeping all the neck meat with the neck. The second cut, for the jowl, can be found by bending the head toward the front leg and cutting where the skin folds.

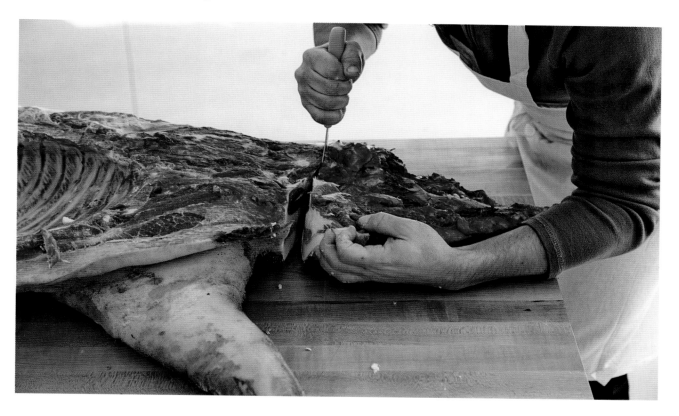

Using the entire length of the blade (the tip of your knife should scrape the table), make the first cut along the back of the skull, starting at the top and drawing toward the spine. When you reach the atlas, push the knife handle down and swing the blade around the back of the spine. Make the second cut by inserting the knife on the underside of the spine, connecting to the first cut, and drawing the knife down through the jowl, again using the full length of the blade. Use the tip of your knife to sever connections in the atlas joint and then lift the head and bend it toward the rear, snapping the spine.

Remove the Tenderloin

The tenderloin is a group of three muscles nestled under the spine. It originates inside the pelvis and tapers as it travels toward the head. Leave it attached only if you plan on cutting porterhouse-style pork loin chops.

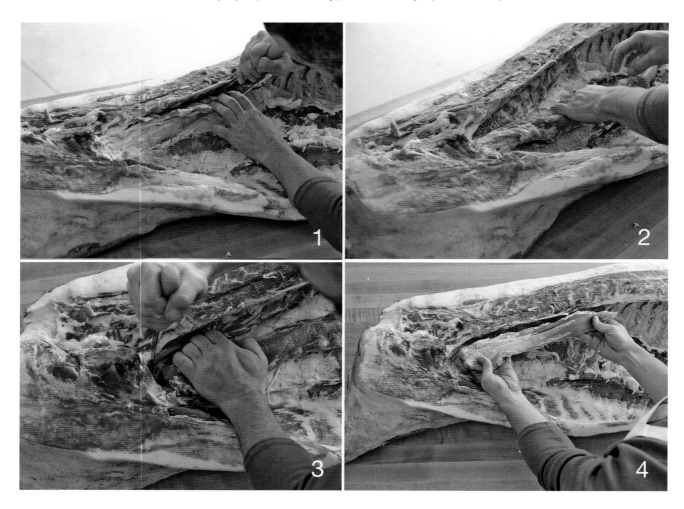

1. TRACE ALONG THE SPINE.

Starting at the last lumbar vertebra, use the tip of your knife to trace the upper edge of the tenderloin, where it connects to the spine. Cut until you reach the finger bones that sit behind the tenderloin. Follow your incision past the spine and along the pelvic bone until you reach the butt end of the tenderloin.

2. PEEL THE TENDERLOIN.

Now use your hand to peel the tenderloin away from the spine and pelvis.

3. RELEASE THE BUTT.

Cut across the butt end of the tenderloin, about 2 to 3 inches deep and through the thick layer of silverskin on the underside. The first couple of times, you may cut too deep and into the sirloin; don't worry – this is part of understanding the boundaries of muscles.

4. FREE THE "FOOT."

Use your knife tip to free the "foot" of the tenderloin, the small muscle that juts away from the base at an angle. Finish removing the tenderloin by peeling the long, tapered muscle away from the spine with the flat of the blade, pressing your knife against the bones to stay close.

Split the Loin from the Sirloin

Technically, the loin primal includes the sirloin, but when breaking down a pig I tend to leave the sirloin attached to the hind leg. It's easier and avoids the later removal of the sirloin from the loin. You can separate the loin and sirloin with a knife or a saw. When using a knife, you'll cut between the last and second-to-last lumbar vertebrae (see the photo in step 4); when sawing, aim for the middle of the last lumbar vertebra. This point is referred to as the separation point. These two points roughly line up with the front (anterior) tip of the pelvis, which is where we want to make the split.

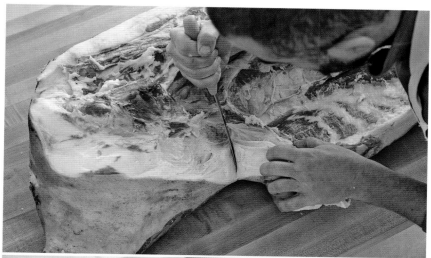

1. SEPARATE THE BELLY FROM THE HIND LEG.

Before you make the split at the separation point, you'll need to separate the belly and flank from the leg. The flank end of the belly attaches to the leg. You can feel the seam by pinching the belly right next to the leg: the leg is firm, from the whole, strong muscles present; the flank end of the belly is soft, thin, and fatty. Start cutting along that seam.

2. FOLD BACK THE BELLY.

Follow that seam toward the loin, stopping a few inches from the separation point. Fold the flank end of the belly over and out of the way of cutting.

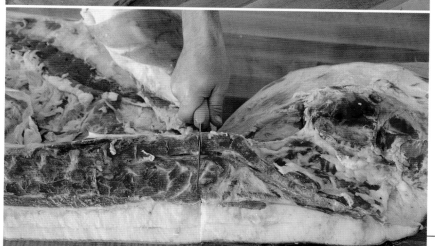

3. CUT BETWEEN THE LAST TWO VERTEBRAE.

Finish the split with a knife by inserting your blade between the last two lumbar vertebrae; you won't cut all the way through, but you should sever some of the connection.

4. **CUT TOWARD THE SPINE.**

Starting on the top (dorsal) side of the back, line up your knife with the separation point and cut toward the spine, making sure your knife is cutting all the way down to the table. When you reach the spine, swing the blade underneath by pressing the knife handle down and continue cutting.

6. **CONNECT THE CUTS.**

On the underside (ventral) of the spine, insert a butcher knife into the cut you just made and connect with the cut that separated the flank end of the belly.

5. **CUT THROUGH THE LOIN MUSCLES.**

Stop when you have cut through the loin muscles down to the bone of the spine and the tip of the knife comes through the other side.

7. **SEVER THE SPINE.**

With all the muscles and skin severed, lift the leg up and bend it toward the head, causing the spine to split at the point where you severed the connections. Or finish the split by sawing in the middle of the last lumbar vertebra.

THERE ARE TWO METHODS of separating the shoulder, the more common being the square-cut method, which gives you a whole shoulder with the loin and side separated together.

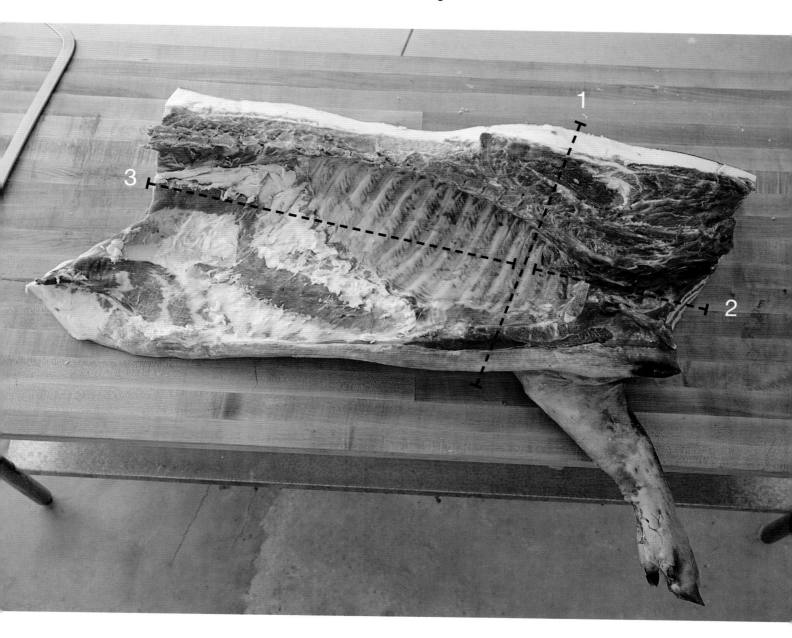

1. SHOULDER.

Separate the whole shoulder from the loin and belly, using the space between the fourth and fifth ribs as your guide.

2. BUTT/PICNIC.

Split the shoulder into the butt and picnic, cutting just below the thoracic vertebrae.

3. LOIN/BELLY.

Separate the belly from the loin, cutting through the boneless portion and then sawing through the ribs.

Separate the Shoulder from the Loin

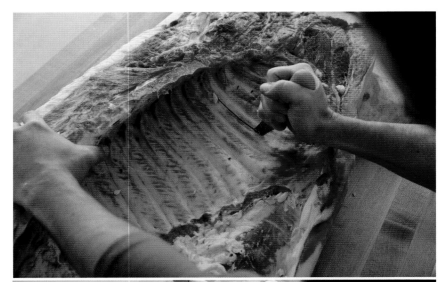

1. **CUT BETWEEN THE FOURTH AND FIFTH RIBS.**
Count out four ribs from the front of the rib cage; don't miss the minute first rib, which may be lost underneath a layer of fat and bloody meat. Separate the shoulder from the loin between the fourth and fifth ribs, in a similar manner to the approach used to separate the loin and sirloin. Insert the knife down to the table and draw the knife toward you between the ribs, cutting through the sternum.

2. **SCORE THE SPINE.**
Find the space between the fourth and fifth thoracic vertebrae (corresponding to the ribs) and score it.

3. **SEVER THE SHOULDER MUSCLES.**
Start above (dorsal to) the spine, making a deep cut in line with the space between the ribs, dragging the knife tip across the table.

4. CUT DOWN TO BONE.

Angle the knife handle as you swing the blade underneath the spine. When you encounter the scapula, try to cut through it. (If you can't cut through the scapula, cut over it and sever the muscles around the spine, then follow the alternate directions in step 5.)

5. SNAP THE SPINE.

With the muscles and scapula separated, fold the shoulder up and toward the loin, separating it. (If you weren't able to cut through the scapula, saw through the spine between the fourth and fifth ribs. Then, saw through the scapula and finish separating the underlying muscles with a knife.)

Split the Shoulder

The whole shoulder can also be split into the Boston butt and picnic subprimals.

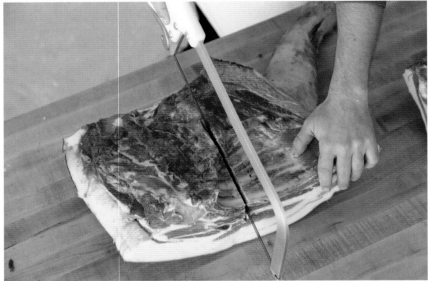

1. SAW ACROSS THE SHOULDER.

Find where the small first rib connects to the spine. Saw across the ribs at that point, roughly parallel to the top of the shoulder. Then, use your breaking knife to cut through the underlying muscle down to the scapula.

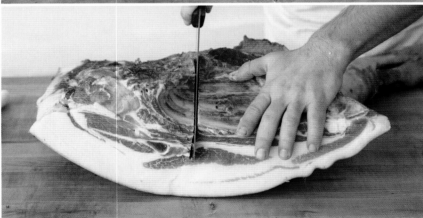

2. FINISH THE SEPARATION.

Saw through the scapula, and then finish the cut with a breaking knife.

Boston Butt (Shoulder)

Split the Loin from the Side

The location for splitting the loin and side will depend largely on how you plan on using the loin. The popular center-cut pork chop, coming from the rib section of the loin, will typically have a tail length that measures 2 to 3 inches from the loin eye; alternatively boning the loin means you can maximize your belly size and leave almost no rib on the loin (and still have back ribs).

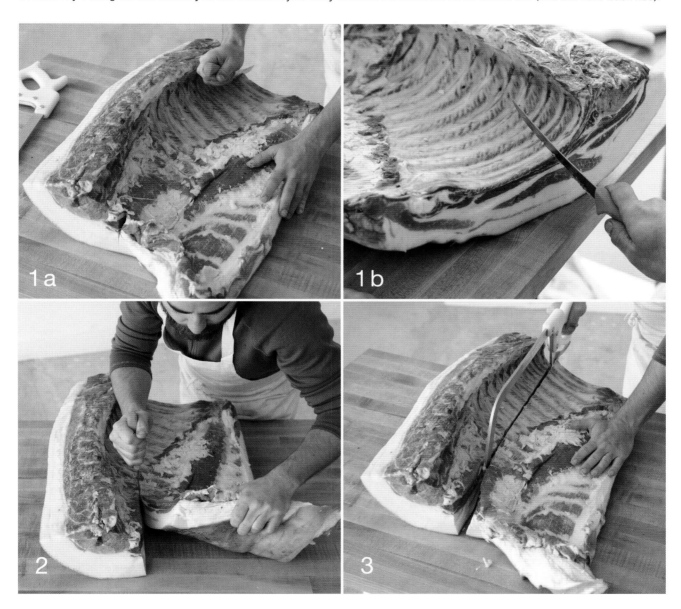

1. MARK THE CUT.

Measure out from the loin eye where to make your cut and mark it on both the shoulder end and the sirloin end.

2. CONNECT THE TWO MARKS.

Cut all the way through the ribless section of belly and score the rib bones.

3. SAW THROUGH THE RIBS.

Switch to the saw and make your way through the rib bones. Switch back to a knife and finish the separation.

Remove the Spareribs from the Belly

The ribs can be left on the belly for bone-in preparations, but more frequently they are removed as spareribs. For ideal spareribs, avoid scraping your knife against the ribs when removing them. Instead, leave about ¼ to ⅜ inch of meatiness on the ribs.

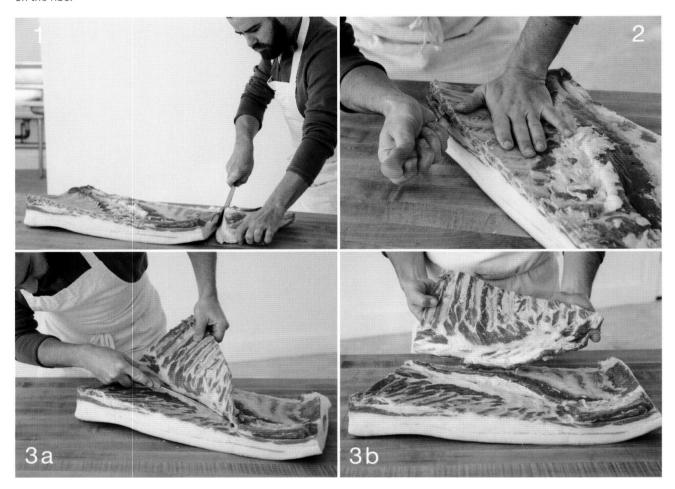

1. SQUARE UP THE CUT.

The sirloin end of the belly is best removed and used for grind, as the fat composition responds differently to cooking than the rest of the belly. Remove it with a cut perpendicular to the upper edge of the belly. Spareribs have a membrane that I highly recommend removing before cooking. Removal is easiest at this point, while the ribs are still attached to the belly. Follow the instructions on page 393.

2. SEPARATE THE RIBS.

Starting at the shoulder end, use the full length of your blade to cut along the underside of the ribs.

3. REMOVE THE RIBS.

Make sure the ribs are released down to their costal cartilage. Pull up on the ribs and cut underneath the cartilage, then the sternum, starting at the smaller end and heading toward the sternum.

Shoulder Option 2: THE SOUTHERN METHOD (UPPER-LOWER SPLIT)

THE SECOND APPROACH to separating the shoulder is geared toward folks who want to maximize the spareribs and plan on splitting the Boston butt and picnic. This involves making a long cut that separates the upper (dorsal) region, which includes the Boston butt and loin, from the lower (ventral) region, which includes the picnic and side.

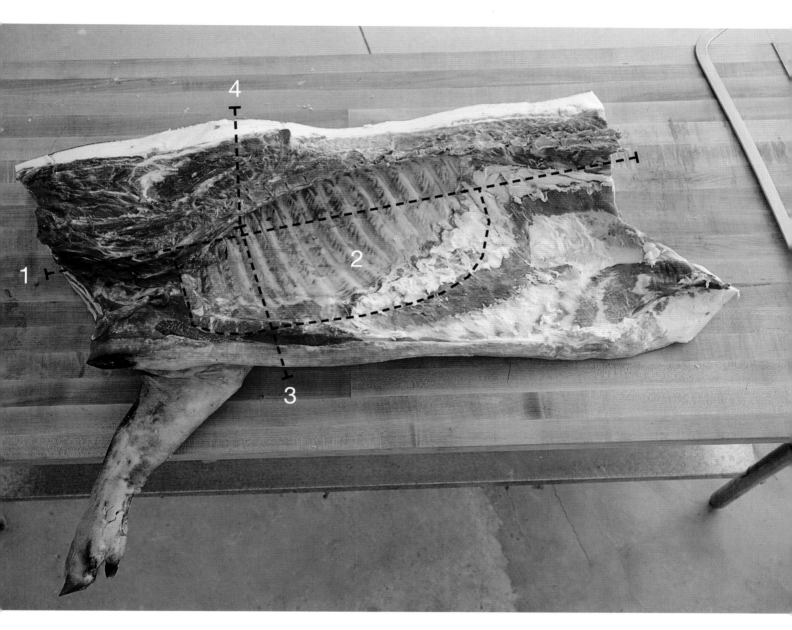

1. PICNIC/BELLY.
Remove the picnic and belly together, sawing through the entire ribcage.

2. SPARERIBS.
Remove the full sparerib rack.

3. BELLY.
Separate the belly from the picnic at a point posterior to the elbow of the ulna.

4. BUTT/LOIN.
Divide the butt and loin by sawing through the spine at the space between the fourth and fifth ribs.

Split the Shoulder

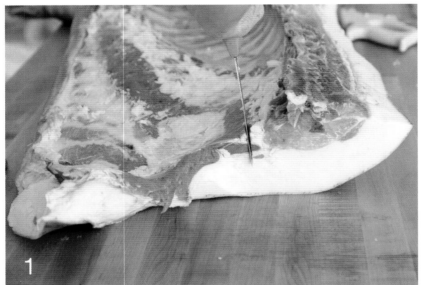

1. MARK THE SIRLOIN END.
Decide on the length of rib tail you want on the loin, commonly around 2 to 3 inches, and make a mark at the sirloin end of the loin.

2. MARK THE SHOULDER.
In the shoulder, find the point where the first rib connects to the thoracic vertebrae.

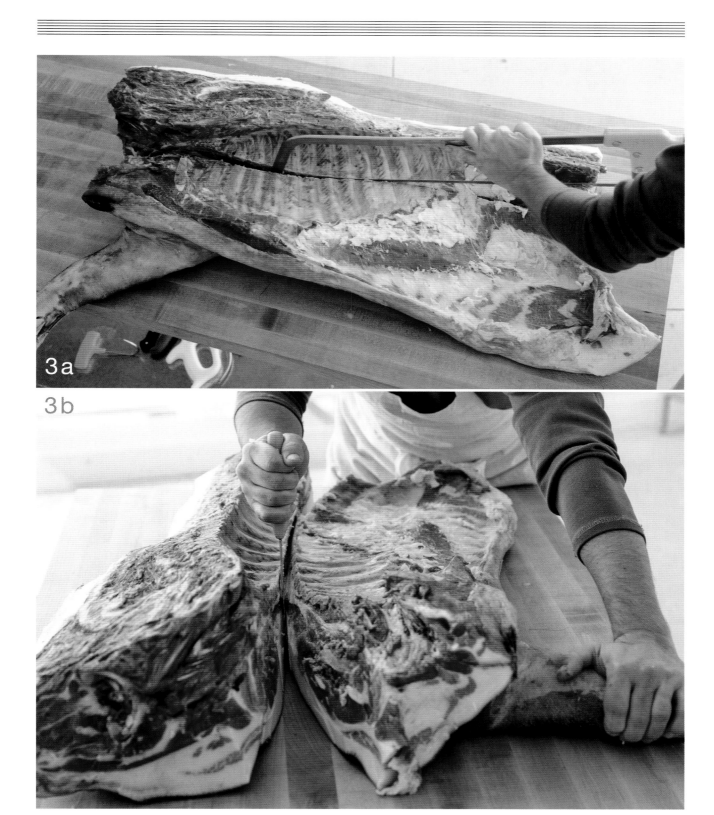

3a

3b

3. SCORE AND SAW.

Score along the shoulder, ribs, and cut through the ribless section of belly. (Stay closer to the spine if you want a wider belly and longer ribs.) Pull out the saw, and saw through the ribs and scapula. Finish the separation using a knife.

Remove the Spareribs

With the picnic and side attached you can remove the entire sparerib rack. Leave a good portion of meat (¼ to ⅜ inch thick) on the ribs when removing them. Remove the membrane, according to the directions on page 393.

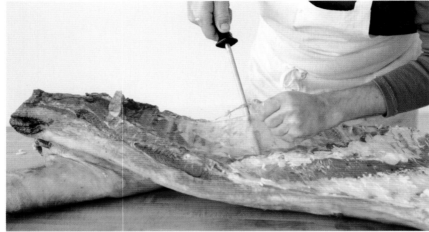

1. PEEL THE MEMBRANE.

While the ribs are still attached, remove the tough membrane that covers them. (Complete instructions are on page 393.)

2. START AT THE SHOULDER END OF THE RIBS.

Slide your knife beneath the beginning of the ribs and cut along the full length.

3. PULL THE BONES BACK.

Cut around the small end of the ribs and beginning of the costal carti-lage. Pull up on the ribs while cutting beneath the cartilage, following it toward the sternum. Work your way down the ribs and, in the shoulder, remove the attached sternum.

Make the Final Cuts

1. SPLIT THE PICNIC FROM THE BELLY.

With the rack of ribs removed, the picnic and side are simple to split. You will notice the thick area where the belly transitions to the shoulder. (It's just behind the "elbow" of the ulna.) Feel for it with your hands, and make a straight cut to separate the two.

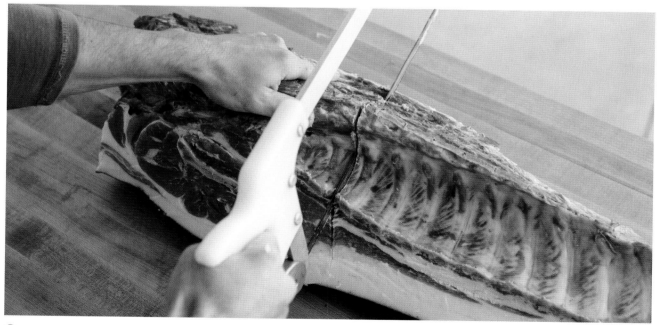

2. SPLIT THE BUTT FROM THE LOIN.

Count out four ribs and saw through the spine between the fourth and fifth ribs, stopping before you hit meat. Finish the separation with a breaking knife.

PORK: FURTHER PROCESSING

Each primal and subprimal removed during the previous process can be further processed into portioned cuts, tied roasts, or other preparations. This section is organized anatomically, starting with the head and working toward the tail.

Bone-In Sirloin Chops

Boneless Loin Roast (Saddle End)

Boneless Sirloin Roast

Tenderloin

Top Round (Inside Round)

Sirloin Tip (Knuckle)

Bottom Round (Outside Round)

Loin

HAM (LEG)

Side (Belly)

Semi-Boneless Ham (Leg)

Pork Steamship Roast

Belly (Side)

Bacon (Uncured)

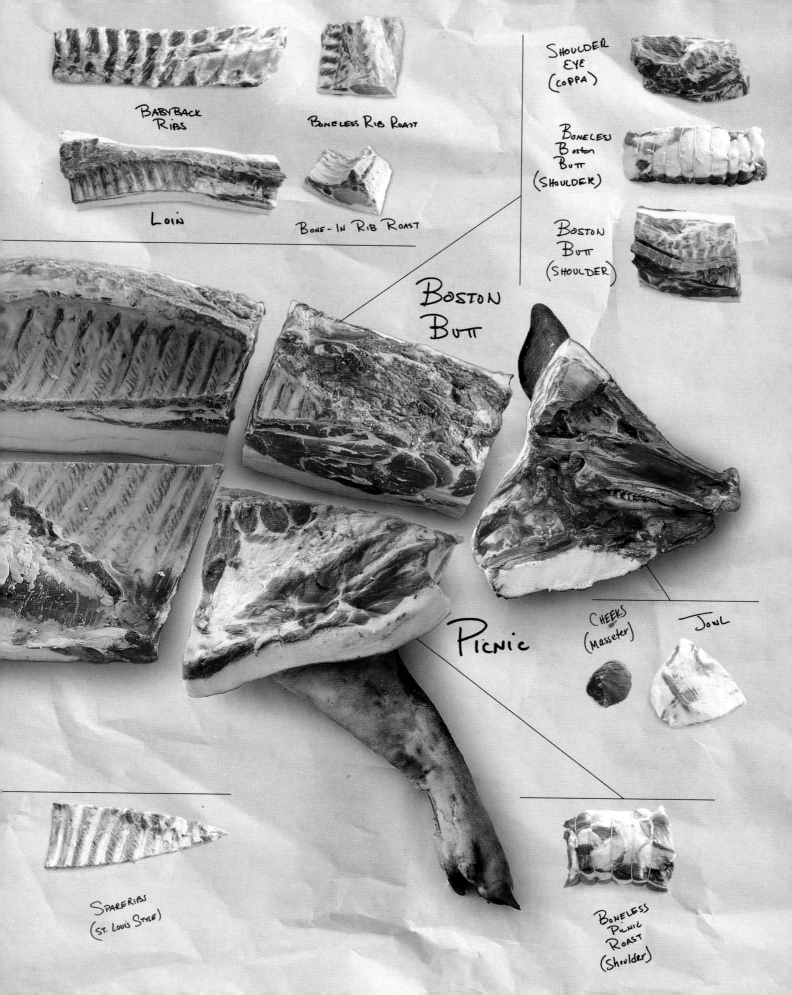

BABYBACK RIBS

BONELESS RIB ROAST

SHOULDER EYE (COPPA)

BONELESS BOSTON BUTT (SHOULDER)

LOIN

BONE-IN RIB ROAST

BOSTON BUTT (SHOULDER)

BOSTON BUTT

PICNIC

CHEEKS (Masseter)

JOWL

SPARERIBS (ST. LOUIS STYLE)

BONELESS PICNIC ROAST (Shoulder)

Diamond hatching on a roast allows for the meat to self-baste with its own fat while also producing delicious cracklings from the skin.

PORK SKIN can be a delectable addition to a roast or other preparations. Prepared right, it is salty, fatty, and crispy: the perfect snack. But it does require some preparation to ensure deliciousness.

SKIN-ON ROASTS Any pork roast that is prepared skin-on will need some special consideration. First, be aware that underneath the skin is normally a thick layer of fat, some of which we want to render into liquid and use to help baste the roast during cooking, adding moisture and flavor. Second, skin reacts to heat a bit differently than muscle does, contracting at a rate that can cause unwanted changes in the shape of the roast. Both of these concerns can be addressed by cutting the skin into a diamond pattern.

Start by cutting lines diagonal to the roast, spacing them 1 to 2 inches apart depending on the overall size of the roast. Try to cut through the entire skin and underlying layer of connective tissue but not too far into the fat and certainly not into the flesh below it. Then rotate the roast 90 degrees and cut another set of lines, maintaining the same spacing.

These cuts will provide pathways for rendered fat to escape while also preventing the skin, which was formerly a solid sheet across the whole roast, from contracting and causing a misshapen result.

REMOVING SKIN In many cases you will not want to leave the skin attached to roasts or cuts, and in these cases it is best to remove the skin in one full sheet, allowing you to use it for other purposes. When removing skin, keep the blade of your knife against the skin the entire time, leaving as much fat with the muscles as possible. This can prove quite difficult, though, because the fat layer changes in thickness and the skin follows the various contours of the muscles. Removing skin from the belly, a generally flat cut, will be far easier than removing the skin from an entire hind leg, where the skin wraps around the whole primal and has no flat area to follow. With the harder cuts start by making smaller strokes with the knife, working toward longer strokes as you get more comfortable with the process. And always keep your blade angled slightly toward the skin.

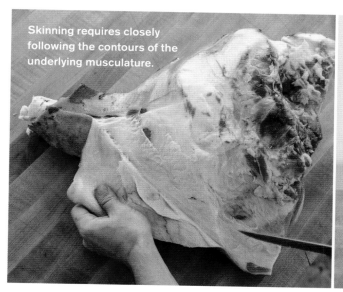

Skinning requires closely following the contours of the underlying musculature.

Fat and skin can be easily separated by working in strips.

Pork skin is thick and tough, allowing for scraping during removal, so keep the blade angled toward the skin. Keeping the skin taut is key: start by releasing enough of the skin to hold on to, then keep it taut while cutting against the skin. Use long knife strokes. I prefer drawing the knife toward me when skinning, but any direction can work.

When skin is used solo it is best scraped of fat, and the best method for fat scraping is flat against a table using a boning knife. Plus, this allows you to save the fat, for which rendering or grinding is a much better use.

Cut the skin in manageable strips or sections. Place the skin fat-side up and next to the edge of the table. Having the skin next to the edge of the table allows your knife to stay parallel with the surface because your knife handle won't be hitting the tabletop. Firmly hold one end while cutting away with your knife, pressing against the skin and angling the blade toward the skin. Turn the skin around and repeat, clearing the opposite side of fat.

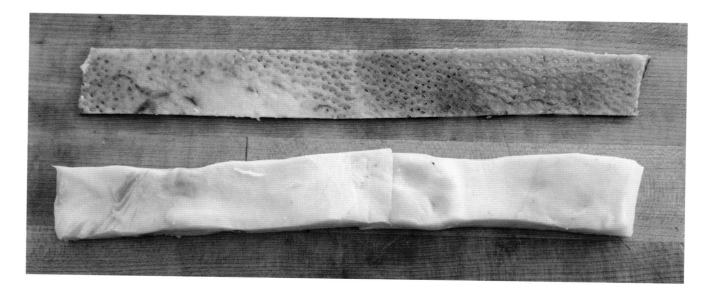

JOWL

The jowls of a pig contain a healthy deposit of fat with some minor streaking of muscle throughout. In Italy and many other countries, they are often cured into *guanciale*, a preparation similar to pancetta. Jowls are riddled with salivary glands, also called parotid glands. You should remove these glands prior to cooking or curing while leaving as much fat and meat on the jowl as possible.

CHEEKS (MASSETER)

Pork cheeks are dense muscles, chock-full of small muscle fibers and connective tissue. They are best prepared using moist-heat methods, converting the collagen into unctuous gelatin. But first they need to be removed and cleaned.

EARS

Pig ears are frequently thrown to the dogs, but they need not be. The cartilage in ears must be broken down through long periods of moist cooking, after which the ears can be fried, whole or sliced. Remove the ears by cutting through the base, where they connect to the skull. Cleaning is minimal: singe away any leftover hair using a torch or open flame on a stove, and then scrub well with a brush.

HEAD

Remove the Jowl

The jowl should be removed without the cheek (masseter), a muscle that is better left for braising.

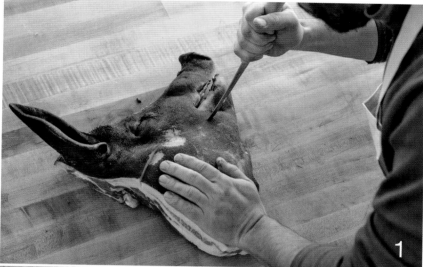

1. TRACE THE JOWL.

Place the split head bone-side down with the jowl facing you. Outline the upper edge of the jowl: start a couple of inches below the eye, curving behind the mouth and ending at the jaw. Cut an inch or two deep to avoid the cheek muscle.

2. PEEL THE JOWL AWAY FROM THE CHEEK.

Slowly cut deeper until you can see the cheek muscle. Pull the jowl away from the cheek, helping with your knife at times. Find the jawbone, starting just below the cheek, and follow it to finish separating the jowl.

Remove the Jowl CONTINUED

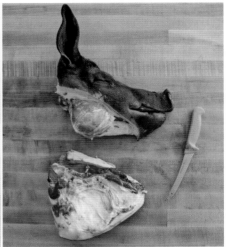

3. CUT ALONG THE JAW.
Find the jawbone, starting just below the cheek, and follow it to finish separating the jowl.

4. FIND THE SALIVARY GLANDS.
Place the jowl skin-side down and notice all the pockets of light brown salivary glands.

5. REMOVE THE GLANDS.
Trim away the glands, keeping your knife flat and above the muscle. Trim as little as possible while ensuring that all the glands are removed.

Remove the Cheek

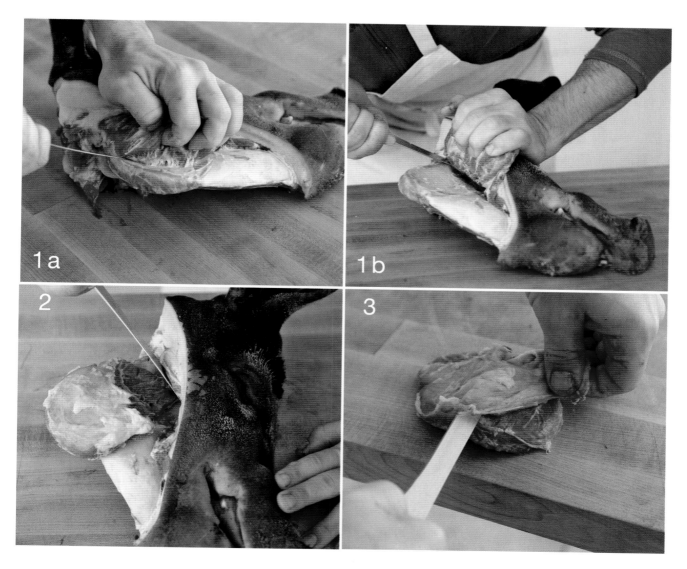

1. START AT THE JAWLINE.

The cheek muscle helps connect the lower jawbone, or mandible, to the skull. Begin removing the cheek at the bottom of the jawbone, pressing the knife tip against the bone and working all the way toward the skull, peeling the muscle away with your fingers as you go.

2. CUT FROM THE TOP.

The skull overlaps the upper portion of the jawbone, and getting in the tight space to remove the full cheek can be tricky. Fully insert your knife into the space underneath the muscle and cut across. Repeat above the muscle. Pull the muscle out of the space while severing the top edge.

3. REMOVE THE SILVERSKIN.

Trim away all the exterior silverskin and any other connective tissue until you are left with just the dark red cheek muscle.

Remove the Cheek and Jowl CONTINUED

CHEEKS (Masseter)

Jowl

Cheek and jowl

BOSTON BUTT and PICNIC

THE SHOULDER IS THE WORKHORSE of the animal so it would follow that it also has the greatest potential for flavor. More often than not, the shoulder is split into its two subprimals: the Boston butt and the picnic. The butt is the more tender of the two — this makes sense, since the picnic is closer to the ground (and the rule is, the closer to the ground the less tender it is) — and is comprised mostly of the notable shoulder roll. The picnic is best utilized for either pulled pork or other slow, lengthy, low-temperature preparations, or boned and used in sausage. Lop off the hock (and smoke it!) and use the trotters to embolden your stock.

Boston Butt

Picnic

Option 1: Boneless

The Boston butt can be boned for roasts or for use in sausage.

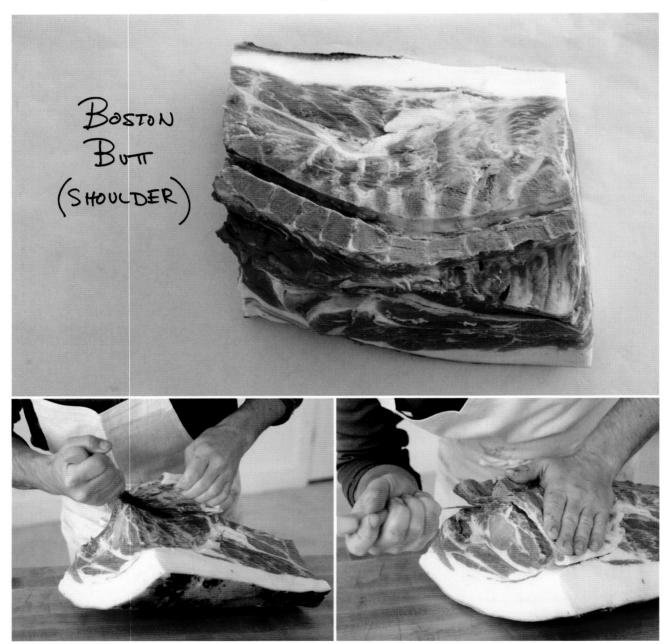

BOSTON BUTT (SHOULDER)

1. RELEASE THE FEATHER BONES.

Start by releasing the feather bones with the butt skin-side down. One by one, work your knife underneath the tip of the bone and then cut toward the spine, keeping your blade angled and against the bone.

2. RELEASE THE RIBS AND SPINE.

Begin working down the ribs with long strokes that travel against the bones. A section of spine steps out where it joins with the ribs. When you reach it, scrape along and out with your blade until you can cut around it. The thoracic vertebrae (those with ribs) should now be released.

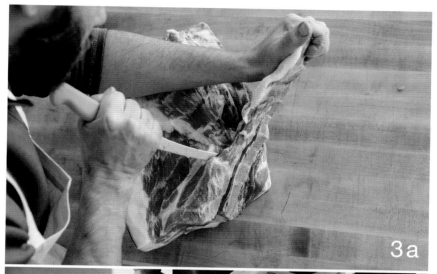

3a

3. CUT AROUND THE NECK BONES.

The neck bones are some of the hardest to extract cleanly but can be removed solely by using the knife tip. Pull up on the released section of the spine and begin working your way down the neck bones, cutting around them one by one. It will take some exploring to learn their shape, and this can only be done hands-on.

3b

4

4. ACCESS THE SCAPULA.

Turn the butt so that you can see the two edges of the scapula: the posterior edge was connected to the loin; the distal edge was connected to the picnic. The face-up side of the scapula is largely flat. Cut in from either edge and follow the entire flat until you've reached the interior edges.

Option 1: Boneless CONTINUED

5

5. TRACE THE UNDERSIDE OF THE SCAPULA.

Follow the flat underside of the scapula, cutting along the posterior edge and staying close to the bone. The tip of your knife should scrape along the scapular ridge.

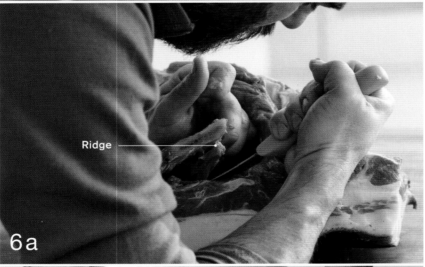

Ridge

6a

6. RELEASE THE RIDGE.

The bony keel of the scapular ridge presents the main challenge in extracting this bone from the shoulder: it curves much more dramatically than that in other species. The ridge also runs diagonally, starting at the anterior distal edge (where the picnic was) and heading across. Cut in where it starts, pulling up on the bone and using the tip of your knife to explore the shape of the ridge until it's released. (This will largely be an exploratory exercise the first number of times.)

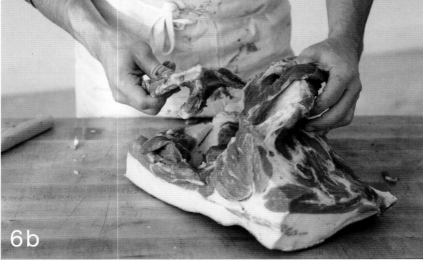

6b

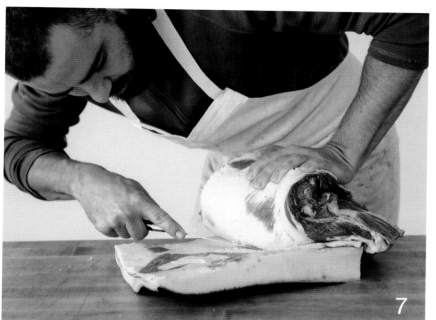

7. REMOVE THE FATBACK.

Trim away any interior connective tissue that was uncovered during the boning process. Remove the fatback as a sheet, peeling the meat away as you cut in long, even strokes. (Leave a bit of fat on if you plan on roasting the butt.) Skin the fatback according the directions on page 371.

Boneless Boston Butt (Shoulder)

A boned and tied butt of pork.

Option 2: Shoulder Roll

The pork shoulder roll (also known as the cottage, cellar, or coppa) is a glorious collection of muscles and fat, providing a more interesting array of textures and flavors than any other part of the carcass. (It is the equivalent to the beef chuck eye roll, which is also delicious.) It includes a diverse selection of muscles — *longissimus dorsi*, *multifidus dorsi*, *complexus*, and *serratus ventralis*, just to name a few — many of which are a continuation of the loin. It's also extremely versatile: it can be roasted, braised, grilled, or cured.

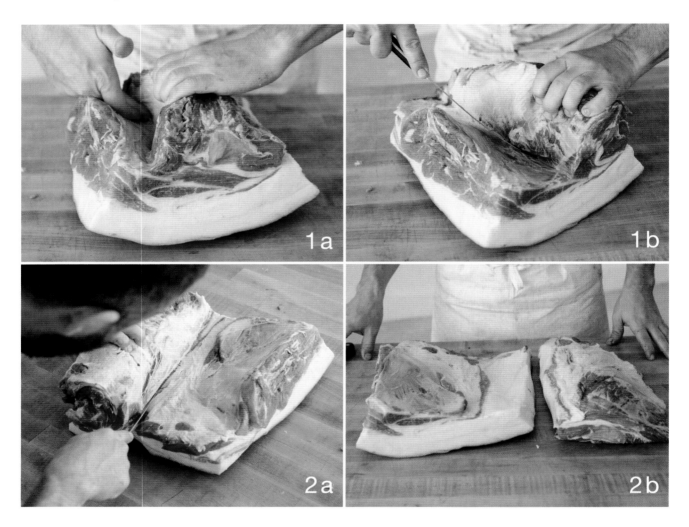

1. FIND THE SEAM.

Follow the first three previous steps for boning a butt. With the butt skin-side down, there is a seam that runs above the scapula. Find it with your fingers, then follow it with your knife, severing the muscles that attach to the scapula.

2. ROLL YOUR EYE.

Follow the seam until you hit the fatback. Then cut along the fatback while rolling the muscle group away until released. (You can leave some fat on the roast for basting.) Remove the scapula, starting with step 5 of the aforementioned butt boning process, then separate the muscle and fatback.

Shoulder (Blade) Chops

In my opinion, these pork shoulder chops are far superior to loin chops, because of the diversity of their muscle and therefore taste. You don't get many of them on an animal. Cut them thick to allow for a slower cooking process than the extremely lean loin chop.

The process is similar to cutting bone-in loin chops (page 398) though I tend to start on the bone side with shoulder chops. Decide whether to cut the chops in one- or two-rib increments. Saw through the chine, then cut through the underlying muscle until you hit the scapula; saw through the scapula, then finish the cut with your knife.

The first couple of shoulder chops from the Boston butt will provide perhaps the best bone-in cuts on the whole carcass.

VARIATION: COUNTRY-STYLE RIBS

Country-style ribs are best prepared bone-in, though they can be boneless, and ideally come from the third through the sixth ribs, which span sections from the shoulder and loin. Thus, these instructions can apply to the loin as well. Basically, it's a chineless shoulder chop cut in half diagonally.

Decide how many ribs you'll need and remove all the ribs as a whole. (Keep in mind that each rib section of meat will give you two country-style ribs. Want six country-style ribs? Remove a three-rib section.) Remove the chine bone (page 394).

Then cut your one-rib chops, sawing through the scapula if you run into it. Place each chop on its side and split into two country-style ribs by cutting corner to corner, passing between the ribs and feather bones.

Boning the Picnic

The picnic can be prepared skin-on, but if you plan on removing the skin, do so before boning out this subprimal. For those not wanting to cook the picnic whole, there is one significant solid muscle group, known as the cushion (*triceps brachii*), that can be separated and tied for a smaller roast.

REMOVE THE TROTTER

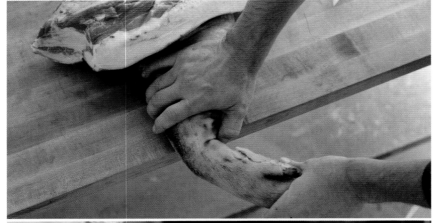

1. FIND THE CREASE.
Place the front trotter over the edge of the table. Bend it back and forth, feeling for the joint with your fingers.

2. CUT AT THE JOINT.
Cut through the skin and ligaments at the crease in the joint and, using the edge of the table as leverage, snap open the joint.

3. FINISH REMOVAL WITH A KNIFE.
If using the knife proves challenging, you can always saw the trotter off at the joint.

REMOVE THE HOCK

1. CUT ABOVE THE ELBOW.

One of the arm bones, the ulna, juts out and forms the elbow. (Note the skeletal diagram on page 349.) The end that joins with the humerus is shaped like a "C," tightly cupping the end of the humerus. Holding the bones together are some tough ligaments that have to be severed to open the joint. Cut above the elbow and down across the joint, making sure to cut down to the bone and through the ligaments.

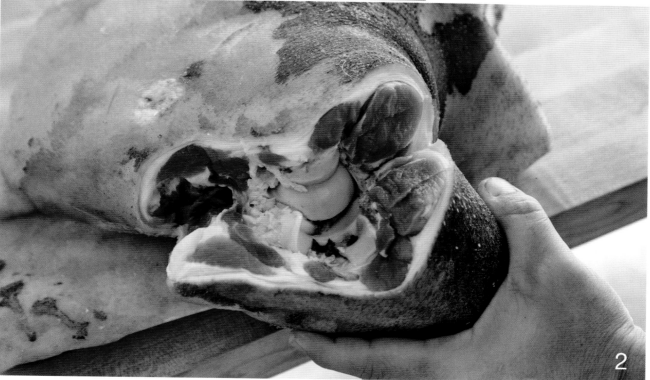

2. OPEN THE JOINT.

Place the hock over the edge of the table and apply some pressure to open up the joint, going back over the ligaments with the tip of your knife if they weren't severed on the first pass. Once the joint is open, finish removing the hock by cutting through the underside. Use a saw if knifing through the joint proves too challenging.

Boning the Picnic CONTINUED

REMOVE THE HUMERUS AND SCAPULA

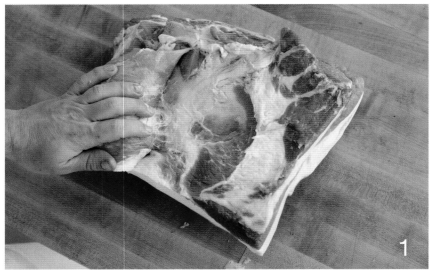

1. SEVER AND SEAM THE BRISKET.

The humerus, or upper arm bone, and a portion of the scapula are left. Place the picnic skin-side down with the humerus facing you. The pork brisket (pectoral muscles) sits atop the humerus with an easily separated seam beneath; to cleanly get to the bone you'll need to cut through them. Cut through the muscles, down to the seam, on a path that roughly follows the humerus bone. Then peel back the brisket at the seam to clear a way to the humerus.

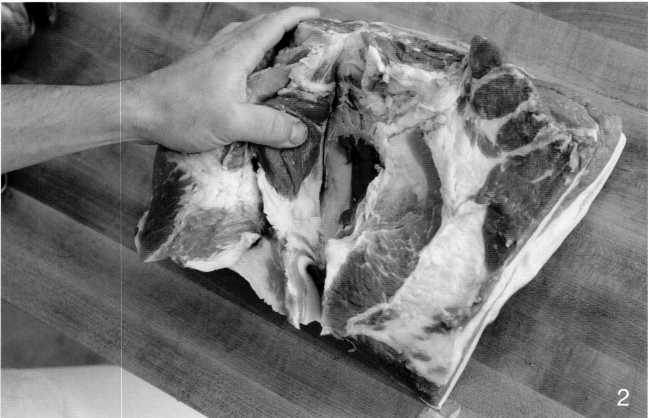

2. CUT DOWN TO THE BONE.

Sever the overlaying muscles to access the bone, following seams if possible.

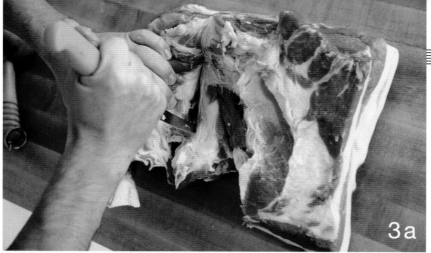

3a

3. RELEASE THE BONES.

Bone both edges of the humerus and the distal end (that was connected to the hock). Work your way under the bone until you can insert your fingers and lift up on the bone. Finish removal by boning the scapular joint and removing the remnants of the scapula.

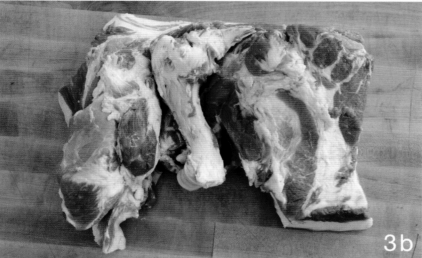

3b

4a

4. ROLL THE ROAST.

Roll, tie, and square off the ends of the picnic for an ideal roast.

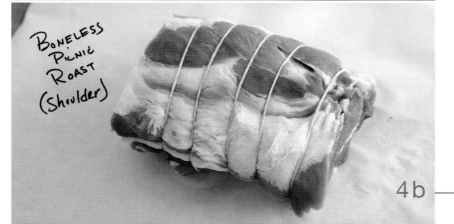

Boneless Picnic Roast (Shoulder)

4b

THE PORK LOIN PROVIDES the ubiquitous pork chop (which, in turn, provides most of the revenue from a pork carcass). But chops need not be the only product prioritized from this delicious primal: back ribs, fatback, and loin roasts all provide unique flavors and offer a diversity of preparations.

BACK RIBS

Back ribs, also known as baby back ribs, are the chineless by-product of a boneless rib loin fabrication. Another type of back rib, button ribs are the finger bones from the lumbar section of the loin. The consequence of a good boning process is that little to no meat is left on the bone, so if you're looking for good back ribs, be sure to leave ¼ inch or so of meat on the bones.

BONE-IN ROASTS

Little beats the presentation and quality of a bone-in loin roast. The loin is separated between the rib (thoracic) and saddle (lumbar) sections for bone-in roasts, and then further portioned from there. Most frequently, the rib loin is used for bone-in chops while the saddle is used for roasts. Bone-in roasts can be prepared with the skin and fat on, but if you plan on removing the skin or fatback, follow the instructions on pages 371 and 381, respectively. In this case, I removed the skin and fatback before proceeding.

The rib loin can also be frenched, a process that removes all meat and connective tissue to produce bare bone ends. (I prefer to leave the delicious intercostal rib meat on.) You'll need to prepare ahead of time for this and leave some extra length on your ribs when you split the loin and belly. Frenching a pork rack is the same process as with lamb, so follow the instructions on page 286 to learn how.

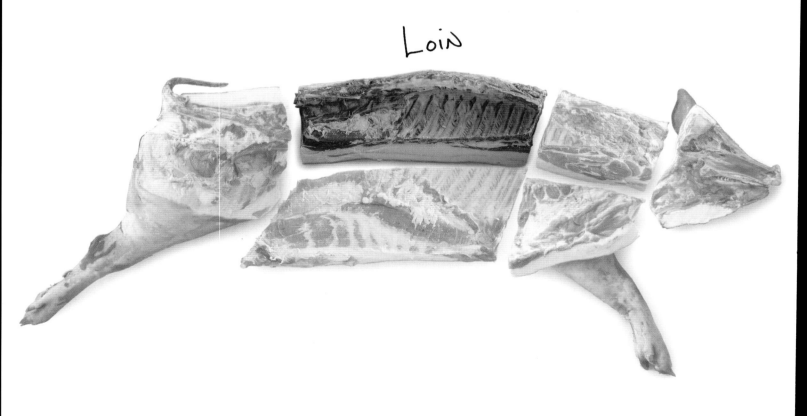

Loin

BONELESS ROASTS

The loin is the most valuable primal on the pig, as well as one of the leanest. The loin primal provides a number of fabrication options, especially if you bone it out. The key to boning a long section of lean meat like a loin is to work with long strokes of the knife while staying close to the bone. Short cuts inevitably mar the surface of muscles, making them less presentable or requiring later trimming and a loss. Use your longest boning knife for this job.

There are two approaches to boning a loin: to chine or not to chine. (Chining is the process of removing the vertebral connection between the ribs or finger bones and the feather bones.) Both work, so choose what suits your style best. If your loin is longer than your saw blade, chine the loin according to the first four steps for fabricating a Portion-Ready Lamb Rack (page 283).

CHOPS: RIB AND LOIN

Loin chops are probably the most popular cut from the loin. Fortunately, you can cut some chops and still have loin left over for roasts or other preparations. There are three kinds of chops: rib, loin, and porterhouse. Rib chops are from the rib section; loin chops are from the saddle; and porterhouse chops are loin chops with the tenderloin left on. The chops can be prepared skin- and fat-on, though if you choose to remove the skin and the fat, do so before cutting the ribs; use the instructions on page 371.

Home fabrication of loin chops is best done between the ribs, so you don't have a lot of choice when it comes to thickness. (You can, of course, bone the loin out and cut boneless chops to whatever thickness you want.) If you want to chine the loin first (page 394), then you can skip the steps that saw or cleave through the vertebrae. Rib or saddle, the process is the same all the way down the spine.

FATBACK

Overlapping the shoulder and loin, and continuing on into the belly, is a thick layer of subcutaneous fat known as fatback. It is characteristically dense, more so than any other fat, and is a great addition to sausage or cured as lardo. There are two layers of fat on the loin: the exterior is the fatback, and the interior is a thinner layer that covers the loin and is separated from the fatback by a thin layer of connective tissue. You can remove one or both layers, though the thinner layer is often left on completely or partially for flavor. If you want to remove the skin from the fat, do so after removal. The following removal process can be done with a bone-in or boneless loin, and can also be applied to the shoulder. A breaking knife or other long knife works well for this process.

TENDERLOIN

On all animals profiled in this book, the tenderloin is the leanest, most tender muscle in the body, resulting in little flavor and a tendency toward a mealy texture. I tend to leave as much fat on as possible, but the cut is customarily fabricated by removing all the exterior fat and connective tissue, the instructions for which are here. The tenderloin is composed of three muscles: *psoas major*, *psoas minor*, and *iliacus* (a muscle that juts angularly away from the head, forming the "foot" of the tenderloin).

BONE-IN
RIB

BONELESS
RIB

CHOPS

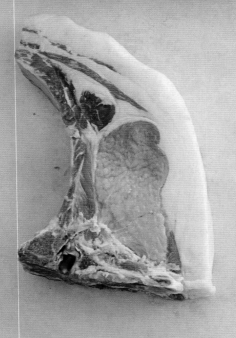

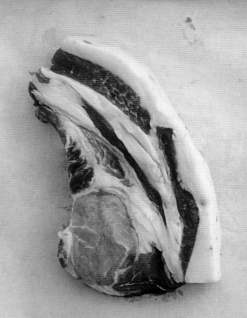

BONE-IN
RIB

BONELESS
RIB

Bone-In
Loin

Boneless
Loin

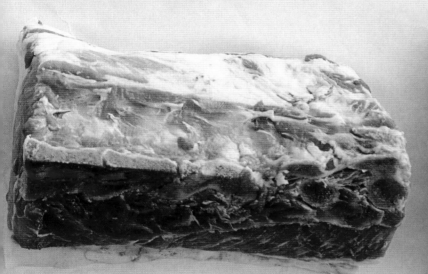

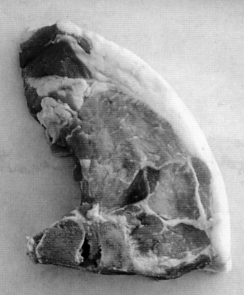

Bone-In
Loin

Boneless
Loin

Removing Baby Back Ribs

Start with a chined bone-in rib loin.

1

2a

2b

1. CUT ALONG THE PROXIMAL EDGE.

As you know, ribs curve. To minimize any damage to the valuable loin, remove the ribs in two steps. Start by releasing the entire edge that was formerly connected to the spine, staying close to the bone.

2. CUT ALONG DISTAL EDGE.

Follow that by cutting along the opposite edge. Pull up on the ribs and sever any remaining attachments between your two incisions.

Removing the Membrane

Button ribs need no cleaning, but before baby back ribs can be cooked, they need to have a membrane removed.

1. LIFT THE MEMBRANE.

Start at either end. On the outside of the ribs — the side that has no meat — insert your honing rod next to one of the bones and underneath the membrane. Slowly pull up and release the end of the membrane.

2. PEEL THE MEMBRANE.

Hold the bones to the table and carefully peel the membrane away. A paper towel can help provide grip on the membrane.

Split the Loin into Rib and Saddle

Assuming that you aren't going to roast the entire loin whole, start by separating the two sections of loin, rib and saddle. Cut through the meat between the last two ribs, making sure to cut down to the bone, and then saw through the spine.

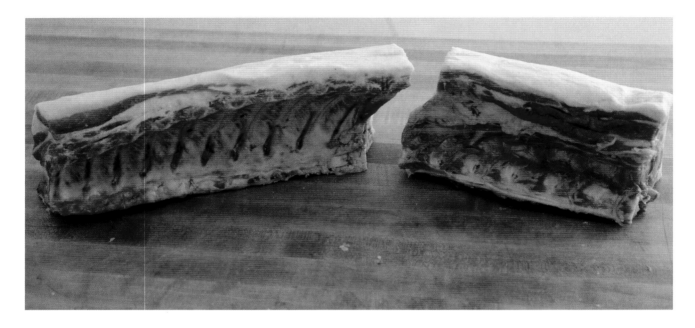

OPTION: REMOVE THE CHINE

Any bone-in roast benefits from having the chine bone removed. This will provide the option to easily remove the feather bones while leaving the ribs or finger bones on. After cooking, bone-in portions can be easily cut without the need to get through bone. Follow the first three steps on page 395 to remove the chine and feather bones.

With chine

Without chine

Boneless Loin Roast, Method 1: Chined

Chining any part of a loin — whether it's the rib, saddle, or the entire length — involves the same fundamental tasks. Here we're using the rib section of the loin.

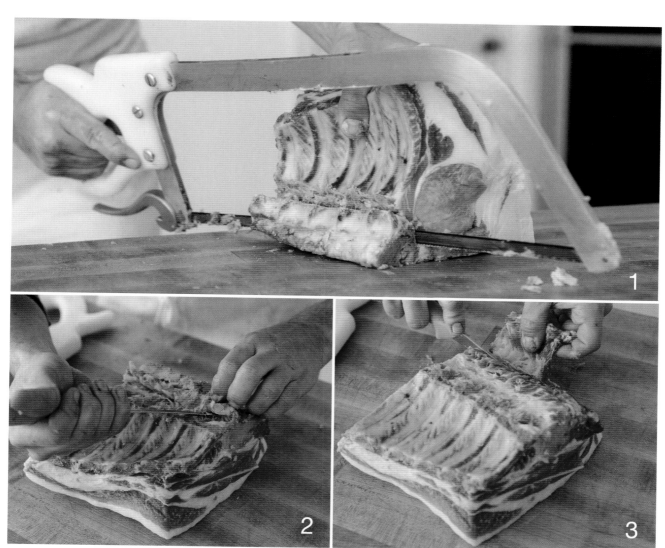

1. SAW OFF THE CHINE.

Place the loin flat on the table, feather bones down. Saw across all the ribs, starting about an inch above where they join with the spine. Keep your blade at roughly a 45° angle.

2. REMOVE THE BUTTONS.

The ends of the ribs are shaped like buttons where they join with the feather bones (the costovertebral joints). Use the tip of your knife to remove them, staying shallow and away from the loin muscle. (This step can be skipped when boning only the lumbar portion of the loin.)

3. REMOVE THE FEATHER BONES.

Starting from the chined end, get under the feather bones with the blade edge and cut along the flat face, pressing against the bone. Remove them with their cartilaginous tips.

4. REMOVE THE RIBS.

Follow the instructions for removing baby back ribs on page 392.

Boneless Loin Roast, Method 2: Unchined

Boning the rib and saddle regions of the loin follow the same principles; in this example we are boning a lumbar region.

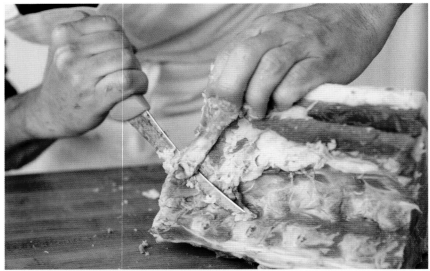

1. REMOVE THE LAST RIB.

(This is for lumbar loins only.) Get under the last rib with your knife and cut down to the junction with the spine. Twist the rib and remove it by hand.

2. SEPARATE THE FINGER BONES.

Removing finger bones can be a bit tricky because they are more embedded in the meat than ribs are. Make a shallow cut along the top edge of the finger bones (or ribs), prying them away with the blade edge. Follow the initial cut and continue down the flat of the bones, removing them as one whole piece (a.k.a. button ribs).

3. NAVIGATE THE RIDGE.

At the base of the ribs, turn your blade outward to get around the step in the spine and, once cleared, angle the blade inward, staying close to the bone.

4. REMOVE THE FEATHER BONES.

Follow the flat face of the feather bones, cutting away from the spine. Stay close to the bone, and then cut through the thick connective tissue at the tip. (For sections of loin much longer than your knife, it may be easier to approach the finger bones from the outside. Cut along the tip of the feather bones, keeping your knife angled slightly at the bone. Continue cutting, in long strokes, until you hit the connection with the ribs and finger bones.)

4a

4b

Cutting Chops

WITH A SAW

1. CUT THROUGH THE MEAT.
Use a breaking knife in one long stroke.

2. SAW THROUGH THE SPINE.
With the feather bones against the table, saw through the vertebrae.

WITH A CLEAVER

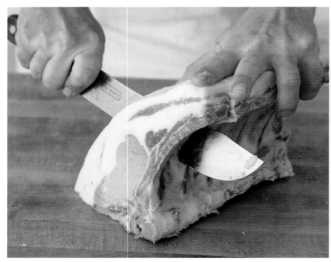

1. CUT BETWEEN THE RIBS.
One by one, cut between each rib or finger bone, making sure to sever the loin muscles all the way down to the spine.

2. CLEAVE THROUGH THE BONE.
Place a cleaver on the bone. Strike the spine of the cleaver using a mallet until you're through the vertebrae.

Removing the Fatback

Decide on the thickness of fat to be left on the loin. (For our example we'll use ½ inch, but choose whatever you want.) Score the fat ½ inch from where it overlaps the meat, cutting 1–2 inches deep along the entire loin. Use this as your guide. Continue cutting the fat in long, smooth motions, following the curvature of the loin muscle and working the entire length of the loin.

Cleaning the Tenderloin

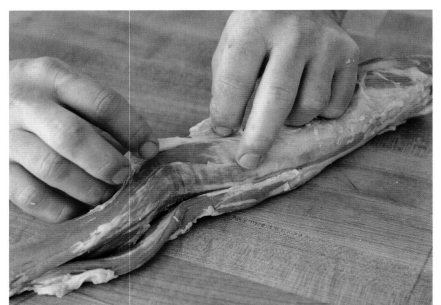

1. REMOVE THE FAT AND CONNECTIVE TISSUE.

Use your fingers to get underneath the obvious layers of fat and connective tissue, using your knife only when needed.

2. RELEASE A STRIP OF SILVERSKIN.

On top of the main muscle is a layer of silverskin that must be removed. The topside grain of the tenderloin runs from the tapered tail toward the large butt end, and the fineness of the grain requires you to work with it, otherwise you are left with a rough surface. Its obvious origin is near the middle of the tenderloin. Slide your knife underneath a portion of the silverskin, blade facing toward the butt. Hold the tenderloin against the table and run your knife along the silverskin and out the butt end, keeping your blade angled just slightly toward the silverskin.

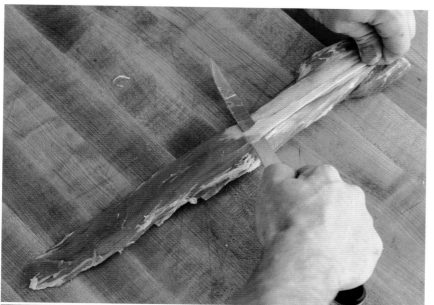

3. SEVER THE STRIP OF SILVERSKIN.

Hold the recently released strip of silverskin taut and cut through its origin. Repeat the process until the entire silverskin is removed. You can also trim the underside, though be forewarned that the grain runs in the opposite direction, starting at the butt end and heading toward the tapered tail.

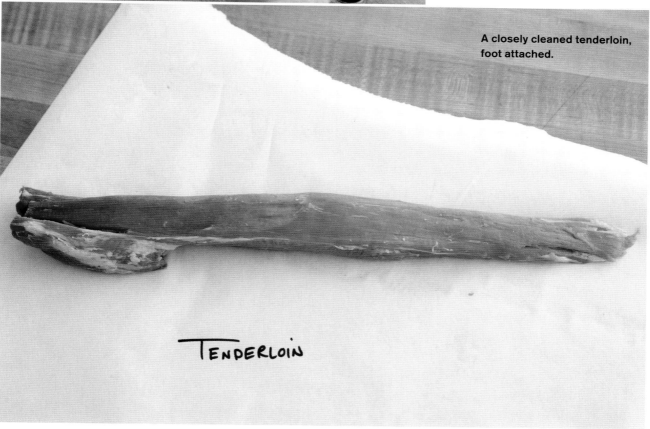

A closely cleaned tenderloin, foot attached.

TENDERLOIN

BELLY

GLORIOUS PORK BELLY, the provider of the most hallowed pork product of them all: bacon. Also known as the side, this primal is often cured, smoked, sliced, then fried as bacon. Though somewhat rare, a bone-in version of bacon can be produced; I find it more useful to remove the spare ribs and treat them as a stand-alone product. Aside from a cured preparation, the belly can also be slow-cooked, ideally over wood coals, to produce bites of fatty, unctuous, smoky, deliciousness.

Side
(Belly)

Skinning the Belly

Pork belly needs little cleaning other than to be skinned and squared up.

BACON
(UNCURED)

1. REMOVE THE SKIN.

Start at one edge and cut between the skin and belly fat, releasing enough skin for you to get a grip on. Keeping the skin taut is key, and long, smooth strokes provide the best result. Hold the skin tight while you cut against it, angling your blade so that it scrapes against the skin. Work your way down one long edge and then across to the other side. (For more on removing and cleaning skin, see page 371.)

2. SQUARE IT OFF.

You can square off your belly before or after you skin it. Cut along the bottom edge, removing the nipples and staying parallel with the top edge; cut both sides, trimming as little as possible, so they are roughly perpendicular to the top edge. Slabs for bacon should be crosscut from the belly.

Removing the Membrane from Spareribs

Spareribs need to only have a membrane removed before ready for cooking. That membrane, called the outer pleura membrane, is tough and unpalatable even after lengthy cooking times. While shown here already separated, the membrane is easiest to remove when the ribs are still attached to the belly.

1. FREE THE MEMBRANE.
Each rib has a blood vessel that runs alongside it; find one that still has blood in it. Press alongside the rib, starting at the base and going up, until the blood comes out. Insert a honing rod into the opening of the blood vessel, going as far as possible while staying between the membrane and the meat. Gently lift up on the rod to begin separating the membrane.

2. PEEL THE MEMBRANE.
Repeat across the ribs, hitting every third rib or so, and then peel the membrane away by hand.

Cutting St. Louis–Style Spareribs

Spareribs can be further fabricated into the popular St. Louis style. Find the bottom edge of the bones, where they join with the costal cartilage, which in turn connects to the sternum. Just above this point, cleave or saw through the ribs and then finish the cut with a knife. For squared spareribs, keep the cut parallel with the sawed edge or lop off a few of the smaller ribs after following the line of cartilage.

LEG

A PORK LEG IS QUITE LARGE, so only under rare circumstances is it cooked whole. If it is to be cooked whole, the best preparation of it is semi-boneless. Otherwise, it's better to bone the entire leg and either separate it into smaller roasts, cutlets, and boneless ham steaks or use it for sausage. This same process applies to a leg with the sirloin still attached, though the pelvic bone will be larger and need more cutting to remove.

SIRLOIN

The sirloin is a collection of gluteus muscles that is often referred to as the rump. It can be left bone-in and roasted whole or boned for other preparations.

HAM
(LEG)

Semi-boneless Pork Leg

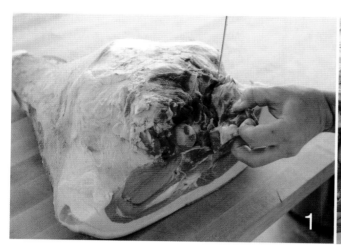

1

2a

1. CUT THROUGH THE BALL-AND-SOCKET JOINT.

Place the leg bone-side up. The first bone to remove is the aitchbone. Its only connection is a ball-and-socket joint with the femur, and removal is aided by pulling the bone away from the meat while cutting around it. Trace the upper edge of the bone with the knife tip, going deeper until you reach the ball-and-socket joint. Cut through the surrounding connective tissue and the ligament that runs inside the joint. (You will feel the pelvis release once the ligament is severed.)

2. NAVIGATE THE BONE.

Pull back on the bone and use the tip of your knife to continue tracing the shape of the bone. The end of the bone is lodged into the meat, connected by a bony protuberance. Cut around it, using the tip of your knife, to finish removing the aitchbone. Trim any meat that was left on the bone and save it for grinding. Remove any sacral and caudal vertebrae by cutting around them.

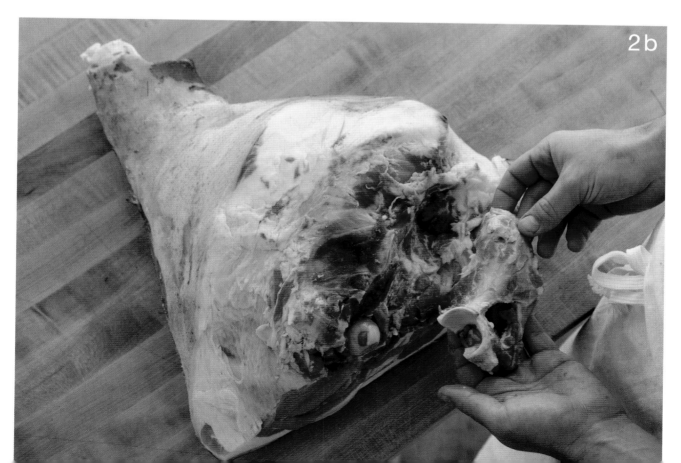

2b

Semi-boneless Pork Leg CONTINUED

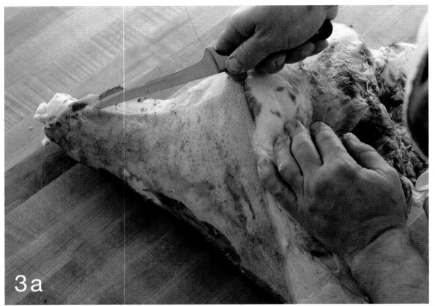

3a

3. REMOVE THE SKIN.

Make a shallow cut through the skin along the inside of the leg and hock. Remove the skin, starting at this incision and working your way around the leg. Trim the fat to a desired thickness. Finish preparing a semi-boneless pork leg by removing any connective tissue or other unpalatable trim.

3b

SEMI-BONELESS
HAM
(LEG)

Boning the Sirloin

1. TRACE THE PELVIS.

Place the sirloin fat-side down. Use the tip of your knife to trace the pelvis, releasing the joint end of the bone first.

2. REMOVE THE PELVIS.

Pull the joint end of the bone away from the meat as you continue to trace its shape, staying close to the bone as it flattens and curves towards the loin end. Finally, cut around the sacral vertebrae to remove them.

3. TIE AND TRIM.

The whole sirloin will make an amazing roast. Trim the exterior fat to a desired level, then tie the entire roast and square off both ends.

Cutting Sirloin Chops

Sirloin chops can be cut from a bone-in or boneless sirloin. The ideal thickness for a sirloin chop is 1.25 inches, which allows for adequate searing but slow finishing.

For boneless chops, follow the instructions for boning the sirloin on page 409. Then tie the entire sirloin in increments that correlate to your chop thickness; cut between the ties to separate the chops.

For bone-in chops, place the sirloin bone-side down. Press one hand flat against the loin end of the sirloin to steady the meat. One by one, use a breaking knife to cut down to the bone. Finish the cut using light saw work.

The size and shape of sirloin chops progresses from the loin end to leg end (left to right).

Remove the Trotter

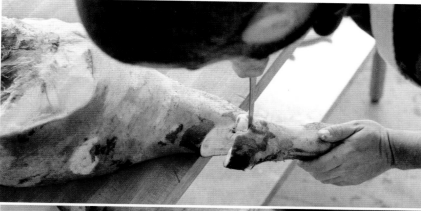

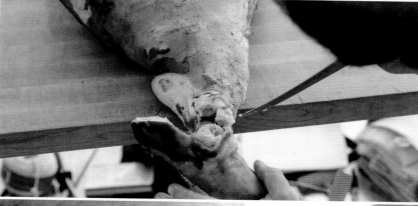

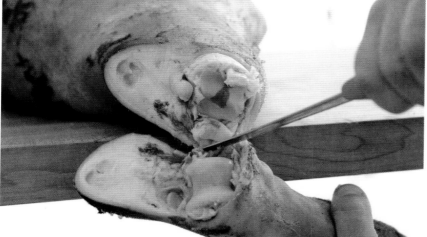

1. CUT ABOVE THE JOINT.
Let the trotter hang over the edge of the table. Bend it back and forth, and use your fingers to feel for the joint, near where the skin creases. The hind shank bone juts out, almost like an elbow. Start above that elbow and then curve downward and across the joint, cutting through the skin, muscle, and ligaments that sit atop the joint.

2. OPEN THE JOINT.
Press down on the trotter, using the table as leverage to open up the joint. Cut through the ligaments using just the tip of your knife if the joint doesn't open up easily when pressure is applied.

3. CUT THROUGH THE UNDERSIDE.
Finish removing the joint by cutting through the underside using a knife. If knife work proves challenging (or boring), saw off the trotter just above the joint.

CROSSCUT HOCKS

Pork hocks, when not boned, are often crosscut and treated as a braising cut, or brined and smoked. Crosscut hocks are best from the center of the bone, where the marrow is concentrated and meat is thickest. Remove both joint ends of the hock, cutting through the meat, sawing through the bone, and finishing with a knife. Cut the remaining hock portion in half or thirds, depending on the desired thickness, using a knife through the meat and sawing through the bone.

Steamship Roast

A pork steamship leg (also known as a handle-on semi-boneless leg) is a presentation approach to preparing a large leg roast. Start with a semi-boneless pork leg.

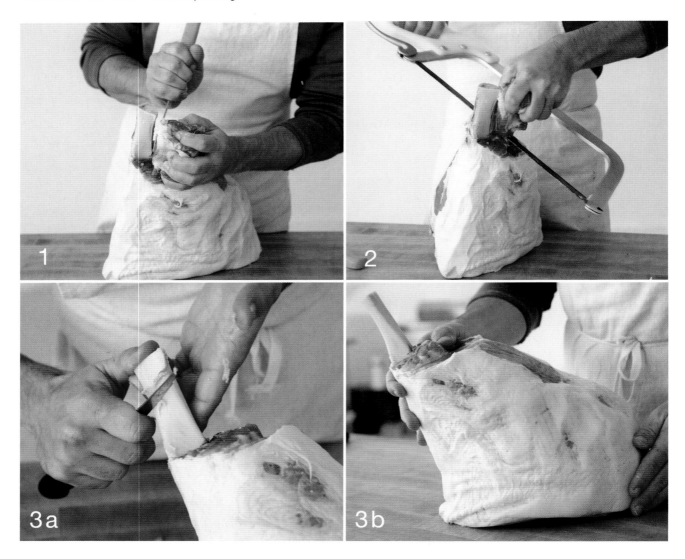

1. REMOVE THE HOCK MEAT.
Just below the stifle joint, cut all the way around the hock and separate the meat below the incision.

2. REMOVE THE FIBULA.
The fibula, one of the two lower leg bones, is fused to the tibia but must be removed for a proper steamship presentation. Saw through the small bone in line with your first cut.

3. FRENCH THE BONE.
French the exposed portion of the tibia, cutting through the bone sheath with the tip of your knife and using your blade or a boning hook to scrape the bone clean of any membrane or meat.

Cutting Ham Steaks

The rear leg of a pig holds many large, lean muscle groups. Crosscut, these can make for delicious bone-in ham steaks. Another option is boneless ham steaks cut from any one of the seamed muscle groups featured in the three-way boneless leg on page 416. For ham steaks, use the longest knife you have. Ham steaks are best cut thick, at least 1 inch, giving you the chance to sear them and finish them slow in the oven.

Start with a semi-boneless leg and remove the sirloin. Place the leg skin-side down and start by squaring off the top (proximal) portion of the leg. Find the ball-and-socket end of the femur. Just below the joint, and perpendicular to the bone, cut through the muscles and down to the bone.

Saw through the bone, and then finish the cut with a knife. The end is now squared off. Decide on the thickness of the ham steaks and measure them off with marks, starting at the square end. Make the first cut on all the ham steaks, staying perpendicular to the bone and even in thickness. Saw through the bone on each cut, and then finish with a knife. Scrape the steaks clean of any bone dust or debris.

Boneless Pork Leg

There are many ways to bone a pork leg, including pocket boning the femur (where you remove the bone without separating the muscle groups). Here I bone the entire leg, including the hock, by following muscle seams. If you want to use the hock, you can remove it whole, with a knife or saw, before boning the femur.

1. CUT INTO THE SEAM.

Start with a semi-boneless pork leg (above). Lay the leg fat-side down. Note the natural seam between the leg tip and inside round. Access the femur from that seam, cutting down to the bone while pulling back the inside round.

2. SEPARATE THE FEMUR.

Continue around the top face of the round, separating the leg tip. Cut around the ball-and-socket end of the bone; lift it up and work your way back toward the stifle joint while cutting on the underside of the bone, using the tip of your knife and staying close.

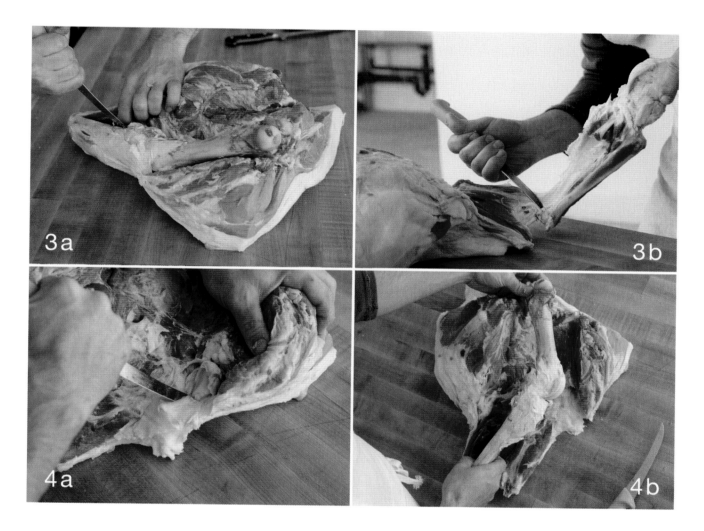

3. BONE THE HOCK.

The lower leg has two bones – the tibia and the fibula – which are fused near the stifle joint. Cut along the inside of the stifle joint and down to the end of the hock bones. Pull the femur away from the meat and, just below the stifle joint, sever the hock muscles down to bone. Peel the muscles away from the bones (or continue to use your knife).

4. SAVE THE BONES.

Find the patella, remove it, and discard. The femur can be saved for stock or cut in two and used as a ham bone for soup. Other bones should be used for stock. The upper and lower leg bones can be separated at the stifle joint by cutting through the joint with the tip of your knife.

Three-Way Pork Leg

The boneless leg can be further seamed into the three main subprimals: inside round, outside round, and leg tip. They share the same leg seams as the primals from their farmyard cousins cattle and sheep. Lay the boneless leg fat-side down. It should look similar to the photo but may not look exactly like it depending on where you removed the femur. Note the indication of seams between the three muscle groups. Separate the groups according to those seams, removing any interior connective tissue, remnants of the sciatic nerve, blood clots, and other unpalatable trim. Areas of the leg not included in these three muscle groups (the heel and the hock muscles) can be used for sausage or grinding.

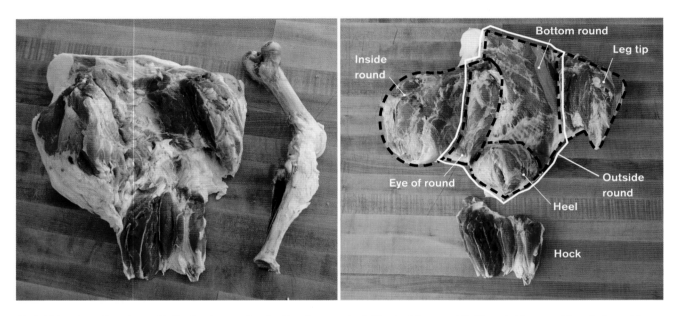

At right is a boneless leg with the hock muscles (bottom) separated. The inside round (left), outside round (center), and leg tip have had their seams opened up and are ready to be separated. The heel (attached at the bottom of the outside round) is partially peeled back to illustrate the seam.

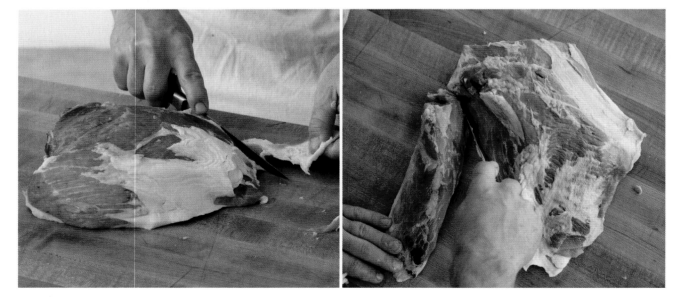

Trimming the inside round.

Separating the two muscle groups of the outside round: the eye of round (left) and bottom round.

CONSTITUENT PARTS OF A PORK LEG

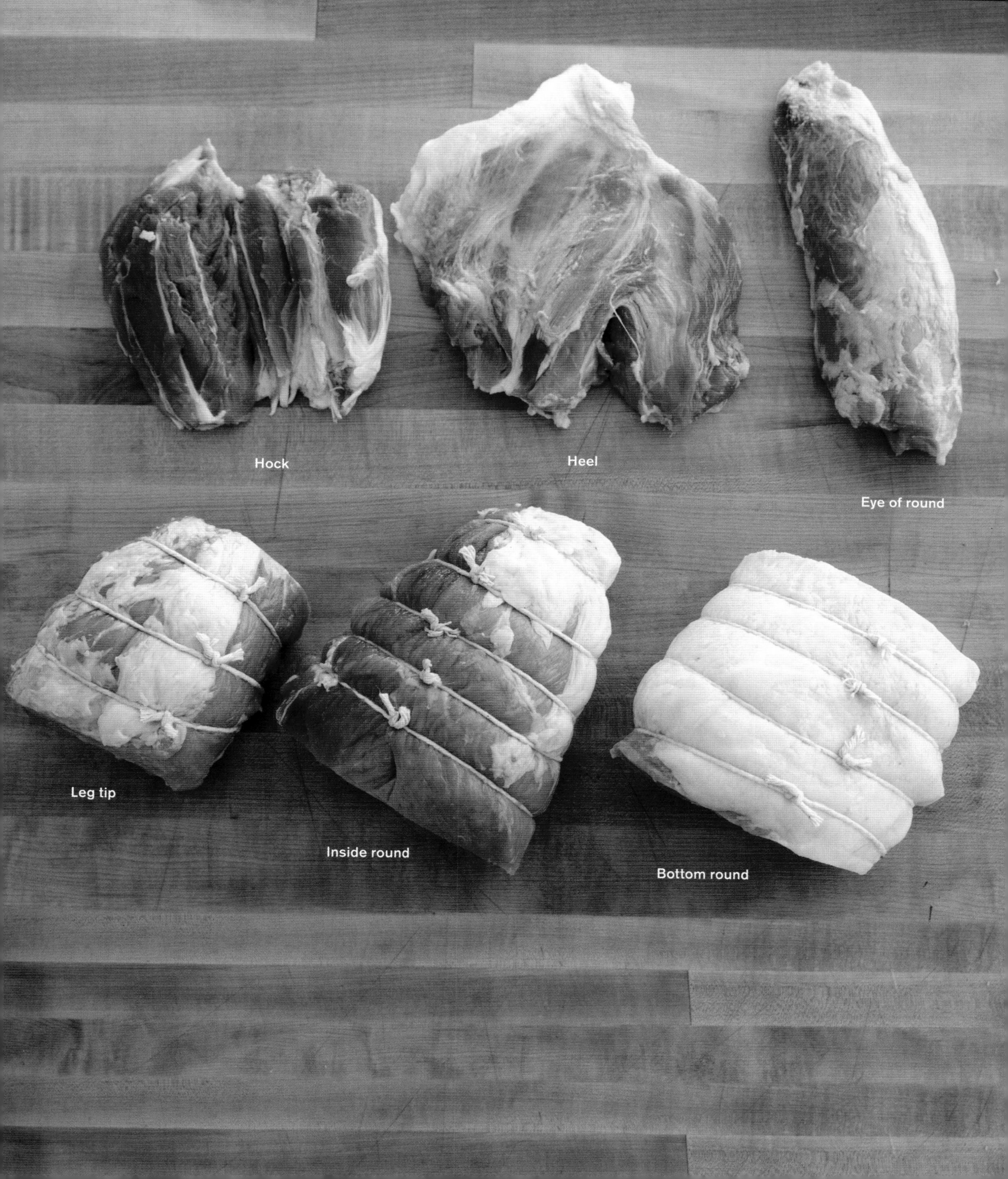

Hock
Heel
Eye of round
Leg tip
Inside round
Bottom round

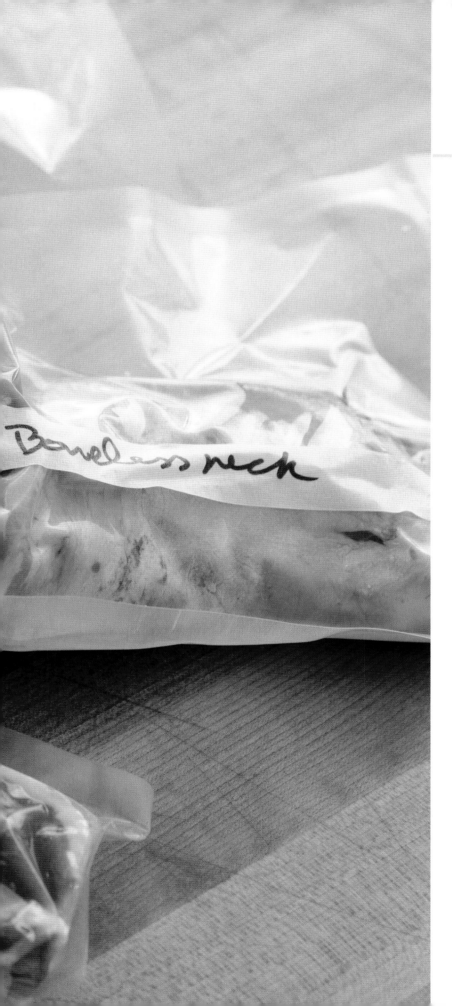

PACKAGING & FREEZING

PROPERLY PACKAGING AND STORING meat is essential for its preservation, enabling a harvest to be used over months or years. Failure to properly wrap and freeze meat products can lead to nutrient and flavor loss, absorption of foreign flavors, dehydration, and other textural changes. The irregular shape of meat products — especially the whole carcasses of small animals like chickens and rabbits — presents challenges when packaging. This chapter covers the main options for overcoming obstacles in packaging and freezing meat, and addresses alternatives for various species.

Freezing

FREEZING FOOD AS A FORM of preservation has been around since prehistory, but for many areas of the world freezing was never an option until the advent of mechanical freezers. (Instead, salt was used extensively.) Today, freezing is the main method of meat preservation. Most home refrigerators come stocked with freezers that provide adequate functionality for short-term storage of meats. But for people looking to store large quantities for long periods, investing in a stand-alone freezer is the most effective solution.

The Effects of Freezing

As soon as foods are exposed to temperatures of 30°F or lower, their moisture begins to change from liquid to solid. This process aids in food preservation because at such cool temperatures the chemical and biological activities that cause spoilage slow down (though they do not cease) and liquid water is now trapped in a solid form, halting the microbes — bacteria, yeasts, and molds — that require water to continue their normal functions and propagation. This also means that the enzymatic activity that causes maturation and decay of the meat is slowed to an inconsequential pace. This lack of enzymatic activity is why freezing meat before resolution of rigor mortis will yield a tough, dry result. (For more on rigor mortis and cold shortening, see page 18.) If conditions are maintained in perpetuity, these effects — halting the activity of microbes and enzymes — maintain the safety of the meat product, though extended periods of freezing will have adverse effects on quality.

Anywhere from 66 to 73 percent of the total mass of meat is water. This means that freezing causes the main component of meat to change from a liquid to a solid state; this shift has the potential to cause considerable damage. There are two types of water within meat:

- **EXTRACELLULAR MOISTURE** is water that exists outside of the cells. It is unadulterated and has few dissolved nutrients or other substances, such as salt, sugar, or minerals.

- **INTRACELLULAR MOISTURE** exists inside cells and contains all the dissolved components required for cells to operate.

When cellular components and other dissolved substances are in water, they crowd the area that water molecules have, getting in the way of bonds forming while the water is freezing. This interruption of the molecules' ability to bind at freezing temperatures requires lower temperatures to achieve a solid state, a process called *freezing point depression*. While the extracellular fluid can freeze at around 30°F, the intracellular water typically freezes at around 14°F. As the temperature of meat drops below freezing, the extracellular fluid begins to form solid ice crystals. Under normal home or commercial freezing environments (temperatures above −40°F), these ice crystals develop quite slowly, growing larger as they form a network.

This is the beginning of freezing any tissue: all the extracellular fluid needs to freeze before the intracellular fluid can freeze. But, as the extracellular ice crystals are forming, they start to leach moisture out of the cells, further concentrating the substances within the intracellular moisture — those same substances that interfere with water's ability to freeze — thus causing the freezing point inside the cells to drop even further and requir-

THE RIGHT FREEZER

Before you harvest meat, make sure that your freezer space is adequate. In most cases, a stand-alone freezer will be the best solution. Stand-alone models provide better temperature accuracy, more space, and slightly shorter freeze times than the freezers provided in conjunction with a refrigerator. This all adds up to better meat preservation. The right size can be hard to estimate — you need enough space for all the meat and then some to help organize and access everything. A helpful estimation is 50 pounds of meat per 2 to 3 cubic feet of freezer space. Chest freezers offer an efficient use of space, though a well-organized stand-up freezer will make the packaged meat more accessible.

ing an even lower temperature to achieve a truly solid, preserved state. Furthermore, the extracellular absorption of water dehydrates the cells, the first of many ways in which freezing damages meat.

The Importance of Rapid Freezing

Water in meat is held within fragile structures like cellular membranes. As the water freezes and expands, the sharp ice crystals cause the cellular walls to rupture. The slower the water freezes, the more expansion occurs and the sharper the ice crystals are, resulting in greater damage to these water-holding frameworks. To make matters worse, the crystals continue to grow after the product appears to be completely frozen, drawing more moisture out of the cells and causing more dehydration and damage.

The ice acts as a plug to the holes made in cellular walls and other structures, but as meat thaws some of the water seeps out of these holes, draining nutrients, pigments, and minerals along with it, resulting in what is called *drip loss*. You may have experienced this: thaw a frozen steak on a plate, and there will be a pool of red liquid collecting around it that you may have thought was blood. It is not: rather, it is formerly frozen liquids full of nutrients and proteins like myoglobin, which provides the red hue. Drip loss is inevitable, and all frozen products will experience it; the amount of drip, and the resultant quality of the meat, is directly linked to the speed of freezing. Drip loss is not the only quality risk from the formation of large ice crystals. These structures will also break emulsions — the stable mixtures of pureed substances, like mayonnaise, hollandaise sauce, or hot dog filling — causing the separation of fats and proteins. This is the reason why hot dogs and other emulsified meat products are not recommended for freezing. Commercial hot dogs destined for the freezer utilize additive ingredients to counteract this effect and stabilize the emulsion.

Because of its effect on the quality of the meat, the speed of freezing is one of the most important factors in preservation. Commercially frozen foods are exposed to extremely low temperatures, –30°F and lower, and high air velocities that promote rapid freezing and thus smaller ice crystals. At home the best option is a stand-alone freezer that provides a quick-freeze or deep-freeze option, or another comparable setting that can be used temporarily when initially freezing items.

> Freezer burn is simply a form of dehydration and oxidation caused by exposure of a frozen product to air.

KILLING PATHOGENS

Slow freezing is detrimental to more than just the cellular structures of the meat. Many bacteria and other microbes, especially those that are single-celled, can be destroyed during freezing; however, because the numbers would be uncertain and the effect not uniform, it is safer to treat freezing as a method of suspending microbial functions rather than destroying them. One exception is trichina, the roundworm sometimes found in pork and game meat that is responsible for the disease trichinosis (see page 29), which can in fact be killed by freezing it. (Proper cooking is another way to kill it.) The U.S. Department of Agriculture recommends that meat be exposed to 0°F, the common home freezer temperature, for at least 106 hours to ensure the death of trichina. The lower the temperature, the less exposure time required for death.

AVOIDING DEHYDRATION

The formation of large ice crystals may be one of the more damaging effects of freezing, but it is not the only risk to consider. Freezer burn is simply a form of dehydration and oxidation caused by exposure of a frozen product to air. It is a consequence of improper temperature maintenance or packaging and can severely reduce the quality and nutrient value of frozen products. While freezer burn most often does not make the product unsafe to consume, it does make it unpalatable.

Dehydration occurs in a frozen product despite the fact that water is held in a solid state. The air within a

Do not freeze whole primals or large cuts that will not be cooked whole.

freezer is innately dry; when it comes into contact with frozen moisture, a process called *sublimation* occurs, in which the solid ice crystals change state to gaseous water vapor without transitioning through the liquid state. Sublimation pulls moisture from the interior of the frozen product to the surface, where it forms large ice crystals and causes the area to permanently dry out, essentially freeze-drying it. The ridges and holes left by the absence of water reflect light, causing light-colored spots on the meat. Furthermore, the air brings oxygen into contact with the product surfaces, causing oxidation and the resultant discoloration to darker or brown tones. (For more on myoglobin and oxidation of meat and fats, see page 14.)

The Tenets of Freezing

Following some baseline recommendations can help you avoid the more detrimental consequences of freezing.

MAINTAIN A CONSTANT TEMPERATURE

First and foremost, keep frozen foods at 0°F or lower and avoid fluctuations in temperature. Fluctuations can result from opening the freezer too often or from inaccurate temperature regulation by the freezer itself. Changes in temperature can cause additional growth spurts in ice crystals, which in turn will contribute to additional cellular membrane damage, drip loss, and moisture loss. If you suspect that your freezer doesn't maintain a consistent temperature, set it lower than 0°F to prevent harmful fluctuations.

DON'T USE FREEZING AS A FORM OF SANITATION

Freezing does nothing to improve the quality of meat; it merely aims to maintain the quality of the product as it was prior to freezing. In fact, meat that has been frozen will be of lesser quality than fresh meat (though with the right approach and equipment, the degradation can be minimized). Therefore, start with the highest-quality product possible. Do not try to save meat that is beginning to spoil by freezing it. Freezing is not a form of sanitation; you cannot expect it to reduce the colonies of microbes that can cause spoilage or food-borne illnesses.

FREEZING ORGANS

Organs — such as the liver, the stomach, and the intestines — are the operational center of an animal's existence. While all muscle tissue bears a likeness, organs are drastically dissimilar from one another due to their highly specialized functions. This makes for a splendid culinary challenge, offering a wide range of textural and flavor possibilities far beyond that of muscle meat.

Freshness with organs is paramount. Organ tissues tend to have higher levels of enzymes to aid in their activity, and for that reason they degrade quickly. Muscle meat may benefit from aging, but organs immediately begin to deteriorate once the systems shut down that have controlled enzymes and maintained balance in the body.

Freezing organs is the best way to halt the degradation of quality, and doing so immediately after slaughter is ideal. This includes all visceral organs, tongues, and marrow bones (which are organs, too). Use an ice-water bath to rapidly bring down the temperature of organs. As soon as they're cold, proceed with packaging and freezing. You can also use an ice glaze to package them. (See page 432 for more information on ice glazing.)

PORTION BEFORE FREEZING

Before you freeze them, portion your meats according to your expected needs — single pounds of ground meat, individual chops, small roasts, and so on. When freezing multiple portions together in one package — say, a few steaks or some preformed burger patties — prevent unappetizing discoloration by making sure the cuts don't touch one another. Avoid having to thaw and refreeze portions, because the adverse effects will become more pronounced with each freezing.

Do not freeze whole primals or large cuts that will not be cooked whole. The larger the item, the slower the freeze, and the higher the risk of cellular damage. Freezing time is squared in relation to thickness — cut a pork chop twice as thick as another, and it will take four times as long to freeze. Smaller packages of meat will freeze more quickly and therefore maintain quality for longer periods of storage. When simultaneously freezing multiple items for the first time, disperse them among the shelves of the freezer to promote quick, even freezing. Use the quick-freeze or deep-freeze option if the freezer has it. Do not stack or crowd items that are being initially frozen; all items need cold airflow to hit every surface to maximize results.

LEANER CUTS STORE LONGER

Fat oxidizes and goes rancid more quickly than lean meat does. Thus, leaner cuts of meat hold their quality longer than fattier cuts. If you expect extended storage, trim exterior fat to the thinnest acceptable amount prior to freezing. Unsaturated fats go through oxidative and deteriorative changes more quickly than saturated ones. Therefore, pork and poultry, which both have relatively high levels of unsaturated fats, will remain in palatable states for shorter periods than beef, lamb, and goat will. Ground meats will spoil sooner than whole pieces of meat because they have more surface area and are therefore more susceptible to oxidation and other chemical changes.

COOKED AND CURED MEATS DON'T FREEZE WELL

Cooked and cured meats freeze poorly compared to fresh, uncooked meat. This is because cooking causes considerable moisture loss; with less moisture in the meat, the efficacy of freezing is also lessened. Furthermore, cured and processed meats have increased levels of salt, a mineral that acts as a catalyst for oxidative reactions, thus increasing the rate of rancidity.

Packaging

PACKAGING IS A LINE OF DEFENSE for your meat, shielding it from the detriments of freezing. If packaged improperly, your meat will lose its battle against ice and oxidation, and will be dry, limp, and unpalatable. With the proper tools and approach, however, preserving meat with little to no perceivable degradation is entirely possible.

The Tenets of Packaging

Packaging for freezing has one objective: to preserve the quality of the product for as long as possible. Each decision about which materials to use and how to use them should be made in response to one question: Will this help preserve my product as long as I need it to? Technology in packaging has developed immensely over the many decades since commercial and home freezing became the norm, but some simple rules based on scientific principles have gone unchanged. And, thankfully, these protocols apply to storage for both fresh and frozen goods.

AIRTIGHTNESS

Packaging intended for the freezer needs to be airtight — both moisture-proof and vapor-proof — in order to have any chance of preserving meat. Permeable materials allow air and moisture through, resulting in freezer burn and quickened rancidity. This is why you never want to freeze meats bought from stores using the traditional "store wrap" — a foam tray with a drip-liner wrapped in plastic — a style of wrapping that intends for air passage in and out, allowing the meat to "bloom," the process in which the myoglobin, the protein responsible for meat's coloration, oxidizes and turns bright red. (For more on oxidation, see page 15.)

EXPULSION OF INTERIOR AIR

While being impervious to the outside air and moisture, packaging should also promote expulsion of as much interior air as possible. Skintight packaging is ideal, with the exterior material touching the surface of the meat wherever possible. This prevents air from reaching the surface. Pliable materials are the best option for nonliquid products, and those with good tensile strength will enable tightening around the meat when air is expelled. Any applicable material should also be easy to remove from the meat after thawing.

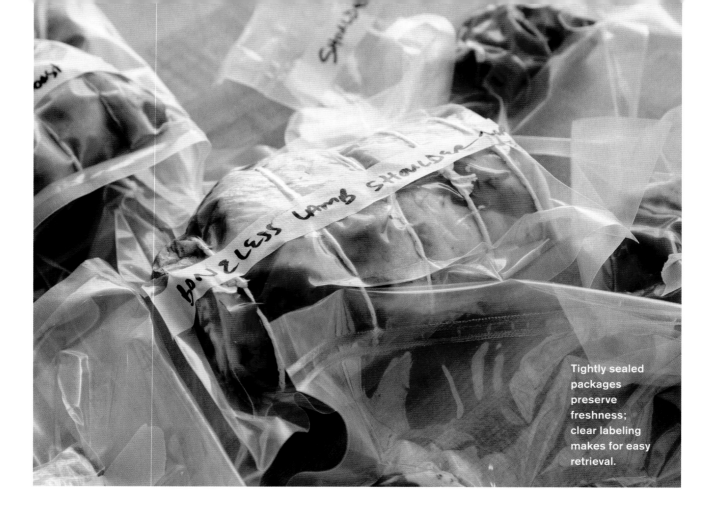

Tightly sealed packages preserve freshness; clear labeling makes for easy retrieval.

DURABLE, FOOD-GRADE MATERIALS

Materials need to be made from an odorless food-grade substance that is tasteless and resistant to transferring any off-flavors. Its impenetrable state should also prevent the transmission of off-flavors between products frozen within the same space. (The last thing you want is for your steak to taste like its plastic container or the whole frozen fish next to it.) Cracking, brittleness, or other forms of detrimental warping should never occur during freezing, when moisture is expanding, or when the product is kept at low temperatures indefinitely. You will likely need to reorganize your frozen products from time to time; therefore, the packaging will need to protect the interior product while also being durable enough to resist the resulting wear and tear. When applicable, package designs that enable convenient storage, stacking, and organization are preferred.

PROPER LABELING

Create labels that you can read easily without having to hold the item in your hand. Otherwise you will be riffling through the entire freezer, picking up item after item, searching for a specific cut. Most meat packaging will be disposable, so the label may simply consist of information written directly on the package with a high-quality permanent marker. But tagging on white or brightly colored backgrounds certainly improves legibility. Temporary labels need to have a freezer-proof adhesive. Reusable containers, used mostly for liquids, should have the space for proper temporary labeling.

AFFORDABILITY

Packaging needs to be affordable, falling into range with your budget. Expensive solutions abound, but you can also find affordable ones. While protecting your product is important, so is profitability. And those with customers must use packaging that appeals to buyers' needs.

Packaging Solutions

Currently there is no single type of packaging that is both perfect and affordable, though some types come very close. The solutions here range from ideal to bare-bones. Assess your priorities by considering both your products and your preferences. Depending on the product to be packaged, the best solution may be to employ multiple techniques.

VACUUM SEALING

Vacuum sealing is by far the single best option for packing perishable meat products. It's the only method that's applicable to every animal species and works in all conditions. Interior air is removed under pressure, pulling the pliable bags tight against the product surface. The bags, made from impermeable food-grade plastic, are sealed permanently through the application of heat and can be used for products in any shape or size. Plus, the process is quick and easy to execute.

Vacuum sealing does have its drawbacks, however. First, investing in a vacuum-sealing machine that can adequately handle your volume may tip the cost scales. Second, edge sealers — the more affordable models designed for home use — do not handle liquids or highly moist products well (this does not include raw meat). Third, home models are also unable to handle extra-large items like a standing rib roast of beef. Fourth, vacuum-sealing bags tend to be more costly than other options, even in bulk, and are prone to being punctured, allowing dry air to flood the interior and leading to freezer burn (though it may be minor). Despite these caveats, vacuum sealing is still the best solution for many processors, especially those dealing with multiple species.

EDGE SEALERS. There are two types of vacuum sealers: edge sealers and chamber sealers. The most common models for home use are edge sealers, with the most popular being the Food Saver brand. The operation requires much simpler machinery than the chamber sealers, making them more affordable and smaller in overall size. Their major drawback is the inability to handle liquids or highly moist products.

Every model will have specific instructions for use, but the concept is the same across the board. Start with a bag, either premade or cut from a roll, that is large enough to fit the product while allowing enough space for sealing. Place the product inside, insert the bag into the machine, and start the vacuum. It will draw out the air, compress the pliable bag around the product, and seal the bag.

CHAMBER SEALERS. The industrial-oriented machines for sealing foods are known as chamber sealers. Rather than suck the air out of the bags, as with edge sealers, chamber sealers utilize a different technique. The bag, with food inside, is placed into the chamber with the opening of the bag laid across a sealing strip. When the chamber is closed, air is pumped out of the chamber, thus creating a low-pressure environment, or vacuum, inside. Once the target pressure is reached, the bag is sealed and the pressure from within the chamber is released. Since the bag was sealed when there was almost no air inside the chamber, once the chamber is opened, the bag compresses tightly around the food. This allows for the bag to contain liquids. Chamber sealers also allow you to control the amount of compression, which is helpful when working with fragile items (rabbit racks) or products that need to hold their shape (cased sausages). Compression also helps to speed up marinating: when meat and liquid are sealed together, the pressure forces the marinade into the meat.

TIPS FOR BETTER VACUUM SEALING

- To cut down on costs, consider joining with other processors and farmers in purchasing a vacuum-sealing machine for shared use.

- To avoid punctures, use additional padding for bone-in butchered parts and other items that have bony or sharp protrusions. You can also use cut-up pieces of vacuum bags or prefabricated bone guards to cover the areas of concern, effectively creating a double layer of plastic, before you seal the bag. For very sharp areas, consider taking a sanitized rasp or file and smoothing the edges.

- While any size bag can be used for any size item, reduce costs by having a selection of bag dimensions, using the smallest option for the product being sealed.

- Hollow items, like a whole chicken, can be filled with plastic wrap to help maintain its shape and expel the air.

- Before filling a bag, fold the edge down about 3 inches (typically the area that gets melted into a seal) in order to avoid soiling it with fats or getting it wet, which will prevent a clean seal.

How to Package Food with an Edge Sealer

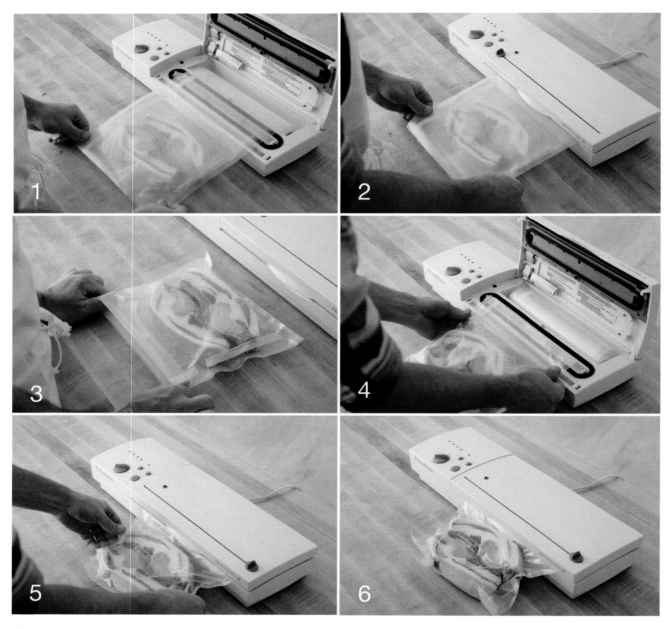

1. Measure out the length of bag needed to accommodate the contents.

2. Cut the bag from the roll and seal one end.

3. Fold down the edge of the bag and place meat inside.

4. Unfold the bag edge and align across the sealing strip with the open end of the bag inside the suction chamber.

5. Close the lid and start the removal of air from the bag.

6. When the indicated level of vacuum is achieved, the unit will permanently seal the bag.

How to Package Food with a Chamber Sealer

1. Fold down the edge of the bag and place meat inside.

2. Unfold the edge of the bag and place in the chamber. The edge of the bag should lay flat against the sealing strip with no wrinkling.

3. Close the lid, which starts the removal of air from within the chamber.

4. The lid opens after the bag is sealed and the chamber returns to normal pressure.

How to Shrink-Wrap a Cut of Meat or a Small Carcass

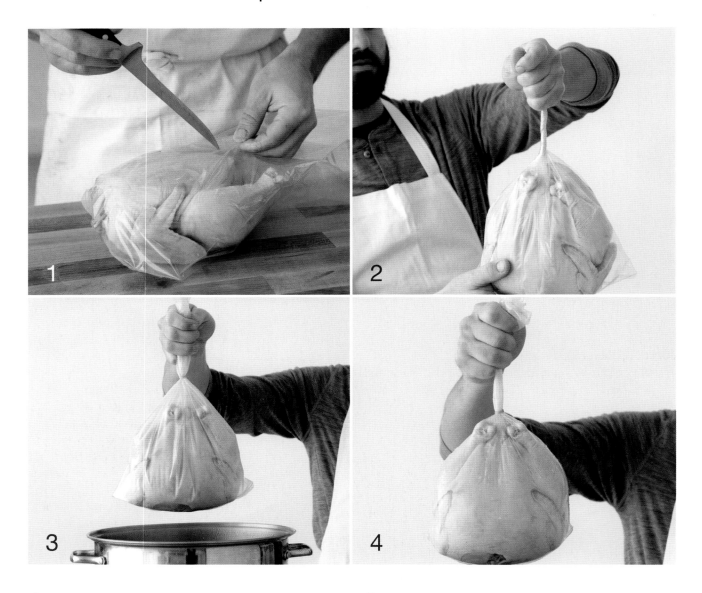

1. Place the meat in the shrink-wrap bag. If you're working with a whole carcass, place it headfirst and then press down on the legs a bit. Poke a small hole in the bag, thus allowing air to escape while the plastic shrinks around the meat.

2. Twist the top of the bag several times and then use hog rings or a tape-based bag sealer to secure it. Expel as much air as you can.

3. Hold the bag by its twisted top and dip it into hot water. Use a utensil — tongs, a strainer, or a spider skimmer all work well — to move the product around while keeping it submerged. (You may find that in the beginning the air in the bag encourages the product to float.) After a few seconds the bag will shrink and expel the air out of the hole.

4. Remove the product and make sure the bag is secured around all areas of the carcass. Avoid submerging the bag multiple times, as water may enter the punctured hole.

SHRINK-WRAP BAGS

Shrink-wrap bags are a great, inexpensive way to package meat and, unlike vacuum sealing, do not require any expensive equipment. They are ideal, and economical, for dealing solely with whole chickens or rabbits, or their constituent parts. The two pieces of equipment are a container large enough to allow you to submerge the entire product and a heat source that can maintain the water temperature around 190°F. The setup is completed with a needle or other sharp object that can poke a small hole through plastic, along with sticky labels for the bags. (The labels will serve two purposes: to provide information and to cover the small hole you will make to let excess air out.) The food-grade bags, which are easily sourced online, need to be thick enough to stretch around hard and soft objects alike without puncturing, so look for ones designed specifically for wrapping the products you work with.

Like any pliable plastic packaging, the shrink-wrap bags can be punctured. But, unlike vacuum-sealed bags, shrink-wrap ones will not flood with air once punctured because the interior is not a vacuum (an environment with a lower gaseous pressure than the atmosphere around it). So, once punctured, a shrink-wrap bag will allow small amounts of air to come in but will hold its shape and maintain its skintight form. This is a huge advantage over vacuum sealing.

Shrink-wrap is not without its own downsides, however. Water temperature can be challenging to maintain: if it's a bit too hot, the bags can overshrink or melt, exposing the product to scalding-hot water; it it's a bit too cold, the bags will only partially shrink and then harden, making a second pass of shrinking difficult even with proper temperature. The bags only shrink so much, requiring you to have a range of bag sizes for products with substantial size differentials. The process is wet and tends to be messy, which is why this type of bagging is often handled outdoors. What's more, the process involves variables that, in order to get right, involve some trial and error, lost bags, and potentially damaged product. Finally, shrink-wrapped products cannot be submerged in ice water, one of the recommended methods of thawing, because the bags are punctured in the process and the opening is never permanently sealed (freezer labels are not water-resistant).

FREEZER-GRADE WRAPPINGS

Manually wrapping items is another effective approach, and there are many kinds of materials available for use by the home processor. The main challenge in manual wrapping is achieving a skintight, airtight package. The most suitable choice is a moisture-impermeable plastic wrap. The elasticity of plastic wrap enables it to conform to various shapes, allowing for a closer fit, thereby expelling more air. This material is often cheaper than any bag option, especially when you buy it in bulk. But there is a trade-off: the need for another layer. You'll need to use freezer or butcher paper in conjunction with plastic wrap to provide the necessary protection against puncturing. Any suitable paper will be two-layered, the exterior being paper and the interior being a thin plastic film to retard air.

This approach works well for small items such as portioned cuts, small roasts, or ground meat, but it is cumbersome and less effective when dealing with whole carcasses or larger items. Also, wrapping with plastic and paper is a two-step approach that takes more time than a single-step solution. For products without a uniform shape, appearance tends to suffer, because the ability to wrap well with paper is challenging. In turn, salability is also affected, potentially from appearance but also because the interior product is obscured by the opaque outer wrapping. In the end, the main advantage to manual wrapping is cost savings in materials (despite the need for two layers) and the lack of equipment needed.

FOILING THE ELEMENTS

Aluminum foil can be an alternate option for the initial layer of wrapping, but it is hard to get a hermetic seal with foil. Plus, it tears easily. But aluminum foil is an excellent choice as an extra measure against oxidizing elements. After wrapping with plastic wrap, cover the cut with a single layer of aluminum foil, carefully sealing the edges. Then proceed with wrapping in paper. This layer of extra protection is often helpful for cuts with sharp protrusions (like bone-in cuts), whole subprimals, primals, or other large items.

Two Methods for Wrapping Meat

THE DRUG-STORE METHOD

1. Place your cut in the middle of the paper.

2. Lift the top and bottom edges and align them above the meat.

3. Starting at the aligned edge, fold the paper down, in 1-inch increments, and crease tightly. Repeat until the paper is tight against the meat.

4. Fold the four sides of the paper inward, pressing out the air and forming a triangular point on both ends.

5. Flip the package over and bring the triangular points together in the middle.

6. Secure the points with freezer-grade tape and label the package.

THE BUTCHER METHOD

1. Place meat near the corner of the paper nearest you.

2. Fold the near corner of the paper over the meat and tuck underneath your cut.

3. Flatten the sides of the paper and crease tightly, expelling the inside air.

4. Fold the sides of the paper up and tightly across the meat, further expelling air.

5. Keeping the paper taut, roll the meat toward the far corner of the paper.

6. Secure the final corner of the paper with freezer-grade tape and label the package.

> Ice conducts heat up to four times faster than liquid does. Because of this, when the exterior of frozen meat thaws first, it actually insulates the frozen center, preventing the warmer exterior temperatures from reaching the interior.

ICE GLAZING

Another effective way of keeping air and other oxidizing elements away from your meat is to lock them in ice, a process called ice glazing. Ice glazing is simple and incredibly effective. Encased in ice, the chance of oxidation is highly unlikely. Some sublimation may occur, but its effect on your edibles will be insignificant. This approach is best for organs and for small carcasses such as poussin, game hens, small game birds. Anything larger tends to be too cumbersome when storing and thawing.

Find a waterproof container that's large enough to hold the items you wish to store while being covered with water. (Carcasses should be frozen individually. Organs, especially those of smaller animals, should be frozen in convenient amounts.) Place the items in the container and cover them with water. Carefully place the container in the freezer and loosely cover it, allowing some air to escape (to avoid the top popping off). Return after 24 hours and tightly seal the container for long-term storage. Ice-glazed items should be thawed like any other frozen cuts, though it is advisable to have something beneath the thawing container, as cracks may have developed in the container during the freezing and storing process.

ZIP-TOP BAGS

Last, and least desirable, is the zip-top plastic bag method. Using a zip-top bag will inevitably cause some freezer burn, the severity of which is dependent on how long you store the frozen carcass. Therefore, use zip-top bags only for short storage times — days to weeks at most — and take the effort to remove as much air as you can. A vacuum cleaner tube (sanitized, of course, and attached to a working vacuum) stuck into the mouth of the bag will remove a majority of the air. You can also use a simple bucket or stock pot full of water, similar to the setup for shrink-wrap bags but without the warmth. Take the bagged product and slowly lower it into the water, being careful not to let the opening of the bag fall underneath the surface of the water. The water will force out a majority of air as the bag descends. Stop at the lowest point at which you can still seal the bag above the waterline. Once the bag is sealed, remove it from the water, dry it off, and store it.

Additional Materials

The preceding solutions can all benefit from some additional materials, depending on what is being packaged. Freezer sheets, or another type of liner, are required when packaging portioned or separate red-meat items together in a package. Foam trays can help provide structure to items, especially loose products like patties or cubed meat and those being stored with manual wrapping. Soaker or purge pads, often accompanying foam trays, will absorb the drip coming off thawing items, helping to prevent any off-flavors that may develop. Ground-meat bags are great for that item and come in sizes demarcated for weight, making it easy to estimate the volume when stuffing them. For all bags, the opening must be sealed; the three options for that are staples, hog rings, and bag tape. The first two are the best all-around options, while the bag tape is really only effective with ground-meat bags.

Get a scale if recording weight is a priority. Prices range dramatically, but something as simple as a bathroom scale will do or, for more precise results, a digital model that has a decimal readout. The scale will need to accommodate the maximum size that you will be processing.

Labels for home use should be large enough to contain such information as contents, weight, and date. For salable items labels may need to include further information such as price, breed, or information about your farm. Most packages of meat cuts are disposable, allowing you to write simple information directly on the package with a high-quality permanent marker (although writing on white or brightly colored backgrounds certainly improves legibility). Temporary labels on reusable containers need to have a freezer-proof adhesive.

Thawing

DO NOT UNDERESTIMATE the importance of proper thawing. You will spend an incredible amount of time on packaging and freezing the bounty from a harvest, and your efforts will be wasted if you thaw the meat incorrectly. Poorly thawed meat is, at best, poorly textured and, at worst, dangerous to consume.

What *Not* to Do

Never defrost a product by microwaving it, sitting it in a warm-water bath, or leaving it to sit out on the counter or anywhere else that has warm ambient temperatures. Microwaving may be a safe way to quickly defrost meat, but the result is a steamed, partially cooked mass that has already undergone major chemical and physical change, none of which is advantageous to texture or taste. Attempting to speed up the defrosting process by exposing the meat to any temperatures above that of your fridge is just dangerous. These temperatures inevitably reside between 40°F and 140°F, the range considered the danger zone. (For more on the danger zone and foodborne illness, see page 25.) This is especially applicable to poultry and offal, though the danger is inherent in any perishable product.

One thing to understand is that thawing meat takes longer than freezing it does; there aren't any suitable shortcuts. The reason for this has to do with the comparative ability of water to transfer heat in different states. Ice conducts heat up to four times as fast as liquid water does. During freezing, the exterior of meat solidifies first, which helps conduct heat out of the meat, quickening the rate of the temperature drop inside. Thawing acts in just the opposite way: the exterior thaws first and, due to less efficient conductivity, actually insulates the frozen center, preventing the warmer exterior temperatures from reaching inside. It is for this reason that the time needed to thaw meat properly is considerably longer than that needed to freeze it.

Proper Thawing

The ideal methods of thawing just take some forethought and time. First, and easiest, is the refrigerator. Anything can be safely thawed in a properly working fridge, but it takes awhile because air doesn't conduct heat well. Thus, the air that surrounds the frozen package is an inefficient coolant. Nonetheless, it works. Leave the meat packaged, set it in a container to catch any drip loss or condensation, and allow at least a day for small items, like steaks and sausage, and up to three or more days for large items like roasts and whole poultry.

The second method, using ice water, illustrates how much more efficient a liquid is than a gas at conducting heat. This is the only suitable shortcut for thawing. Use a container about four times the size of the frozen product. Leave the product in its package and submerge it in cold water. (I also add some plates on top to keep the item submerged, but anything with weight will work if you find the product bobbing up for air.) Add lots of ice and leave the water trickling. The ice-cold temperature will actually improve the thawing time while also retarding microbial growth. The trickling flow of liquid will keep a mild circulation in the water bath, helping keep cold water against the frozen product at all times. Using this method, you can defrost small items in as little as 30 minutes and large items in a couple of hours (or more depending on how large).

ACKNOWLEDGMENTS

THERE ARE SO MANY PEOPLE that I feel blessed to have worked with, during my career as a butcher and the creation of this book. I don't have space to thank them all, so here's the "short" list, starting with my dear editor, Carleen. She had no idea what she was stepping into when she contacted me the first time, God bless her. Even so, she's owned this project from the onset and stood behind it, and me, thus allowing it to maintain more of my original vision than I ever expected. It simply wouldn't be the project it is without her guidance and support. It's been a true joy working with her. (I think it's toughened her up a bit, too.)

Carolyn Eckert may be the warmest, most dedicated, and talented designer I've worked with. I had immediate confidence in her ability to produce designs that aesthetically and functionally worked (even for me!). What a relief. Her positivity is infectious. And the folks at Storey Publishing, who surprised me all along the way with their openness and sincere dedication to publishing books that educate. Their decision to turn this project into two volumes (this book and its companion, *Butchering Beef*) was visionary. They listen to their authors, and really hear them. And, of course, no book published in this modern era could be what it is without the talented and dedicated work of a photo retoucher. Hartley Batchelder, your work on this project was indispensable.

Joe Keller and his assistant, Jeff, were the perfect choice for photographing this project. It takes a certain kind of person to stand in a cold rain all day capturing images of slaughtering without losing his sense of humor. Not only are his images wonderful, but he is as well.

William P. Barlow (a.k.a. Billy the Butcher) was generous enough to volunteer his help on consecutive days of our photo shoot. A chef, butcher, consummate ice cream maker, and dear friend, Billy's understanding of food and his brilliance with it as a medium is stunning, and I'm indebted to him for innumerable areas of education within this industry.

I don't think we would have ever survived our first day of slaughter photography without the help of Greg Stratton. Proficient, efficient, and easy to work with, Greg's knowledge and foresight helped us avoid numerous slip-ups as the rest of the crew was focused on "getting the shot." I am truly in his debt.

Eric Shelley surprised me from the moment I met him; his depth of knowledge is extraordinary and his generosity in sharing it even more so. His compassion for animals is evident in the way he talks and teaches, and is demonstrated when you see him handling livestock.

I owe my foundational knowledge to his talent as an instructor and mentor. Clint Lane — the Robin to Eric's Batman — is also a fantastic teacher. He makes it look too easy.

Speaking of those who make it look too damn easy, let's talk about Jose Morquecha, butcher and pasta chef at Blue Hill at Stone Barns. His innate abilities are staggering, and I've benefited immensely from the opportunities I've had to work at his side. Watching him cut is like watching the slow, fluid strokes of a painter, though before you know it, it's over and you're scratching your head, wondering how he finished so quickly. His generosity, in general but also with his butchering wisdom, is why I'm indebted.

There are so many talented butchers cutting meat today — too many to mention — and they all inspire and push me to better understand the utility of these animals. I can learn something from each one of them, and look forward to gleaning such knowledge when our paths cross.

I truly appreciate the opportunity created by Severine von Tscharner Fleming, leader of a fabulous and influential collective of young farmers called the Greenhorns. Look 'em up.

We are all indebted to this country's farmers, small and large, who provide us with the necessary nourishment we crave. Connect with some in your area and show them the respect they deserve, and find out first hand where some of your local food is grown and raised.

A special thanks to the manufacturers of excellent apparel and equipment that enable us butchers to perform at the levels that we do, especially the companies that were featured in this project: Rabbit Wringer, R. Murphy Knives, and Duluth Trading Co.

My mother instilled this love for and fascination with food at an early age; I was "making" dinner for my family before I could read and write. That allure with what we consider sustenance has never stopped. The greatest gift of that is simple: I know how to cook. I've lived a happier, healthier life because of it, thanks to her. Her love and my father's support have given me the confidence to continue expanding my comfort zone through risks and adventure.

My daughter, Moxie Blaze, challenges me to better understand myself while providing joy and love every day. She was born right when this project began. Whoa. Amazingly enough, I was able to write these books from home while also helping to raise a newborn, thanks to my loving wife, Lola. She understood the challenge of balancing these priorities and helped run the house when I was under the gun. Hence, this project, as with any book, is a burden largely shared by the partnership (in so many ways). I couldn't have done this without her.

BIBLIOGRAPHY

BOOKS

Chambers, Philip G., Temple Grandin, Gunter Heinz, and Thinnarat Srisuvan. *Guidelines for Humane Handling, Transport and Slaughter of Livestock.* FAO, RAP Publication, 2001.

FAO. *Guidelines for slaughtering, meat cutting, and further processing.* FAO, 1991.

Heinze, Gunter, and Peter Hautzinger. *Meat processing technology for small- to medium-scale producers.* FAO, RAP Publication, 2007.

Lawrie, R. A. *Lawrie's Meat Science.* CRC Press, 1998.

McCracken, Thomas O., Robert A. Kainer, and Thomas L. Spurgeon. *Spurgeon's Color Atlas of Large Animal Anatomy: The Essentials.* Wiley-Blackwell, 1999.

McGee, Harold. *On Food and Cooking: The Science and Lore of the Kitchen.* Scribner, 2004.

Myhrvold, Nathan, Chris Young, and Maxime Bilet. *Modernist Cuisine: The Art and Science of Cooking.* The Cooking Lab, 2011.

North American Meat Processors. *The Meat Buyer's Guide.* North American Meat Processors. 2011.

Romans, John R., William J. Costello, Wendell C. Carlson, and Marion L. Greaser. *The Meat We Eat.* Prentice Hall, 2000.

ARTICLES

Akester, A. R. *Structure of the glandular layer and koilin membrane in the gizzard of the adult domestic fowl.* Sub-Department of Veterinary Anatomy, University of Cambridge, 1985.

Ali, M. S., H. S. Yang, J. Y. Jeong, S. H. Moon, Y. H. Hwang, G. B. Park, and S. T. Joo. *Effect of Chilling Temperature of Carcass on Breast Meat Quality of Duck.* Poultry Science Association Inc., 2008.

Berry, Joe G. *Home Processing of Poultry.* Oklahoma State University.

Bliven, Lynn, Tatiana Stanton, and Erica Frenay. *On-Farm Poultry Slaughter Guidelines.* Cornell University Cooperative Extension, 2012.

Boles, Jane Ann, and Ronald Pegg. *Meat Color.* Montana State University.

Bonhotal, Jean, Lee Telega, and Joan Petzen. *Natural Rendering: Composting Livestock Mortality and Butcher Waste.* Cornell University, 2002.

Bowker, B. C., A. L. Grant, J. C. Forrest, and D. E. Gerrard. *Muscle metabolism and PSE pork.* American Society of Animal Science, 2000.

CDC, FDA, USDA. *Foodborne Illnesses Table: Bacterial Agents.* USDA, 2004.

Du, Wayne. *Pork Safety and Quality: Feed Withdrawal Prior to Slaughter.* Queens Printer for Ontario, 2005.

Fanatico, Anne. *Small-scale Poultry Processing.* NCAT, 2003.

FDA, Keith A. Lampel (ed). *Bad Bug Book — Foodborne Pathogenic Microorganisms and Natural Toxins.* FDA, 2012.

Goodsell, Martha, and Dr. Tatiana Stanton. *A Resource Guide to Direct Marketing Livestock and Poultry.* Cornell University, 2010.

Grandin, Temple, Ph.D. *Recommended Animal Handling Guidelines & Audit Guide: A Systematic Approach to Animal Welfare.* American Meat Institute Foundation, 2010.

Grimes, Lauren M., and Chris R. Calkins. *Mapping Tenderness of the Serratus Ventralis.* Animal Science Dept., University of Nebraska–Lincoln, 2008.

Hamilton, D. N., M. Ellis, K. D. Miller, F. K. McKeith, and D. F. Parrett. *The effect of the Halothane and Rendement Napole genes on carcass and meat quality characteristics of pigs.* Journal of Animal Science, 2000.

Hulot, F., and Ouhayoun J. *Muscular pH and Related Traits in Rabbits: A Review.* World Rabbit Science Vol. 7(1), 1999.

Kim, G. D., J. Y. Jeong, S. H. Moon, Y. H. Hwang, G. B. Park, and S. T. Joo. *Effects of muscle fiber type on meat characteristics of chicken and duck breast muscle.* Division of Applied Life Science, Gyeongsang National University.

King, D. A., M. E. Dikeman, T. L. Wheeler, C. L. Kastner, and M. Koohmaraie. *Effects of Cold Shortening and Cooking Rate on Beef Tenderness.* Cattlemen's Day, 2002.

McKee, S. R., and A. R. Sams. *Rigor Mortis Development at Elevated Temperatures Induces Pale Exudative Turkey Meat Characteristics.* Poultry Science Vol. 77, 1998.

Purchas, Roger W., Dean L. Burnham, and Stephen T. Morris. *On-Farm and Pre-Slaughter Treatment Effects on Beef Tenderness.* Massey University, NZ, 2001.

Savell, J. W., S. L. Mueller, B. E. Baird. *The Chilling of Carcasses.* International Congress of Meat Science and Technology, 2004.

Scallan, Elaine, Robert M. Hoekstra, Frederick J. Angulo, Robert V. Tauxe, Marc-Alain Widdowson, Sharon L. Roy, Jeffery L. Jones, and Patricia M. Griffin. *Foodborne Illness Acquired in the United States — Major Pathogens.* Emerging Infectious Diseases Vol. 17(1), 2011.

Snyder, Jr., O. Peter, Ph.D. The Microbiology of Cleaning and Sanitizing a Cutting Board. Hospitality Institute of Technology and Management, 1997.

Stanton, Dr. Tatiana. *Marketing Slaughter Goats and Goat Meat.* Cornell University.

Van Laanen, Peggy. *Safe Home Food Storage.* Texas Agricultural Extension Service.

Wulf, Duane M. *Did the Locker Plant Steal Some of My Meat?* South Dakota State University.

WEBSITES

Gamble, H. Ray. *Trichinae: Pork Facts — Food Quality and Safety.* USDA, Agricultural Research Service (http://www.aphis.usda.gov/vs/trichinae/docs/fact_sheet.htm)

Grandin, Temple. *Recommended Stunning Practices.* (http://www.grandin.com/humane/rec.slaughter.html)

GLOSSARY

ABDOMEN | the area between the thoracic cavity and sacral region; the belly

ABDOMINAL | relating to the abdomen

ABOMASUM | the fourth, and final stomach, of the rumen before food is passed into the small intestines

ACETABULUM | the socket of the hip where the ball of the femur attaches

ACHILLES TENDON | the tendon connecting the shank muscles to the hock or cannon bones

ACTIN | one of the two main contractile proteins within muscle fibers; one of the proteins responsible for the effects of rigor mortis

ADIPOSE TISSUE | commonly known as fat

AGING | a process by which natural chemical reactions occur under controlled conditions in order to counteract the effects of rigor mortis and increase flavor and tenderness within meat

AITCHBONE | half of the pelvis after a carcass has been split into sides

ANTERIOR | relating to the front of an object; occurring closer to the head of the animal; the opposite of posterior

BALL OF FEMUR | the proximal end of the femur bone that attaches to the pelvis at the acetabulum

BILE SAC | see Gallbladder

BLADEBONE | see Scapula

BLOOMING | the act of myoglobin oxygenation that results in the cherry red, or reddish, color of meat

BOVINE | of or relating to the species of cattle

BREAK | to segment a carcass into primal cuts

BUNG | the anus of the animal

CANNON BONE | the lowest portion of the leg of an ungulate, connecting the hock or shank and the hoof; bears most of the animal's weight when walking; is used for raising the animal after stunning during exsanguination

CAUDAL | relating to, or occurring near, the rear end of the animal

CAUL FAT | a lacework of fat and connective tissue extending over the stomach and intestines; often used during aging to inhibit excessive moisture loss

CERVICAL | relating to the neck of the animal; occurring near the neck

CERVICAL VERTEBRAE | the bones in the neck

CHINE BONES | the bones of the vertebrae after a carcass is split into sides

COD FAT | fat occurring in the groin area of steer

COLLAGEN | the main protein occurring in connective tissue, bones, cartilage, and skin

CONNECTIVE TISSUE | see Fascia

CROP | a temporary storage pouch in a bird's esophagus where food stops before passage to the proventriculus

DISTAL | occurring away from the center of the carcass; the outer point; the opposite of proximal

DORSAL | occurring near the back of the animal; situated closer to the back

DRESSING PERCENTAGE | the percentage of the live weight that is left over after slaughter, exsanguination, and evisceration; the carcass weight divided by the live weight then multiplied by 100

ESOPHAGUS | a muscular tube that passes food from the mouth to the stomach

EXSANGUINATION | the draining of blood from a living animal; death from bleeding

FASCIA | sheaths of fibrous tissue, mainly composed of collagen, that connect and wrap organs, muscles, and other parts of the body

FEATHER BONES | the spinous process of the vertebrae after a carcass has been split into sides

FEMUR | the largest bone in the body, connecting proximally to the pelvis and distally to the tibia; informally known as the leg bone

FIBULA | the smaller of the two lower leg bones; bears little weight compared to the tibia, and in some cases is fused with the tibia

FINGER BONES | see Transverse Processes

FINISH | the final fattening of an animal before slaughter; the amount of fat on an animal when it's slaughtered

FOODBORNE | occurring in, on, or around food; contracted through the consumption of food

FOREQUARTER | the anterior half of a beef or pig carcass

FORESADDLE | the anterior half of a sheep, goat, or veal carcass

FREEZER BURN | the drying out of meat through sublimation and oxidation of fats while in a frozen state

GALLBLADDER | an organ, attached to liver, that is responsible for bile production and excretion; requires removal before the liver can be considered edible

GAMBREL | see Achilles tendon; also known as gambrel cord

GASTROLITHS | small stones swallowed by poultry and stored in the gizzard that support digestion; also known as gizzard stones

GELATIN | the resultant protein after hydrolysis of collagen

GIBLETS | the collective offal of poultry; usually includes gizzard, liver, neck, and heart

GIZZARD | the heavy-muscled first stomach of poultry responsible for breaking down food, with the assistance of gastroliths, before passing food to the small intestine; also known as the ventriculus

HINDQUARTER | the posterior half of a beef or pig carcass

HINDSADDLE | the posterior half of a sheep, goat, or veal carcass

HOCK JOINT | the joint connecting the hock to the cannon bones or trotter; the joint at the distal end of the tibia

HUMERUS | the upper arm bone; connects ventrally with the radius and ulna and dorsally with the scapula

ILIUM | the anterior bone of the pelvis; one of the three bones that fuse together to form the pelvis or pelvic girdle; commonly known as the pin bone after a carcass is split into sides

INSENSIBLE | unconscious and unresponsive to pain; incapable of feeling, especially suffering

INTERCOSTAL | occurring between the thoracic rib bones

ISCHIUM | the posterior bone of the pelvis, also forming the dorsal edge of the pelvis; one of the three bones that fuse together to form the pelvis or pelvic girdle

KOILIN | a dense, abrasive-resistant and inedible material that forms the inner lining of the gizzard

LATERAL | of or near the side; the opposite of medial

LEAF FAT | the abdominal fat deposits from inside the abdomen of a pig

LUMBAR | the area between the thoracic and sacral regions; the lower back

LUMBAR VERTEBRAE | the portion of the spine that comes after the thoracic vertebrae and before the sacral vertebrae; in lamb, goats, and pigs there are seven, while on bovines there are six

MEDIAL | near the middle or center; the opposite of lateral

METMYOGLOBIN | the oxidized state of myoglobin responsible for the brown hue of meat

MONOGASTRIC | a digestive system that contains a single stomach; humans, pigs, and dogs are all examples

MYOGLOBIN | the protein responsible for oxygen delivery from the blood stream to the muscle fibers; occurs in various states depending on bonding with oxygen or other elements, resulting in the various states of muscle coloration; the default, deoxygenated state, is responsible for the purplish hue of meat that has not yet been exposed to air

MYOSIN | one of the two main contractile proteins within muscle fibers; one of the proteins responsible for the effects of rigor mortis

NECKBONES | see Cervical Vertebrae

OFFAL | the edible organs from the thoracic and abdominal regions of the body

OMASUM | the third stomach of the rumen and one of the sources of tripe

OSSIFICATION | the hardening and calcification of cartilage that happens as an animal ages; used to judge the age of some species

OXYMYOGLOBIN | the oxygenated state of myoglobin, responsible for the "bloomed" red color of meat

PATELLA | commonly known as the kneecap

PATHOGEN | a dangerous microorganism capable of causing disease

PAUNCH | an informal term for the stomach or rumen of an animal

PECTORAL | the region occurring across the lower portion of the thoracic ribs and sternum; the breast or chest region

PELVIC | relating to the pelvis or around the area of the pelvis

PELVIS | the fused combination of three bones: the ilium, ischium, and pubis; informally known as the hip bones

PH | the level of acidity within a solution

PIN BONE | see Ilium

PIZZLE | the penis of an animal

PLUCK | the thoracic viscera; the heart, lungs, and usually attached trachea

POLLED | an animal without horns, either naturally or by removal

POPE'S NOSE | see Pygostyle

PORCINE | a pig, or relating to a pig

POSTERIOR | the rear end of an object; occurring nearer to the tail or rear of something; the opposite of anterior

POSTMORTEM | after death

POTABLE | able to be safely consumed, as in clean drinking water

PREEN GLAND | see Uropygial Gland

PRIMAL CUTS | the initial results from breaking down a carcass; the largest segments of a carcass that precede further breakdown into subprimals and portioned cuts

PROVENTRICULUS | the first stomach of the avian digestive system, which receives food from the crop before passing it to the gizzard; also known as the "true stomach"

PROXIMAL | occurring near the center or attaching closer to the center; the opposite of distal

PSE | an acronym meaning pale, soft, and exudative meat; meat with too low a pH, occurring due to stress and excitement just prior to slaughter, making meat soft, mealy, and unable to properly hold moisture; occurs most frequently in pigs

PUBIC SYMPHYSIS | the cartilaginous midline between both sides of the pubis

PUBIS | the smallest of the three bones that fuse to form the pelvis, occurring nearest the groin

PURGE | moisture loss from fresh or cooked meats

PYGOSTYLE | the posterior portion of poultry vertebrae; the fatty tail; also known as the "pope's nose"

QUADRUPED | an animal with four legs

RADIUS | one of the two lower arm bones, connecting proximally to the humerus; on most animals is fused with the ulna

RETICULUM | the second stomach of the rumen, occurring between the true rumen and omasum

RIGOR MORTIS | The chemical change after death in which actin and myosin enter into permanent and opposing states of contraction (ionic locking) causing a toughness in the muscles that lasts until the muscle structures deteriorate during aging.

RUMEN | the first of four stomachs in ruminants; informally, the four-chambered stomach responsible for digestion in ruminants

RUMINANT | an animal with a rumen

SACRAL VERTEBRAE | the fused portion of the vertebrae occurring between the lumbar and caudal sections; also known as the scrum

SCAPULA | the flat shoulder bone that connects distally to the humerus; also known as the shoulder blade

SHANK | the lower portion of an arm or leg, occuring above the hoof or foot

SIDE | to split a full carcass laterally in two; one half of a carcass that has been split in two laterally

SPINOUS PROCESSES | the bony extensions of the vertebrae

STERILE | containing no living organisms

STERNUM | the ventral-most bone of the chest, connecting both sides of the thoracic ribs; also known as the breastbone

STIFLE JOINT | the joint between the femur and tibia; also known as the knee joint

SUBPRIMAL CUTS | the breakdown of primal cuts into defined, subdivided sections; usually a collection of muscles separated through natural seams; can be further broken down into portioned cuts

SWEETBREAD | the thymus gland in adolescent animals; sometimes refers to the pancreas, which is also known as the "belly sweetbread"

SYNOVIAL FLUID | a lubricating fluid found amid joints to prevent friction; most frequently found in the stifle joint of beef

THAW RIGOR | the effects of rigor mortis occurring in a thawing carcass that was frozen before the onset and resolution of rigor mortis

THORACIC | the area between the cervical and lumbar regions; occurring in or around the chest region

THYMUS GLAND | a gland occurring in the neck that atrophies as the animal ages past adolescence; the main source of sweetbreads

TIBIA | the main lower leg bone, connecting proximally to the femur and distally to the hoof, foot, or cannon bones

TRANSVERSE PROCESSES | the lateral bony extensions of the lumbar vertebrae; also known as finger bones

ULNA | one of two lower arm bones, connecting proximally to the humerus; often occurring fused to the radius

UNGULATE | a hooved animal

UROPYGIAL GLAND | a gland occurring just anterior to the pygostyle that excretes an oily substance, applied through the act of preening, that helps in avian cleaning and maintaining feather condition

VARIETY MEATS | see Offal

VENTRAL | offering under or near the bottom of an animal or object; the opposite of dorsal

VERTEBRAE | the bones that make up the spine

VISCERA | the collective organs inside an animal

WEASAND | see Esophagus

RESOURCES

APPAREL
Duluth Trading Co.
Belleville, Wisconsin
800-505-8888
www.duluthtrading.com

BUTCHER BLOCKS
Green Mountain Woodworks
Talent, Oregon
866-535-5880
www.greenmountainwoodworks.com

John Boos
Effingham, Illinois
888-431-2667
www.johnboos.com

KNIVES
Friedr. Dick Corp.
Farmingdale, New York
800-554-3425
www.dick.de/en

R. Murphy Company, Inc.
Ayer, Massachusetts
888-772-3481
www.rmurphyknives.com

Victorinox Swiss Army Inc.
Monroe, Connecticut
800-442-2706
www.victorinox.com

Wenger NA
Orangeburg, New York
800-267-3577
www.wengerna.com

POULTRY EQUIPMENT
Anyone Can Build a Whizbang Chicken Plucker
http://whizbangplucker.blogspot.com
Source of the book by the same title by Herrick Kimball

Cornerstone Farm Ventures
Norwich, New York
607-334-9962
www.cornerstone-farm.com

Fleming Outdoors
Ramer, Alabama
800-624-4493
www.flemingoutdoors.com

Planet Whizbang
http://whizbangbooks.blogspot.com
Source of Anyone Can Build a Whizbang Chicken Scalder by Herrick Kimball

Stromberg's Chicks & Game Birds
Pine River, Minnesota
800-720-1134
www.strombergschickens.com

PROCESSING EQUIPMENT
Butcher and Packer Supply Company
Madison Heights, Michigan
248-583-1250
www.butcher-packer.com

LEM Products Direct
West Chester, Ohio
877-336-5895
www.lemproducts.com

MeatProcessingProducts.com
Incline Village, Nevada
877-231-8589
www.meatprocessingproducts.com

QC Supply
Schuyler, Nebraska
800-433-6340
www.qcsupply.com

Walton's, Inc.
Wichita, Kansas
800-835-2832
www.waltonsinc.com

SHARPENING
Edge Pro Inc.
Hood River, Oregon
541-387-2222
www.edgeproinc.com

FOODBORNE ILLNESS INFO
Bacteria and Viruses
U.S. Department of Health & Human Services
www.foodsafety.gov/poisoning/causes/bacteriaviruses

Estimates of Foodborne Illness in the United States
Centers for Disease Control and Prevention
www.cdc.gov/foodborneburden

Foodborne Illnesses & Disease Fact Sheets
Food Safety and Inspection Service, USDA
www.fsis.usda.gov/fact_sheets/Foodborne_Illness_&_Disease_Fact_Sheets

Partnership for Food Safety Education
www.fightbac.org

Safe Minimum Cooking Temperatures
U.S. Department of Health & Human Services
www.foodsafety.gov/keep/charts/mintemp.html
If you like your food overcooked and dry

Trichinae Fact Sheet
Animal and Plant Health Inspection Service, USDA
www.aphis.usda.gov/vs/trichinae/docs/fact_sheet.htm
Includes freezing times, written by H. Ray Gamble

INDEX

honing rod/knife steel, 42, 43, 44, 56–57, 97
horned sheep and goats, 204
hot boning, 95
housing, humane, 81
hydrolysis of collagen, 8
hygiene. *See* cleanup; personal hygiene

I

ice glazing, 432
inedible(s)
    disposal of, 96–97
    removing, 60, 176
    separating, 60–61
infections, foodborne pathogens and, 24
infectious bacteria, 27
insensibility, 80–82
intercostal meat, 287

J

Jaccard meat tenderizer, 51
joint(s)
    atlas, 103, 120, 161, 185, 200, 218, 233, 257, 349, 353
    ball-and-socket, 189, 308, 407
    boning and, 64, 66
    break, 200, 201, 215, 216
    costovertebral, 395
    hock, 118, 156, 171, 172
    humerus-radius, 265
    popping, 189, 192
    scapular, 266, 268, 387
    separating, 66
    shank, 201
    spool, 200, 202
    stifle, 146, 154, 185, 191, 310, 318, 320, 414, 415
jowls, pork, 93
    about, 372
    removal of, 373–374, 376

K

keel bone, 149, 150
kidneys, 94, 136
    pig, 325, 344, 352
    processing of, 93, 137
    rabbit, 161, 174, 176, 177
    sheep, 200, 228
    visceral fat and, 10
killing cones, 102, 104, 106
kitchen scale, 48
knife accessories
    honing rod/knife steel, 42, 43, 44
    scabbard, 44, 45
knives. *See also* boning knife
    blade characteristics, 40
    flexibility of, 40, 41
    shape of, 40
    sharpening, 44, 45, 56–57
    types of, 40–42, 97, 98
knots, making, 74–77
    butcher's knot, 74–75
    packer's knot, 76–77
koilin, 126

L

labeling, 424, 432–433
lamb. *See also* leg and sirloin, lamb; sheep and goat butchering
    carcass species and, 231
    cuts, portioned, 254–255
    cut sheet, 73
    leg of
        boneless, 307–312
        semi-boneless, 309
    leg roasts, 318
leaf lard, removal of, 352

leg and sirloin, lamb
    about, 301
    bone-in/bone-out leg, 311
    boneless leg of lamb, 307–312
    boneless sirloin, 302–304
    semi-boneless leg of lamb, 309
    sirloin chops, 305–306
    steaks, 320–321
    subprimal roasts, 313–318
liver, 94
    chicken, 136
    rabbit, 176, 177
    sheep, 226
loin. *See also* tenderloin
    lamb
        about, 290
        boneless, including tenderloin, 291–295
        roasts and chops, 298–300
        saddle chops, 300
        split loin primals, 297
        supreme loin roast, 296
    pork, 388–401
        about, 233, 235
        baby back ribs, removing, 392
        butt split from, 367
        chops, 389, 398
        common cuts, 348
        fatback, removal of, 399
        membrane removal, 393
        sirloin and, 355–356
        splitting, rib/saddle, 394

M

mad cow disease, 31
mallet, 43, 51
mammary gland, 210
marinades, alkaline, 69
masseter. *See* cheeks
meat(s). *See also* "blooming" meat
    aging time, 20
    animal stress and, 18
    color and freshness, 16
    cured, 16
    damage to, 60, 81, 88–89
    ground, 36, 37
    intercostal, 287
    tenderizer, 51
    tenderness, 8, 12
meat grinders, 50–51
meat lugs, 51
mechanical plucking, 113, 114, 127
mechanical tenderization, 68–69
membrane
    fascial, 211
    removal of, ribs, 287–288, 393, 404, 405
metmyoglobin, 14, 15
microbes, 24
muscle(s).
    about, 5
    actin and myosin, 6, 7, 18, 19
    color of, 14–16
    connective tissue and, 7–9
    denuding, 62
    fats and, 9–12
    fibers, 6–7, 12–13
    grain of, 58, 60, 68
    into meat, 17–21
    structure, 6, 7, 53
    tenderness, 8
    types of, 11, 12
mutton, 202
myoglobin
    cured meats and, 16
    three states of, 14–15
myosin, 6, 7, 18, 19

N

nitrile gloves, 48, 49
norovirus, 28, 29
nose-to-tail use, 323, 347

O

offal. *See also* gland(s); inedible(s); viscera
    chicken, 117, 136–137
    edible, 93–94, 161, 174
    freezing, 422
    packaging, 179
    pig, 325
    rabbit, 178
    sheep and goat, 200, 229
organs. *See* offal; *specific organ*
ossification, 245
oxymyoglobin, 15, 16

P

packaging. *See also* wrapping meat
    about, 419
    ice glazing, 432
    labeling and, 424
    lamb, 232
    offal, 179
    pork, 348
    poultry, 136, 140
    rabbit, 179, 182
    tenets of, 423–424
    vacuum sealing, 424–427
    zip-top bags, 432
packer's knot, 76–77
pancreas, 60, 94
paracord, 88
parasitic worms, 29
pathogens
    about, 27
    contamination and, 24–27
    freezing and, 421
    infections and, 24
    killing with heat, 36
    preventing spread of, 31–34
    types of, 27–31
peeling the bone, 66–67
personal hygiene
    hand washing, 31–33
    sanitary clothing, 33–34
picnic, pork. *See* Boston butt and picnic
pig butchering. *See* pork butchering
pig ears, 372
pig slaughtering
    about, 323
    age of pig, 329
    anatomy, 324
    bleeding the animal, 85, 86, 332–333
    equipment, 324
    evisceration, 90–92, 338–344
    hair removal, 328–329
    hoisting the animal, 87
    overview of, 330
    preparation for, 329
    scalding, 236–328, 334–335
    scraping carcass, 336–337
    securing/stunning animal, 331
    setup for, 324
    skeletal structure/viscera, 325
    splitting/cooling carcass, 96, 345
    stunning and sticking, 324
pig sticking knife, 97, 98
pinfeather removal, 115
pinning knife, 97, 98
pistol grip, 54
pithing (debraining), 83, 110–111
pizzle, 339
pluck (thoracic viscera)
    pig, 344
    rabbit, 177